GEOMETRY, PERSPECTIVE DRAWING, AND MECHANISMS

Don Row
University of Tasmania, Australia

Talmage James Reid
University of Mississippi, USA

NEW JERSEY • LONDON • SINGAPORE • BEIJING • SHANGHAI • HONG KONG • TAIPEI • CHENNAI

Published by

World Scientific Publishing Co. Pte. Ltd.
5 Toh Tuck Link, Singapore 596224
USA office: 27 Warren Street, Suite 401-402, Hackensack, NJ 07601
UK office: 57 Shelton Street, Covent Garden, London WC2H 9HE

British Library Cataloguing-in-Publication Data
A catalogue record for this book is available from the British Library.

GEOMETRY, PERSPECTIVE DRAWING, AND MECHANISMS

ISBN-13 978-981-4343-82-4
ISBN-10 981-4343-82-X

Printed in Singapore.

GEOMETRY, PERSPECTIVE DRAWING, AND MECHANISMS

Dedicated to our wives, Maggie and Bonnie.

Preface

I have admired the elegance and simplicity of combinatorial geometries from the day I first met them at Lakehead University's "Research Week in Geometry" in 1970, and their lack of influence in the teaching and content of undergraduate geometry courses has increasingly puzzled me. I hope that my approach will help to remedy this lack, and that its blend of theory and application to the everyday world will catch the reader's imagination. He or she may feel a little of the excitement that surrounded geometry during the Renaissance as the development of perspective drawing gathered pace, or more recently as engineers sought to show that all the world was a machine. The same excitement is here still, as enquiring minds today puzzle over a random-dot stereogram or the interpretation of an image painstakingly transmitted from Jupiter.

The book attempts to give a sound basis for a variety of undergraduate courses, to provide a basis for a geometric component of graduate teacher training, and to provide background for those who work in computer graphics and scene analysis. It provides a source from which refresher courses for High School and Elementary School teachers may be drawn.

Chapters 1 to 4 form a self-contained development of the geometry of extended Euclidean space. Chapters 5 to 8 use this geometry to systematically clarify and develop the art of perspective drawing and its converse discipline of scene analysis. The behavior of bar-and-joint mechanisms and hinged-panel mechanisms is analyzed in Chapters 9, 10, and 11, again this analysis is based only on the results of the first four chapters. In Chapter 12 we introduce spherical polyhedra and apply scene analysis to drawings of these and associated objects in Chapter 13. In Chapter 14 we relax the axioms of Chapter 2, giving an introduction to matroids that covers their fundamental concepts and is motivated by their underlying existence in important mathematical structures.

The treatment is structured so that the majority of its content is accessible to students without specialist mathematical grounding. It contains many exercises and constructions that link the work closely to practical experience. In particular, fourteen problems and their solutions have been selected to demonstrate the scope of geometric inquiry. I have used a range of computer programs to demonstrate their place in geometry, in particular the relationship between the growing availability of computing power and its use in virtual reality construction.

Anyone teaching from this book will find it easy to tailor a course to suit their particular needs. For example, Chapters 1 to 6, together with the results of Chapter 8, comfortably fills a one-semester undergraduate course that covers basic projective geometry and perspective drawing. Someone with a mechanical bent may prefer to substitute Chapters 9 and 10 in place of Chapters 5, 6, and 8. For a student wanting only planar geometry, Chapters 1 to 3, and 9 can be covered in one semester. A course analyzing the common structure of geometries, graphs and matroids is given by Chapters 1 to 4, 11 (excluding the third and fourth sections), and 14. This material provides a sound geometric background for those wishing to further study matroids (see [49]). These are just some of many sensible selections.

Don Row

Contents

Chapter 1

COMBINATORIAL FIGURES

From the moment of birth, all of us examine our surroundings with fierce intensity, coming to terms with our shared physical world and developing a *modus operandi* for coping with its requirements. We develop a shared understanding of the vocabulary used to describe this world. At least, by and large, our answers to simple questions are in agreement. For example, a question such as: "Is this cat black?" usually receives the same answer regardless of who is asked. For our purposes we need to suppose that our schooling has given us a reasonably common interpretation of words such as "point", "flat", and "straight line".

In this chapter we examine examples of sets of points by drawing them or modeling them and singling out some of their simple features. We will concentrate initially on just two questions about a set of points, namely "Which subsets of the points are on a straight line?" and "Which subsets of the points are flat?". These questions may be simple, but as we show in subsequent chapters, they lead to the heart of the structure of our world, and motivate the axiomatic development to be given in Chapter 2.

We call any set of points that "lies on a straight line" a *linear set*, and say that its members are *collinear* points. We call any set of points that is "flat" a *planar* set, and say that its members are *coplanar* points. Thus we are asking, in the two questions above, "Which are the linear subsets?" and "Which are the planar subsets?".

When we know the answer to these questions for a certain set of points, we have a figure. More precisely, a *combinatorial figure* is a set of points together with any convenient description of exactly which subsets are linear subsets, and exactly which subsets are planar subsets. The adjective *combinatorial*, often omitted, is used in this context to emphasise that we are only interested in questions of collinearity and coplanarity.

1.1 Drawing figures

We can describe figures in many ways. In this section, we use a straight-edge and pen or pencil to make drawings, and agree on notation to indicate those subsets of points that are linear and those that are planar. We represent a point by a small circle or a dot. We use a thin straight line in a drawing, drawn through circles or dots, to indicate a linear set of points. We can always draw a straight line through just two points but often omit it when it seems to give no useful information, and merely clutters up the drawing. In Figure 1.1 the the first and second drawings are each of the same 2-point figure, the third drawing is of three non-collinear points, and the fourth is of a figure of three collinear points. The sixth and seventh drawings are of the same 5-point figure, but the fifth drawing is not.

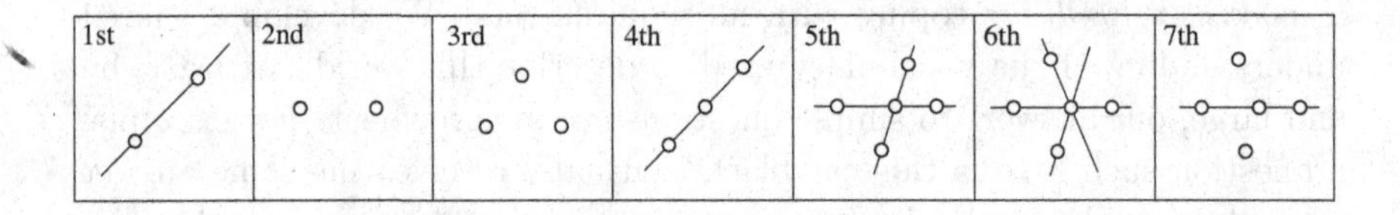

Fig. 1.1 Notation for linear sets of points

Definition 1.1.1. *A combinatorial* **triangle** *is a figure of three non-collinear points. Each point is a vertex of the triangle (see the third drawing of Figure 1.1).*

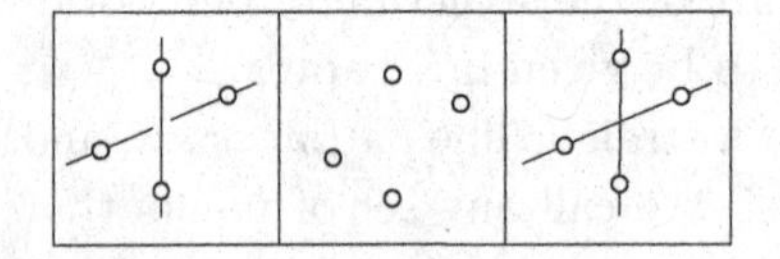

Fig. 1.2 A tetrahedron and two planar figures

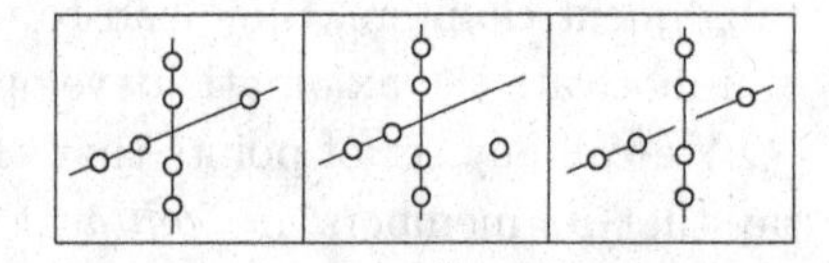

Fig. 1.3 Three 7-point figures

As we know, non-planar figures also exist. For example:

Definition 1.1.2. *A combinatorial* **tetrahedron** *is a figure of four non-coplanar points. Each point is a vertex of the tetrahedron.*

We often draw a tetrahedron as in the first drawing of Figure 1.2, the gap in one of the straight lines indicating that the two do not meet. We may draw a figure of four coplanar points as in either of the last two drawings of Figure 1.2. The second drawing is preferred to the third, but either indicates that the four points of the figure are coplanar. In a similar spirit, the first two 7-point figures drawn in Figure 1.3 are meant to be planar, but the third figure is not.

Exercise 1.1.1. Describe differences between the figures of Figure 1.4.

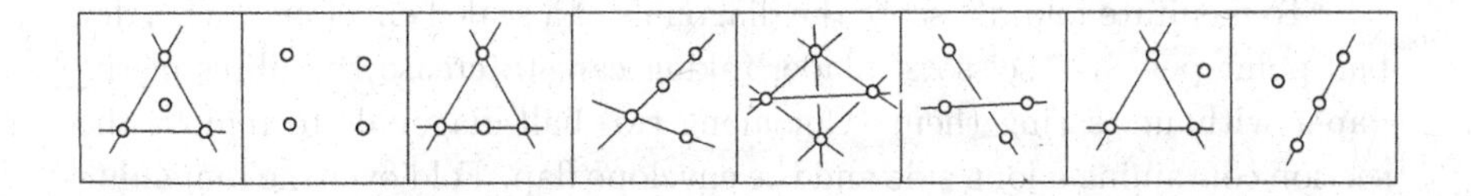

Fig. 1.4 Eight 4-point figures

Shading a drawing, as in Figure 1.5, can help an artist indicate which sets of points are planar and which are not. Figures can be drawn in many ways. Each drawing in Figure 1.6 is of the same figure. We prefer drawings that are attractive and efficiently show the linear and planar subsets of points.

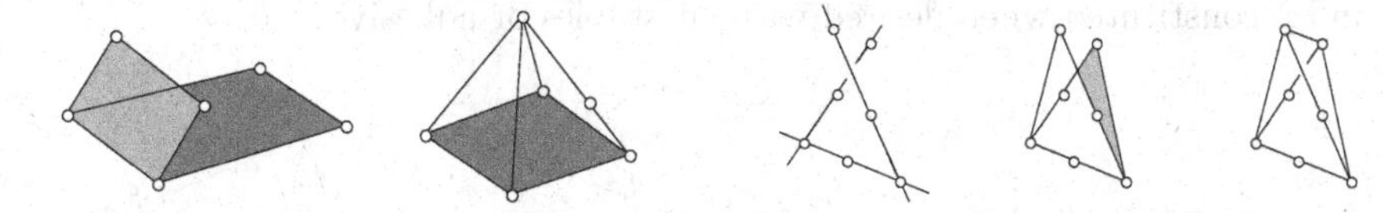

Fig. 1.5 Using shading

Fig. 1.6 Different drawings of the same figure

Construction 1.1.1. Draw a figure of four points, three of them being collinear. Draw a figure of five points, three on one straight line and three on another straight line. Draw a figure of six points, three being collinear, and one of these three belonging to a planar set of four points.

One of the difficult tasks of geometry is the recognition of the same figure in different guises. This mirrors the everyday problems of understanding that occur when we read an instruction leaflet accompanying a product, when we study a painting, or when we struggle with the architectural plans of a proposed house. These recognition problems were the genesis of non-metric geometry, and one aim of this book is to trace their solutions.

1.2 Modeling figures

Another method of specifying a figure is by a model, and in this section, we examine some models and, where possible, construct them.

Construction 1.2.1. Using an envelope, make a model whose points are the vertices of a tetrahedron, by following the instructions below of Charles W. Trigg [58] (see Figure 1.7).

"To facilitate folding, score the diagonals of a sealed envelope with a dry ball-point pen or a scissors' blade, taking care to crease the fibres of the paper without tearing them. Cut along two half-diagonals to remove the section containing a long side and the envelope flap. Fold over the remaining portion along half-diagonals and crease firmly. Fold back along the same lines and crease firmly again. (Avoid envelopes on which the diagonals fall along a sealed seam.) Bring dc onto ab, flatten the envelope to form ef and crease firmly. Fold back along ef and again crease firmly. In the diagram, e' indicates the point on the lower side directly under e. Separate e and e' until efe' is straight. Fold around efe' until d meets a. Tuck d under ab and press up on b and c until the ends dc and ba coincide. The rigid model with no open edges thus produced can be collapsed for storage or carrying and reconstituted when desired without staples or adhesives."

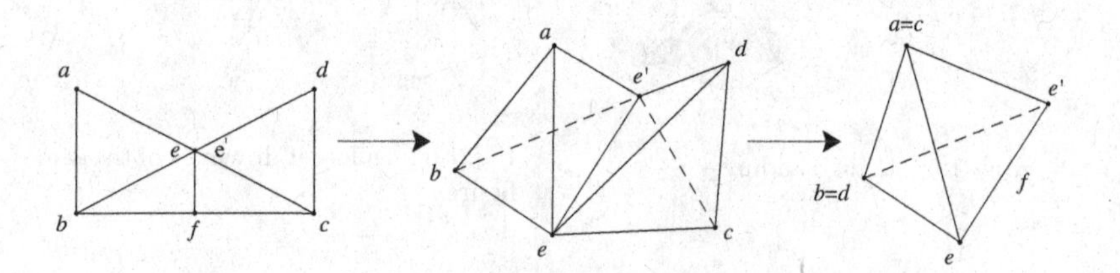

Fig. 1.7 Making a model of a tetrahedron

Construction 1.2.2. Make a model of Figure 1.6 in which each point is represented by a bead-headed pin stuck into a model tetrahedron.

Models undoubtedly make some figures easy to visualize but are less convenient to carry about than are drawings. It seems to be more difficult to make a convincing drawing of a non-planar figure than of a planar figure. This may be because a drawing of a planar figure is in fact a model of it.

In practice we blur the distinction between "a figure" and "a model, or drawing, of a figure" - even though a figure has many models and drawings. This is common practice. How often do we say "Oh, look at this mountain!", when we really want someone to look at a photograph of the mountain. The claim that a figure exists can be tested by constructing a model.

Definition 1.2.1. *A* **construction** *of a model is a recipe, telling us how to make it (or, in the case of a planar figure, telling us how to draw it).*

Can the list of instructions which make up the recipe be successfully carried out? Construction 1.2.1 suggests that labeling is helpful to complete a construction. We usually use numbers or lower-case Roman letters as labels for points and upper-case Roman letters to stand for sets of points.

Construction 1.2.3. Draw a point 1. Draw two straight lines through 1. Draw a point 2 on the first line and a point 3 on the second line. Draw a straight line through 2 and 3, and a point 4 on this line. Draw a point 5 on the straight line through 1 and 2. Draw a point 6 on the straight line through 4 and 5 and on the line through 1 and 3.

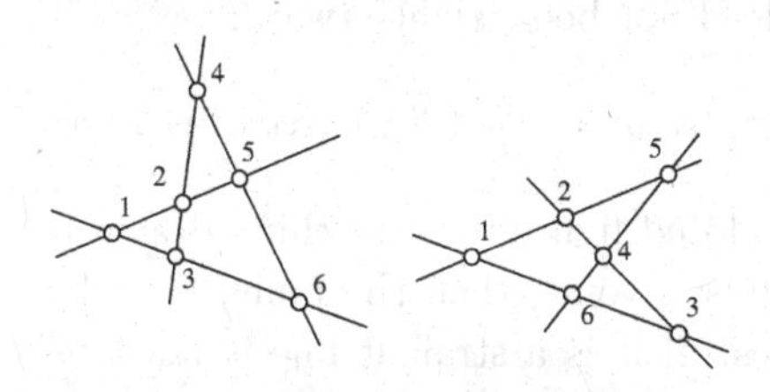

Fig. 1.8 Two examples arising from Construction 1.2.3

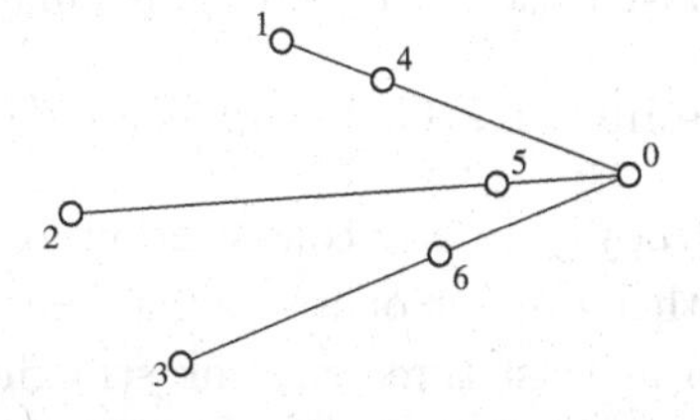

Fig. 1.9 A result of Construction 1.2.4

Construction 1.2.3 is clear while Constructions 1.2.4 and 1.2.5 are not. We call straight lines that share a common point *concurrent.*

Construction 1.2.4. Draw a figure consisting of a triangle with vertex set $\{1,2,3\}$ and a triangle with vertex set $\{4,5,6\}$ so that the straight line through the points 1 and 4, the straight line through the points 2 and 5, and the straight line through the points 3 and 6, are concurrent.

If we start Construction 1.2.4 by first choosing the six points, then usually the three required straight lines will not be concurrent. One way to successfully carry out the construction is by first choosing a point, then drawing three straight lines through this point, and then choosing points 1 and 4 on one of the straight lines, points 2 and 5 on another, and points 3 and 6 on the remaining straight line. In this way we avoid the need either to draw one straight line through each of three existing points, or to find a point that is on three existing straight lines. We need to avoid these two steps in order to be sure that a recipe for drawing a figure can be followed. Let us see whether or not this problem is something to worry about.

Construction 1.2.5. Try to draw seven coplanar points so that seven straight lines can be drawn, with each of the straight lines through three of the points, and each of the points on three of the straight lines. If unsuccessful, then try to draw just six straight lines as above.

We give such figures a special name.

Definition 1.2.2. *A* **closed figure** *is a figure with each point on three or more straight lines, and each straight line containing three or more points.*

Neither figure drawn in Figure 1.8 is closed. If it exists, then the first figure mentioned in Construction 1.2.5 would be closed. We now show that construction of closed figures may indeed not be straightforward.

Lemma 1.2.1. *A claim about the existence of a closed figure requires proof.*

Proof. In any construction for the closed figure the last thing drawn is either a point or a straight line. If it is a point, then the point must be on at least three existing straight lines. If it is a straight line it must be through at least three points of the figure. We have agreed that either of these steps requires some justification. □

Add to the figure drawn in Figure 1.9 by drawing a straight line through 1 and 2 and a straight line through 4 and 5, and drawing the point 9 of intersection of these two lines. Draw the intersection point 7 of a straight line through 2 and 3 and a straight line through 5 and 6, and the intersection point 8 of a straight line through 3 and 1 and a straight line through 6 and 4, as in the first figure in Figure 1.10. Should we draw a straight line through the points 7, 8, and 9? Are the three points collinear? The second ten-point figure in Figure 1.10 seems to be the result of a successful attempt to draw such a straight line. It is a closed figure and is named a *Desargues figure*, after the 17th Century French architect and geometer Girard Desargues. Do our attempts at drawing it conclusively answer the question of its existence?

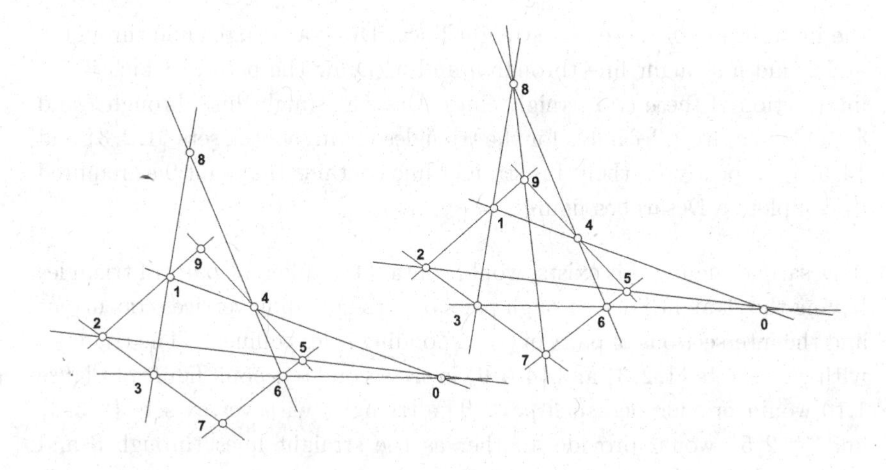

Fig. 1.10 A possible Desargues figure

Are the lines straight as viewed under a microscope? The imperfect nature of our drawing materials leaves us unsure.

Our attempts to construct Desargues figures, and further experiments in Construction 1.2.6, suggest that the following statement may be true.

Conjecture 1.2.1. *Let* $\{1, 2, 3\}$ *and* $\{4, 5, 6\}$ *each be the vertex set of a triangle so that the lines joining* 1 *and* 4, *joining* 2 *and* 5, *and joining* 3 *and* 6, *are concurrent. Label the point of intersection of the straight line through* 1 *and* 2 *and the straight line through* 4 *and* 5 *by* 9, *the point of intersection of the straight line through* 2 *and* 3 *and the straight line through* 5 *and* 6 *by* 7, *and the point of intersection of the straight line through* 3 *and* 1 *and the straight line through* 6 *and* 4 *by* 8. *Then the points* 7, 8, *and* 9 *are collinear.*

Lemma 1.2.2. *The truth of Conjecture 1.2.1 would be justification for the existence of a Desargues figure.*

Proof. Draw a point 0. Draw three more non-collinear points 1, 2, and 3. Draw a straight line through points 0 and 1. Draw a straight line through points 0 and 2. Draw a straight line through points 0 and 3. Draw a point 4, collinear with 0 and 1. Draw a point 5, collinear with 0 and 2. Draw a point 6, collinear with 0 and 3. Draw a straight line through 2 and 3 and a straight line through 5 and 6. Draw the point 7 which is the intersection of these two straight lines. Draw a straight line through 1 and 3 and a straight line through 4 and 6. Draw the point 8 which is the intersection of these two straight lines. Draw a straight line through 1 and 2 and a straight line through 4 and 5. Draw the point 9 which is the intersection of these two straight lines. Draw a straight line through 7 and 8. If Conjecture 1.2.1 holds for the triangles having vertex sets $\{1, 2, 3\}$ and $\{4, 5, 6\}$ respectively, then this straight line contains the point 9 as required to complete a Desargues figure. □

A Desargues figure, if it exists, would contain ten different pairs of triangles having the straight lines through pairs of corresponding vertices concurrent and the intersections of pairs of corresponding sides collinear. The triangles with vertex sets $\{1, 2, 3\}$ and $\{4, 5, 6\}$ in the second ten-point figure of Figure 1.10 would provide one such pair. The triangles with vertex sets $\{8, 3, 6\}$ and $\{9, 2, 5\}$ would provide another as the straight lines through 8 and 9, through 3 and 2, and through 6 and 5 are concurrent at the point 7; while the points of intersection of pairs of corresponding sides are the three collinear points 0, 4, and 1. We may carry out some more construction attempts of Desargues figures with computer-aided drawing in the hope that more accuracy will give us encouragement to press our investigations with the reward being a successful proof of Conjecture 1.2.1.

Construction 1.2.6. Using a computer drawing program such as The Geometer's Sketchpad or GeoGebra, follow the instructions given in the proof of Lemma 1.2.2 to draw a Desargues figure.

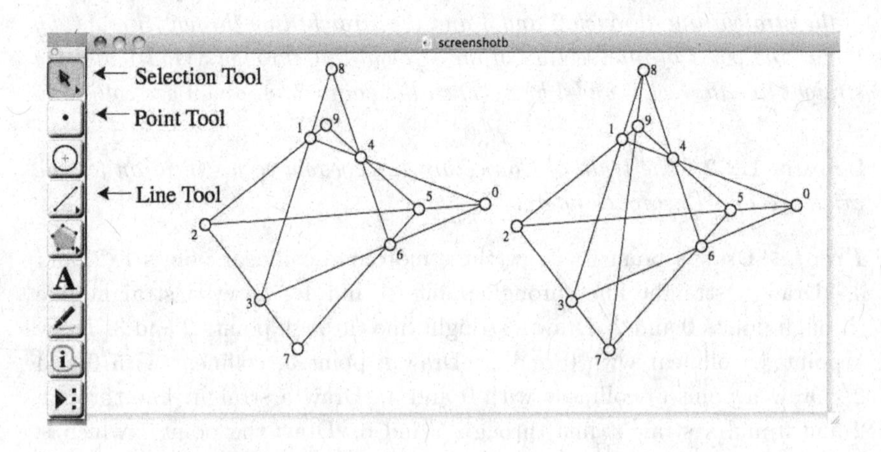

Fig. 1.11 Screen of an open **Sketch**

For those using The Geometer's Sketchpad the process is as follows. Figure 1.11 shows the toolbox on the lefthand side of an open sketch and indicates the selection, point, and line tools of the toolbox. The line tool connects two points by clicking on one point and dragging the mouse to the other point. In addition, it is helpful to label points by using the label submenu of the display menu at the top of the program screen (not shown). Place four points 0, 1, 2, and 3 wherever you wish. Produce a line through 0 and 1, a line through 0 and 2, and a line through 0 and 3. Use the undo menu item to correct any mistakes. Use the point tool to click at a point on the line through 0 and 1 to give a point 4. Repeat to obtain a point 5 on the line through 0 and 2, and a point 6 on the line through 0 and 3. The selection tool can be used to move the diagram around if two lines intersect off of the screen. Construct a line through 2 and 3, and a line through 5 and 6. Choose the point tool and click on the intersection of these lines to give a point 7. Repeat the process to obtain 8 and 9 in turn. Draw a line through 7 and 8, and perhaps through 9. Choosing an aesthetically pleasing final version, print it and check that the so-called lines are straight (certainly not true on a screen which is not flat). Does this experiment strengthen your view of the truth of Conjecture 1.2.1? The truth or falsity of this conjecture is crucial in governing the physical behavior of our world and in determining the geometry we develop to describe the world. We meet it next in Chapter 3.

The following problem arises frequently in perspective drawing, as we see in the section on practical drawing methods in Chapter 5. A little experimenting may suggest a solution based on the truth of the above conjecture, but we must wait until Construction 3.5.4 in Chapter 3 and Theorem 4.4.3 in Chapter 4 before we can guarantee the validity of a solution.

Problem 1.2.1. *Let two straight lines and a point be drawn. If the two lines would meet off the page, devise a construction, using only a pencil and straight-edge, for a straight line that contains the given point and would be concurrent with the two given straight lines.*

Another type of figure that will be central to our investigations of the world around us is a polygon. We define it below, and see that it leads us again into the realm of closed figures.

Definition 1.2.3. *A* **combinatorial** n**-gon**, *sometimes called a polygon, is a figure of* n *points* $\{1, 2, \ldots, n\}$, *together with a cyclic listing* $1, 2, \ldots, n$ *of the points. Each point is a vertex of the* n*-gon and the straight lines* *through* 1 *and* 2, *through* 2 *and* 3, ..., *through* n *and* 1 *are the* n *sides of the* n*-gon. We denote the* n*-gon by* $12\ldots n$.

If the number n is even, then, for each $i = 1, 2, \ldots, \frac{n}{2} - 1$, the vertices i and $\frac{n}{2} + i$ are opposite to one another, and the side through i and $i+1$ and the side through $\frac{n}{2} + i$ and $\frac{n}{2} + i + 1$ are opposite to one another.

We may think of a triangle as a 3-gon and we therefore use abc as a notation for the triangle with vertex set $\{a, b, c\}$. We often refer to a 5-gon as a *pentagon*, and to a 6-gon as a *hexagon*. In a hexagon 123456, the sides through 1, 2 and through 4, 5 are opposite to one another, as are the sides through 2, 3 and through 5, 6, and the sides through 3, 4 and through 6, 1.

If a figure is an n-gon, then we sometimes draw straight lines to indicate its sides - even though they may have only two points of the figure on each. For example, we draw the 4-gon 1234, the 4-gon 1324 and the 4-gon 1243 as in the first three figures of Figure 1.12.

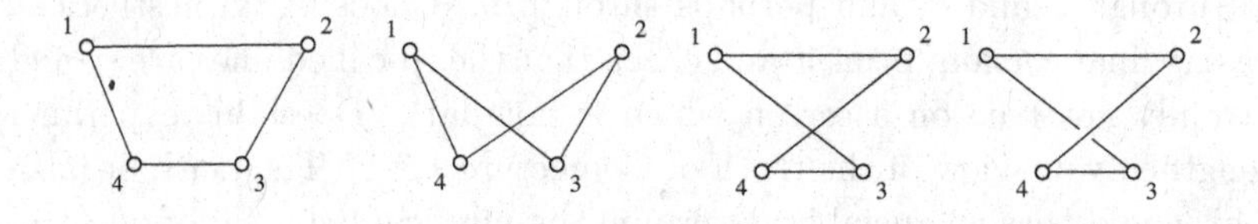

Fig. 1.12 Some 4-gons

An n-gon may not be planar, as we see in the last figure in Figure 1.12.

Exercise 1.2.1. Is it possible that some of the six figures in Figure 1.13 are "shadows" cast by a non-planar 4-gon model? Which of the six may be shadows? Why? It may be helpful to make a wire model and examine it.

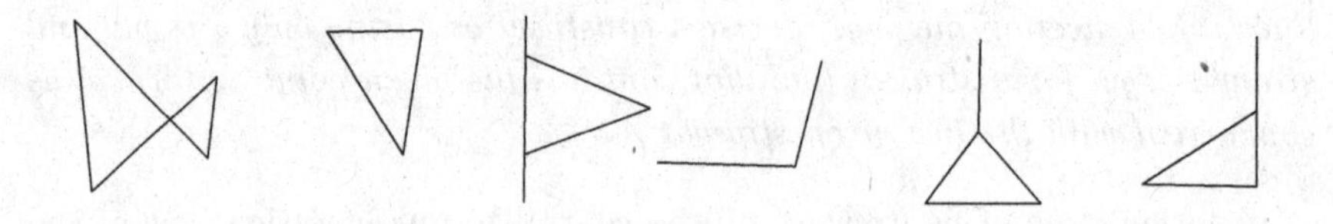

Fig. 1.13 Figures that may be shadows of a non-planar 4-gon

We may construct a hexagon 123456 so that the vertices 1, 3, and 5 are collinear and the vertices 2, 4, and 6 are collinear. Will the three intersections of pairs of opposite sides of such a hexagon be collinear? The example in Figure 1.14 seems to be the result of a successful attempt to draw such a figure. But, as is the case with a Desargues figure, we notice that it is a closed figure and, from Lemma 1.2.1, a claim of its existence would need some justification. If it does exist, then we will name the resulting nine-point figure a Pappus figure, after the 4th Century Greek geometer Pappus of Alexandria. Figure 1.14 suggests that any example of a Pappus figure would be planar. Our attempts at constructing examples of a Pappus figure, and further experiments using The Geometer's Sketchpad, suggest that the following statement may be true.

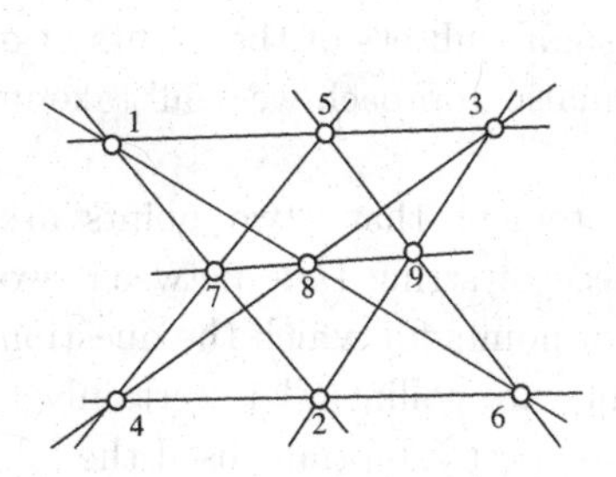

Fig. 1.14 A possible Pappus figure

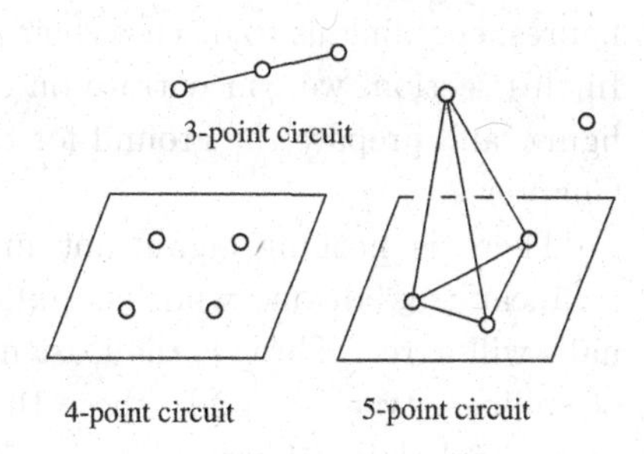

Fig. 1.15 Types of circuits

Conjecture 1.2.2. *Let* 123456 *be a hexagon so that the points* 1, 2, *and* 3 *are collinear and the points* 2, 4, *and* 6 *are collinear. Label the point of intersection of the side through* 1 *and* 2 *and the side through* 4 *and* 5 *by* 7, *label the point of intersection of the side through* 2 *and* 3 *and the side through* 5 *and* 6 *by* 8, *and label the point of intersection of the side through* 3 *and* 1 *and the side through* 6 *and* 4 *by* 9. *Then the points* 7, 8, *and* 9 *are collinear.*

Exercise 1.2.2. Prove that the truth of Conjecture 1.2.2 would be justification for the existence of a Pappus figure.

Exercise 1.2.3. Assuming that the Pappus figure in Figure 1.14 exists, find a hexagon in the figure that has the collinear points 5, 6, and 9 as the three intersections of pairs of its opposite sides.

Construction 1.2.7. Use a computer-drawing program to attempt a construction of a Pappus figure, in an analogous way to our attempt to construct a planar Desargues figure in Construction 1.2.6. Choosing a pleasing arrangement of points, print your drawing, and check the lines for straightness. As a result of your experiments, do you have a view on the likelihood of Conjecture 1.2.2 being true?

1.3 The circuits of a figure

Our agreed meanings for words such as *collinear*, and *coplanar* are still based completely on our intuitive idea of "point", "straight line", and "flat". At the moment we are merely hoping to obtain a working understanding of figures, enabling us to discuss their properties without too much confusion. In this section, we concentrate on certain small subsets of the points of a figure, and prepare the ground for the axiomatic approach we shall take in Chapter 2.

There is nothing significant in the statement that "two points are collinear", as anyone who has pulled a piece of string taut between two nails will agree. Three is the least number of points for which the question of collinearity arises and "these three points are collinear" is certainly a non-trivial claim. There is some evidence [41] that Egyptians used the following method to ensure that the base lines of their structures were straight. Two markers were floated at the same height in a long channel of water and a post sighted and marked along a line of sight through these markers. This procedure gave three points of a straight horizontal line, and repeated application gave further points of the line.

The claim that "three points are coplanar" is not likely to arouse controversy. It's truth is the reason that three-legged stools are common. On the other hand, four is the least number of points for which the question of coplanarity arises and "these four points are coplanar" is a non-trivial and interesting claim.

We hope that knowing the answers to such claims about small subsets of points in a figure will enable us to answer questions about linearity and planarity of larger subsets. Specifically we hope that a subset of points of a figure is linear if each of its 3-point subsets is linear, and that a subset of points of a figure is planar if each of its 4-point subsets is planar. This hope motivates the following definition.

Definition 1.3.1. *A* **circuit** *of a figure is a set of three collinear points of the figure, or a set of four coplanar points of the figure (no three of which are collinear), or a set of five points of the figure (no four of which are coplanar and no three of which are collinear) (see Figure 1.15).*

For example, the figure in Figure 1.6 has three 3-point circuits, two 4-point circuits, and one 5-point circuit. We hope that the set of circuits concisely summarizes all the linear and planar relationships amongst the points of a figure. If so it would give us a means, other than by drawing and modeling, of specifying a figure. The example of Figure 1.6 could thus be described by labeling its points $1, 2, 3, 4, 5, 6$, and 7 and writing down its set of circuits, $\{\{1,2,3\},\{3,4,5\},\{5,6,7\},\{1,2,4,5\},\{3,4,6,7\},$ $\{1,2,4,6,7\}\}$. The following exercises give some practice in recognizing the circuits of figures. In Chapter 2 we use this notion of circuit as a basic tool for analyzing and constructing examples in our geometry.

Exercise 1.3.1. Label the points of each of the eight figures in Figure 1.4 by labels $1, 2, 3$, and 4 and list all the circuits of each figure.

Exercise 1.3.2. Assuming that the attempt at drawing the Desargues figure in Figure 1.10 was successful, list all its circuits.

Exercise 1.3.3. Find a 5-point figure having as many 3-point circuits as possible. Find a 5-point figure having as many 4-point circuits as possible.

Exercise 1.3.4. List the circuits of each of the four figures in Figure 1.16.

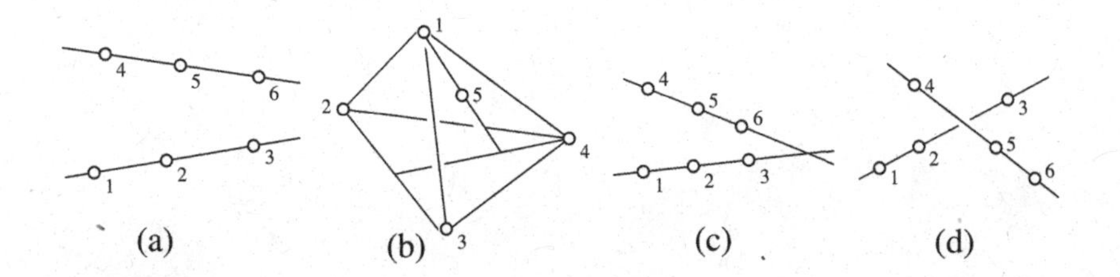

Fig. 1.16 Four figures

Construction 1.3.1. Draw a figure of eight points, say $\{0,1,2,3,4,5,6,7\}$, having no three-point circuits, having $\{0,1,3,5\}$, $\{0,2,3,6\}$, $\{1,4,5,7\}$, and $\{2,4,6,7\}$ amongst its four-point circuits, and having no five points coplanar. You may feel like using a computer drawing program with shading abilities for this construction.

You may feel that a box-like appearance is appropriate for the figure of Construction 1.3.1. Drawing segments rather than lines may aid in clarity, as we saw in Figure 1.5. To draw one "face of a box" select **Segment** rather than **Line** from the **Line** menu. Click on a point, drag, click on another point, drag, click on another point, drag, click on a fourth point, drag and click on the first point. This gives a 4-gon. It may be shaded, as in the second figure in Figure 1.5, emphasizing the coplanarity of the four vertices of the "face". Restricting shading to only one or two 4-gons may achieve the desired clarity.

For those using The Geometer's Sketchpad the process for shading a polygon is as follows. Choose the **Selection** tool, hold down the **Shift** key and click on the vertices *in correct cyclic order*. Release the **Shift** key. From the **Construct** menu choose **Polygon interior**. The shading of the polygon may be varied in depth by choosing the **Color** menu in the **Display** menu and selecting the preferred depth of shading in by using the **Other** submenu.

Summary of Chapter 1. We introduced combinatorial figures and agreed on notation for drawings of them. We drew and modeled figures, discussing the notion of their construction. The construction of closed figures seems to be particularly difficult. We introduced the idea of a circuit of a figure, and listed the circuits of examples.

Chapter 2

COMBINATORIAL GEOMETRIES

In Chapter 1 we gained some expertise in describing and constructing combinatorial figures. In the process, the question of the existence of certain figures arose. In order to answer this and other questions about figures, we show that each combinatorial figure is an example of a structure known as a combinatorial geometry. We then take advantage of a simple axiomatic definition of combinatorial geometries. From these axioms we systematically develop a body of knowledge about all combinatorial geometries, knowing that each theorem provides information about combinatorial figures in particular. The axioms are chosen with two purposes in mind. First they are easy to use and understand, and second they give a theory of geometries that captures the essential properties of figures. In determining the directions in which we push development, we are continually guided by questions and experiences of the world in which we live. Our purpose in this chapter is to derive fundamental properties of combinatorial geometries, and to show how these properties strengthen our intuitive understandings of figures.

2.1 The definition of a combinatorial geometry

In this section, we give an axiom system for combinatorial geometries and then prove that each combinatorial figure is a combinatorial geometry.

Definition 2.1.1. *A* **combinatorial geometry G** *is an ordered pair* $(E, \mathcal{C})$ *consisting of a nonempty set* E*, whose elements are called points, and a collection* $\mathcal{C}$ *of subsets of* E *satisfying the following four conditions:*
Condition C1: *Each member of* $\mathcal{C}$ *is either a three-point set, a four-point set, or a five-point set.*
Condition C2: *If* C_1, C_2 *are members of* $\mathcal{C}$ *and* $C_1 \subseteq C_2$, *then* $C_1 = C_2$.

Condition C3: *(Elimination Condition) If C_1 and C_2 are distinct members of $\mathcal{C}$ and $e \in C_1 \cap C_2$, then there is a member C_3 of $\mathcal{C}$ such that $e \notin C_3 \subseteq C_1 \cup C_2$.*
Condition C4: *Each five-point subset of E contains a member of $\mathcal{C}$.*

If **G** is the combinatorial geometry $(E, \mathcal{C})$, then we say that **G** is a *geometry on E* and E is the *set of points of* **G**. We often write $E = E(\mathbf{G})$ and $\mathcal{C} = \mathcal{C}(G)$ and omit the adjective *combinatorial* so that "figure" stands for "combinatorial figure" and "geometry" stands for "combinatorial geometry". We check in detail that the collection of circuits of each figure (Definition 1.3.1) satisfies the requirements of the set $\mathcal{C}$ in Definition 2.1.1. This is the last appeal made to an intuitive proof in place of formal proof. Our intuitive understanding of our surroundings remains paramount in suggesting the properties of combinatorial geometries that we should investigate.

Theorem 2.1.1. *Each combinatorial figure is a combinatorial geometry.*

Proof. By their definition, the circuits of a combinatorial figure clearly satisfy Conditions **C1**, **C2**, and **C4**. We now investigate the truth of Condition **C3**. Suppose that C_1 and C_2 are distinct circuits of the figure and the point e is in $C_1 \cap C_2$. If $|C_1| = |C_2| = 3$, then the only three possibilities are shown in Figure 2.1. In each case $(C_1 \cup C_2) - \{e\}$ contains a three-point circuit or a four-point circuit (see Figure 1.15).

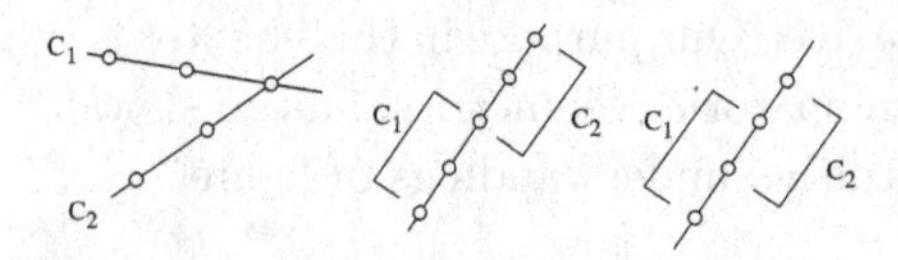

Fig. 2.1 Possible unions of two distinct 3-point circuits that meet

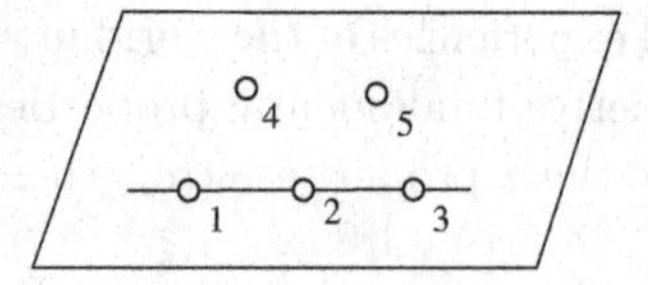

Fig. 2.2 A three-point circuit and a four-point circuit

If $|C_1| = 3$, $|C_2| = 4$, and $|C_1 \cap C_2| = 2$, then the unique possibility, in which all five points of $C_1 \cup C_2$ are coplanar, is shown in Figure 2.2. Here C_1 consists of the three collinear points and C_2 consists of four points with no three of the points being collinear. Then $(C_1 \cup C_2) - \{e\}$ contains either a three-point circuit or a four-point circuit.

If $|C_1| = |C_2| = 4$ and $|C_1 \cap C_2| = 3$, then again all five points of $C_1 \cup C_2$ are coplanar and, in each of the two possibilities, $(C_1 \cup C_2) - \{e\}$ is a set of four coplanar points and consequently contains a circuit.

In all other cases, $|(C_1 \cup C_2) - \{e\}| \geq 5$. Then Condition **C4** ensures that the set $(C_1 \cup C_2) - \{e\}$ contains a circuit. □

Definition 2.1.2. *Let* **G** *be a combinatorial geometry. We call each member of* $\mathcal{C}(G)$ *a* **circuit** *of the geometry.*

This definition and Definition 1.3.1 agree when **G** is a figure and so we continue to speak of the circuits of a figure without ambiguity. A **finite** combinatorial geometry is one whose set of points is a finite set.

2.2 Lines and planes of a combinatorial geometry

We now build on the foundation of Definition 2.1.1, deriving various useful properties of combinatorial geometries, knowing that each property will consequently hold for combinatorial figures. We start by making precise our intuitive understanding of lines and planes in a geometry.

Definition 2.2.1. *Let* **G** *be a combinatorial geometry. For any two points* $a, b \in E(\mathbf{G})$*, the set* $\{x \in E(\mathbf{G}) : \{a, b, x\} \in \mathcal{C}(G)\} \cup \{a, b\}$ *is a* **line** *of the geometry. We denote this set by* $a \vee b$ *and speak of "the line ab". A line is* **non-trivial** *if it contains at least three points. A line is trivial if it contains only two points.*

We see immediately what this means if the combinatorial geometry is a combinatorial figure. If a and b are any two points of the figure, then a third point x of the figure is on the straight line through a and b if and only if $\{a, b, x\}$ satisfies the requirements of Definition 1.3.1. This is equivalent to $\{a, b, x\}$ being a circuit of the geometry as mentioned above. Thus x is on the straight line containing a and b if and only if it belongs to the line $a \vee b$ specified by Definition 2.2.1. Thus any set consisting of all points of the figure that are on a straight line drawn through two points of the figure is a line of the figure, and conversely. It is a trivial line of the figure if it contains only the two points. It is a non-trivial line of the figure if it contains three or more points.

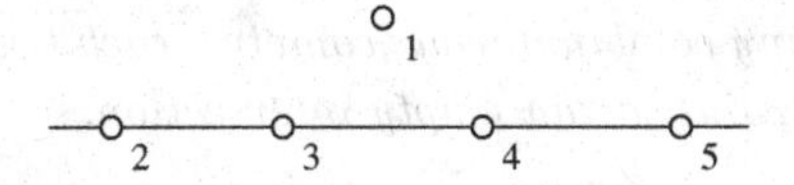

Fig. 2.3 A five-point figure

Example 2.2.1. The circuits of Figure 2.3 are the sets $\{2,3,4\}$, $\{2,3,5\}$, $\{2,4,5\}$, and $\{3,4,5\}$. Using Definition 2.2.1 we obtain the sets $\{1,2\}$, $\{1,3\}$, $\{1,4\}$, $\{1,5\}$, and $\{2,3,4,5\}$ as the lines of the figure. The line $2 \vee 3 = \{2,3,4,5\}$ is non-trivial, the other lines are trivial.

We also notice in Figure 2.3 that $2 \vee 3 = 2 \vee 4 = 2 \vee 5 = 3 \vee 4 = 3 \vee 5 = 4 \vee 5 = \{2,3,4,5\}$. The particular choice of the points used in Definition 2.2.1 does not seem critical. We now see that this is the case in general.

Theorem 2.2.1. *If x and y are any two points of a line L of a combinatorial geometry, then $L = x \vee y$. In other words, a line can be "named" by any two of its points.*

Proof. Let $L = a \vee b$ and c be any other point of L. It follows from $c \in a \vee b$ that $\{a,b,c\}$ is a circuit. We next show that $a \vee b = a \vee c$. Let $z \in a \vee b$. If z is a, b, or c, then $z \in a \vee c$ either by definition or because $\{a,b,c\}$ is a circuit. Suppose that z is distinct from these three elements. Then $\{a,b,z\}$ is a circuit. The Circuit Elimination Condition **C3** guarantees that $(\{a,b,c\} \cup \{a,b,z\}) - \{b\}$ contains a circuit. Thus $\{a,c,z\}$ is a circuit. Hence $z \in a \vee c$. It follows that $a \vee b \subset a \vee c$. It follows from interchanging the roles of b and c in the above argument that $a \vee c \subset a \vee b$. Thus $a \vee b = a \vee c$. Hence c can replace b as a naming point of L. It follows from $a \vee b = b \vee a$ that c can also replace a as a naming point for L. Thus x and y can successively be inserted as naming points for L, if they were not already used as naming points. □

Corollary 2.2.1. *In any combinatorial geometry, each two points belong to exactly one common line.*

Exercise 2.2.1. List all the lines of the first figure of Figure 1.3. List all the lines of the last figure of Figure 1.3.

We now prove that the intersection of any two lines is exactly as our intuition might lead us to expect.

Lemma 2.2.1. *In any combinatorial geometry, each two distinct lines have either one common point, or an empty intersection.*

Proof. If distinct points a and b both belong to lines A, B, then $A = a \vee b = B$. □

Example 2.2.2. Figure 2.4 contains three figures, each with point set labeled $\{1, 2, \ldots, 7\}$. If $A = 1 \vee 2$ and $B = 4 \vee 5$, then, in the first figure, $A \cap B = \{3\}$. But in each of the last two figures $A \cap B = \emptyset$.

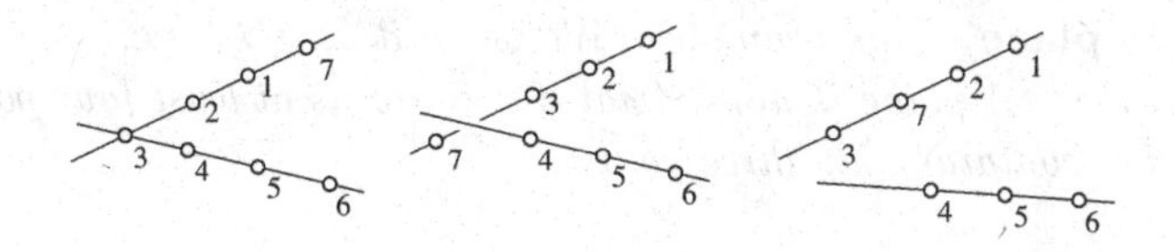

Fig. 2.4 Intersections of pairs of lines

Exercise 2.2.2. In the first figure in Figure 2.4, list the intersection of $1 \vee 2$ and $4 \vee 5$, of $1 \vee 2$ and $5 \vee 6$, of $4 \vee 6$ and $3 \vee 7$, of $1 \vee 5$ and $4 \vee 7$, and of $1 \vee 2$ and $2 \vee 3$.

Next we extend the meaning of linearity to combinatorial geometries. The following theorem characterizes linear sets in terms of 3-point circuits.

Definition 2.2.2. *A set of points of a combinatorial geometry is* **linear** *if it is contained in a line of the geometry or is a single point. The points of a linear set are said to be* **collinear**.

Theorem 2.2.2. *A set A of points of a combinatorial geometry is linear if and only if each three-point subset of A is a circuit.*

Proof. Suppose that A is linear. If $\{a, b, c\} \subseteq A$, then A is a subset of some line L. Then $L = a \vee b$. Hence $\{a, b, c\}$ is a circuit by Definition 2.2.1. Thus each three-point subset of A is a circuit.

Conversely, suppose that each three-point subset of A is a circuit. If A has fewer than three-points, then A is linear. If A has at least three-points, then, for fixed points a and b in A, and for all other points x of A, $\{a, b, x\}$ is a circuit. By Definition 2.2.1, $A \subseteq a \vee b$ so that A is linear. □

We have reduced the general question of whether a set of points is linear to questions only about three-point circuits. Our intuitive understanding of the circuits of a figure in Chapter 1 coincides with their precise meaning as circuits of a particular type of geometry. Theorem 2.2.2 ensures that, in the case of figures, the above definitions of linear and collinear coincide with our intuitive understanding of the terms. We next make precise our intuitive understanding of a plane of a combinatorial geometry.

Definition 2.2.3. *Let* $\mathbf{G}$ *be a combinatorial geometry. For any three non-collinear points* $a, b, c \in E(\mathbf{G})$, *the set* $\{x : x \in C \subseteq \{a, b, c, x\}, C \in \mathcal{C}(G)\} \cup \{a, b, c\}$ *is a* **plane** *of the geometry. We write it as* $a \vee b \vee c$, *and speak of "the plane abc". A plane is non-trivial if it contains at least four points. It is trivial if it contains only three points.*

Example 2.2.3. In Figure 2.5, the planes $1 \vee 2 \vee 3 = \{1, 2, 3, 5, 6\}$ and $1 \vee 3 \vee 4 = \{1, 3, 4, 6\}$ are non-trivial, while the plane $1 \vee 2 \vee 4 = \{1, 2, 4\}$ is trivial.

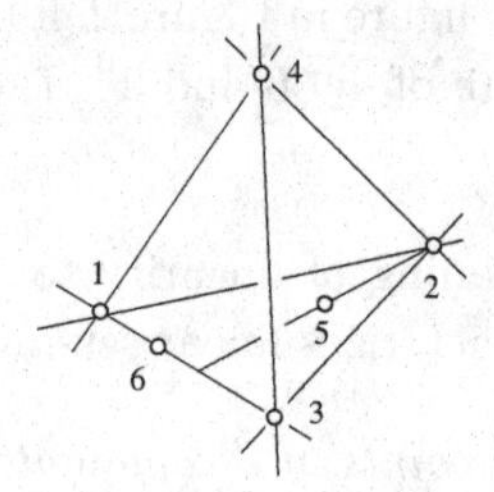

Fig. 2.5 A figure containing both trivial and non-trivial planes

If a, b, c were allowed to be collinear in Definition 2.2.3, then, by Condition **C2** of Definition 2.1.1, each circuit C satisfying $C \subseteq \{a, b, c, x\}$ would be a 3-point set. Hence $a \vee b \vee c$ would reduce to the line $a \vee b = b \vee c = c \vee a$.

We next prove that the choice of any three non-collinear points of a plane is adequate for its definition (for lines, see Theorem 2.2.1). The points a, b, and c are called "naming points" for a plane $a \vee b \vee c$ in a geometry .

Theorem 2.2.3. *If* x, y, *and* z *are any three non-collinear points of a plane* P *of a combinatorial geometry, then* $P = x \vee y \vee z$. *In other words, a plane can be "named" by any three of its points that are non-collinear.*

Proof. We claim that given three naming points for a plane, any point of the plane that is not on a line determined by two of the naming points may be used to replace the third naming point when describing the plane. Let $P = a \vee b \vee c$. The theorem will hold after the claim is proved as follows. There exists a line named by two of the elements of $\{a, b, c\}$ that does not contain x. Suppose that $b \vee c$ is this line without loss of generality. Then $a \vee b \vee c = x \vee b \vee c$. The point y is either not on $x \vee b$ or is not on $x \vee c$. Suppose the latter without loss of generality. Then $x \vee b \vee c = x \vee y \vee c$. Finally, z is not on $x \vee y$ so that $x \vee y \vee c = x \vee y \vee z$. Thus $P = a \vee b \vee c = x \vee y \vee z$.

Let d be a point of the plane P other than a that is not on $b \vee c$. We show that $a \vee b \vee c = b \vee c \vee d$ to complete the proof of the claim. It follows from $d \in (a \vee b \vee c) - (b \vee c)$ that there exists a circuit C contained in $\{a, b, c, d\}$ that contains both a and d. Hence all points of $\{a, b, c, d\}$ are contained in both $a \vee b \vee c$ and $b \vee c \vee d$. Let w be a point of $a \vee b \vee c$ other than a, b, c, or d. Then there exists a circuit D contained in $\{a, b, c, w\}$ that contains w. It follows from applying Condition **C3** that there exists a circuit E contained in $(C \cup D) - \{a\} \subset \{b, c, d, w\}$. Evidently, $w \in E$ as d is not on the line $b \vee c$. Thus $w \in b \vee c \vee d$. Hence $a \vee b \vee c \subset b \vee c \vee d$.

Let w be a point of $b \vee c \vee d$ other than a, b, c, or d. Then there exists a circuit F contained in $\{b, c, d, w\}$ that contains w. If $d \notin F$, then $w \in b \vee c \subset a \vee b \vee c$ which is what we wish to show. Suppose that $d \in F$. It follows from applying Condition **C3** that there exists a circuit G contained in $(C \cup F) - \{d\} \subset \{a, b, c, w\}$. Evidently, $w \in F$ as a, b, and c are not collinear. Thus $w \in a \vee b \vee c$. Hence $b \vee c \vee d \subset a \vee b \vee c$. Thus $a \vee b \vee c = b \vee c \vee d$ and the claim holds. □

Corollary 2.2.2. *In any combinatorial geometry, each three non-collinear points belong to exactly one plane.*

Example 2.2.4. In Figure 2.5, $1 \vee 2 \vee 3 = 1 \vee 2 \vee 5 = 1 \vee 2 \vee 6 = 1 \vee 3 \vee 5 = 1 \vee 5 \vee 6 = 2 \vee 3 \vee 5 = 2 \vee 3 \vee 6 = 3 \vee 5 \vee 6$, while $1 \vee 3 \vee 4 = 1 \vee 4 \vee 6 = 3 \vee 4 \vee 6$.

Exercise 2.2.3. List all the planes of the first figure in Figure 2.4. List all the planes of the second figure in Figure 2.4.

Lemma 2.2.2. *In a combinatorial geometry, if two points of a line are in a plane, then so are all points of the line.*

Proof. Let x and y be two points of a line that are in a plane $a \vee b \vee c$ of the geometry. There is some point z of $a \vee b \vee c$ not collinear with x and y, as otherwise $a, b, c \in x \vee y$. Thus $a \vee b \vee c = x \vee y \vee z$. Moreover, if $w \in x \vee y$, then either $w \in \{x, y\}$ or $\{x, y, w\}$ is a circuit. This guarantees that $w \in x \vee y \vee z$. Hence $x \vee y \subset x \vee y \vee z = a \vee b \vee c$. □

Lemma 2.2.3. *In any combinatorial geometry, let p be any point, and L any line not containing p. Then there is exactly one plane of the geometry containing both p and L. We denote this plane by $p \vee L$ or $L \vee p$.*

Proof. If $L = a \vee b$, then $P = a \vee b \vee p$. □

The next few results show that the intersections of pairs of planes and lines are satisfyingly simple.

Lemma 2.2.4. *In any combinatorial geometry, the intersection of two distinct planes is either a line of points, a single point, or is empty.*

Proof. If distinct points a and b each belong to both planes P and Q, then, from Lemma 2.2.2, $a \vee b \subseteq P \cap Q$. If a point $c \notin a \vee b$ is in both P and Q, then $P = a \vee b \vee c = Q$; a contradiction. Hence $a \vee b = P \cap Q$. □

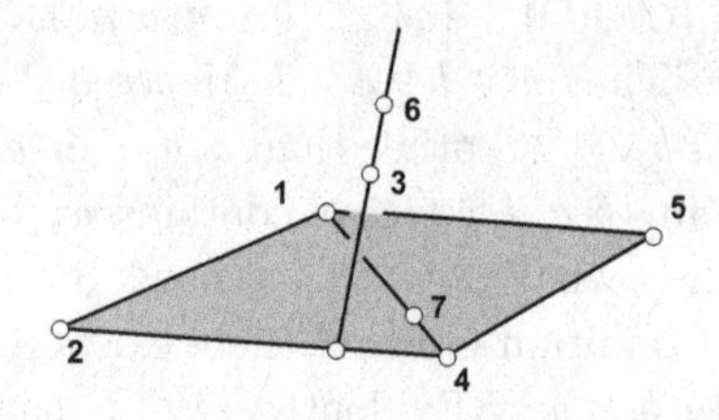

Fig. 2.6 Intersections of pairs of planes

Example 2.2.5. In Figure 2.6, if $P = 1 \vee 2 \vee 3$, $Q = 2 \vee 3 \vee 4$, and $R = 5 \vee 6 \vee 7$, then $P \cap Q = 2 \vee 3$, $P \cap R = \emptyset$, and $Q \cap R = \{6\}$.

Lemma 2.2.5. *In any combinatorial geometry, a plane and a line have the line itself, a single point, or the empty set as intersection.*

Proof. If a, b each belong to a line L and a plane P, then $L = a \vee b$ and Lemma 2.2.2 guarantees the result. □

We used our treatment of lines as a guide in discussing planes. The proofs were more complicated because of the need to consider both three- and four-point circuits, but the approach was the same. Just as we did for linearity, we extend the idea of planarity to combinatorial geometries and thereby characterize planar sets in terms of three- and four-point circuits.

Definition 2.2.4. *A set of points of a combinatorial geometry is* **planar** *if it is contained in a plane of the geometry, if it is linear, or if it is a singleton. The points of a planar set are said to be coplanar.*

Theorem 2.2.4. *In any combinatorial geometry, a set A of points is planar if and only if each four-point subset of A contains a circuit of the geometry.*

Proof. First suppose that A is planar. If $\{a, b, c, d\} \subseteq A$, then either $\{a, b, c\}$ is a three-point circuit or, by Theorem 2.2.3, $A \subseteq a \vee b \vee c$. In this second case, by Definition 2.2.3, the set $\{a, b, c, d\}$ also contains a circuit.

Conversely, we suppose that each four-point subset of A contains a circuit. If each three-point subset of A is a circuit, then, by Theorem 2.2.2, the set A is linear and is therefore planar. If A contains a subset $\{a, b, c\}$ of three non-collinear points, then, by assumption, each other element d in A belongs to a circuit contained in $\{a, b, c, d\}$ and so $A \subseteq a \vee b \vee c$. □

The agreement of our intuitive understanding of the circuits of a figure in Chapter 1 with their precise meaning as circuits of a particular type of geometry, in combination with Theorem 2.2.4, ensures that the above definitions of planar and coplanar coincide with our intuitive understanding of the terms in the case of figures.

Definition 2.2.5. *Two lines of a combinatorial geometry are* **skew** *if their union is not planar otherwise we say that they are coplanar.*

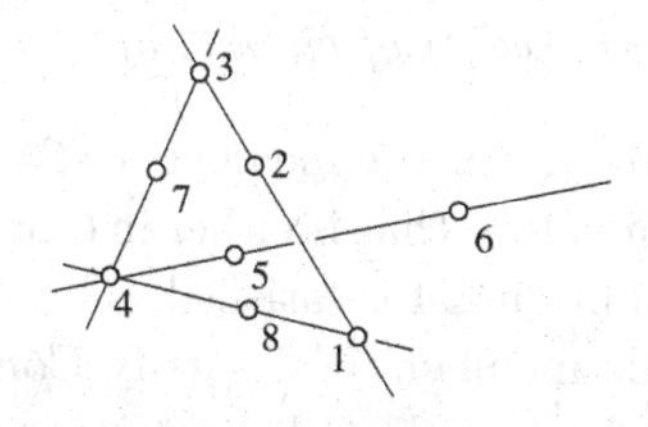

Fig. 2.7 Skew and coplanar pairs of lines

In Figure 2.7, the lines $1 \vee 2$ and $7 \vee 8$ are coplanar (as they belong to $1 \vee 3 \vee 4$), while $1 \vee 2$ and $4 \vee 5$ are skew (as 1245 does not contain a circuit).

Definition 2.2.6. *Let* $\mathbf{G}$ *be a combinatorial geometry. If* $E(\mathbf{G})$ *is either a single point or a line, then* $\mathcal{G}$ *is a* **linear geometry**. *If* $E(\mathbf{G})$ *is a single point, a line or a plane, then* $\mathbf{G}$ *is a planar geometry. In the only other possibility* $E(\mathbf{G})$ *is not planar, and* $\mathbf{G}$ *is a non-planar geometry.*

These terms agree with our previous intuitive use of them for any combinatorial geometry that is a figure. All examples of geometries thus far are figures. We would like to know whether there are geometries which are not figures - or whether our axiomatic description encompasses figures exactly.

Exercise 2.2.4. Let $E = \{1, 2, \ldots, n\}$ and $\mathcal{C}$ consist of all three-point subsets of E. Prove that **C** is the set of circuits of a linear geometry on E. If it is a figure, then we could call it a figure of n collinear points in general position.

2.3 Two families of combinatorial geometries

This section contains two related techniques for constructing examples of combinatorial geometries. They offer the intriguing possibility of the existence of geometries that are not figures. One technique is suggested by our verification of Theorem 2.1.1. The technique enables us to construct a planar geometry by merely choosing a collection of three-point sets whose pairwise intersections are small.

Proposition 2.3.1. *Let E be any nonempty set. If $\mathcal{K}$ is a collection of three-point subsets of E such that $|C_i \cap C_j| \leq 1$ whenever C_i and C_j are distinct members of $\mathcal{K}$, then the set $\mathcal{C} = \mathcal{K} \cup \{C \subseteq E : |C| = 4, C$ contains no member of $\mathcal{K}\}$ is the collection of circuits of a planar geometry on E.*

Proof. Condition **C1** holds for the members of $\mathcal{C}$. Each five-point subset of E contains a four-point subset that is either in $\mathcal{C}$ or contains a three-point member of $\mathcal{C}$. Hence Condition **C4** is satisfied. The choice of the four-point sets in $\mathcal{C}$ ensures that the members of $\mathcal{C}$ satisfy Condition **C2**. If C_1 and C_2 are two distinct members of $\mathcal{C}$, each containing the point e, then, for each of the possibilities for $|C_1|$ and $|C_2|$ we have $|(C_1 \cup C_2) - \{e\}| \geq 4$ and so $(C_1 \cup C_2) - \{e\}$ contains a four-point set which contains a member of $\mathcal{C}$. Thus $(E, \mathcal{C})$ satisfies all four conditions of Definition 2.1.1 and is therefore a geometry. Theorem 2.2.4 ensures that this geometry is planar. □

Exercise 2.3.1. Let $E = \{1, 2, \ldots, n\}$ and $\mathcal{C}$ consist of all four-point subsets of E. Prove that $\mathcal{C}$ is the set of circuits of a planar geometry on E. Is it a figure? If it were we could call it a figure of n coplanar points in general position.

Exercise 2.3.2. Let $E = \{0, 1, 2, \ldots, 6\}$ and let $\mathcal{K} = \{\{0, 1, 3\}, \{1, 2, 4\}, \{2, 3, 5\}, \{3, 4, 6\}, \{4, 5, 0\}, \{5, 6, 1\}, \{6, 0, 2\}\}$. Suppose that $\mathcal{C} = \mathcal{K} \cup \{C \subseteq$

$E : |C| = 4, C$ contains no member of $\mathcal{K}\}$. Prove that $\mathcal{C}$ is the set of circuits of a planar geometry on E.

Definition 2.3.1. *We call the geometry of Exercise 2.3.2 the* **Fano plane,** *or planar Fano geometry, after the Italian geometer Gino Fano.*

Even though we failed in our attempts at Construction 1.2.5, we did construct a figure containing all but one of the three-point circuits of the Fano plane. We now use Proposition 2.3.1 and the next proposition to show that both the failed and successful versions exist in the world of combinatorial geometries. Any example obtained by the method of Proposition 2.3.1 leads to more examples if we successively delete members of $\mathcal{K}$. After each deletion the reduced set is the collection of three-point circuits of a geometry. The following proposition specifies the circuits of this geometry.

Proposition 2.3.2. *Let E be any nonempty set and suppose that $\mathcal{K}$ is a collection of three-point subsets such that $|C_i \cap C_j| \leq 1$ whenever C_i and C_j are distinct members of $\mathcal{K}$. Then $\mathcal{C} = (\mathcal{K} - \{C_1\}) \cup \{C \subseteq E : |C| = 4,\ C$ contains no member of $\mathcal{K}\} \cup \{C_1 \cup \{x\} : x \in E - C_1\}$ is the collection of circuits of a planar geometry on E. We say that this geometry is obtained by relaxing the circuit C_1.*

Proof. The set $\mathcal{K} - \{C_1\}$ is a collection of three-point subsets $C_i \subseteq E$ such that $|C_i \cap C_j| \leq 1$, for all distinct i, j. It therefore satisfies the requirements of Proposition 2.3.1. Thus we need only prove that $\{C \subseteq E : |C| = 4,\ C$ contains no member of $\mathcal{K}\} \cup \{C_1 \cup \{x\} : x \in E - C\}$ consists exactly of the four-point subsets of E that do not contain any member of $\mathcal{K} - \{C_1\}$.

Suppose that C_0 is a four-point subset of E that contains no member of $\mathcal{K} - \{C_1\}$. Then either C_0 contains no member of $\mathbf{K}$ and so is in $\{C \subseteq E : |C| = 4, C$ contains no member of $\mathcal{K}\}$, or $C_0 \supseteq C_1$ and so C_0 is in $\{C_1 \cup \{x\} : x \in E - C_1\}$.

Conversely, suppose first that C_0 belongs to $\{C \subseteq E : |C| = 4, C$ contains no member of $\mathcal{K}\}$. Then either C_0 contains no member of $\mathcal{K}$ or $C_0 \supseteq C_1$ and C_0 is in $\{C_1 \cup \{x\} : x \in E - C_1\}$. Second suppose that C_0 were in $\{C_1 \cup \{x\} : x \in E - C_1\}$ but not in $\{C \subseteq E : |C| = 4,\ C$ contains no member of $\mathcal{K} - \{C_1\}\}$. In this case, C_0 would be equal to $C_1 \cup \{x\}$ for some x and would contain some member $C_i \neq C_1$ of $\mathcal{K}$. But then $|C_1 \cap C_i| > 1$, a contradiction. □

Definition 2.3.2. *The* **non-Fano** *plane is the planar geometry obtained from the Fano plane by relaxing the circuit $\{1, 5, 6\}$.*

We can be certain that the non-Fano plane is a figure, and we may suspect that the Fano plane is not, but we cannot yet prove this.

Definition 2.3.3. *Let* $E = \{0, 1, 2, \ldots, 9\}$ *and let* $\mathcal{K} = \{\{0, 1, 4\}, \{0, 2, 5\}, \{0, 3, 6\}, \{1, 2, 9\}, \{4, 5, 9\}, \{2, 3, 7\}, \{5, 6, 7\}, \{1, 3, 8\}, \{4, 6, 8\}, \{7, 8, 9\}\}$. *Then* $\mathcal{C} = \mathcal{K} \cup \{C \subseteq E : |C| = 4, C$ *contains no member of* $\mathcal{K}\}$ *is the collection of circuits of a planar geometry on* E. *We call this the* **planar Desargues geometry** *(see Figure 1.10).*

Definition 2.3.4. *The combinatorial geometry obtained from the planar Desargues geometry by relaxing the circuit* $\{7, 8, 9\}$ *is called the* **planar non-Desargues geometry**.

Our experiments in Chapter 1 lead us to the tentative hypothesis that the planar Desargues geometry is a figure and the planar non-Desargues figure is not a figure, but this is by no means certain. We can characterize the class of combinatorial geometries obtainable by the methods of Proposition 2.3.1 by an upper bound on the number of points in each line.

Lemma 2.3.1. *A combinatorial geometry may be obtained by the technique of Proposition 2.3.1 if and only if it is planar, and each line contains at most three points.*

Proof. First suppose that a geometry $\mathbf{G}$ is obtained using Proposition 2.3.1. Any four-point subset of $E(\mathbf{G})$ contains a circuit and Theorem 2.2.4 guarantees that $\mathbf{G}$ is planar. A point x, not in $\{a, b\}$, belongs to a line $a \vee b$ if and only if $\{a, b, x\}$ is a circuit. But, in any geometry obtained by the technique of Proposition 2.3.1, there is at most one three-point circuit containing both a and b. Therefore the line $a \vee b$ contains at most three points.

Conversely, suppose that $\mathbf{G}$ is planar and each line contains at most three points. Then no two three-point circuits of the geometry share more than one common point, and the collection of the three-point circuits of the geometry satisfies the requirements asked of $\mathcal{K}$ in Proposition 2.3.1. From Theorem 2.2.4, we have that each four-point subset of E not containing a three-point circuit is itself a circuit. □

If we use four-point sets in the role that three-point sets played in the discussions above, then we obtain similar results. The resulting construction technique generates examples of non-planar geometries.

Proposition 2.3.3. *Let E be any nonempty set. If $\mathcal{K}$ is a collection of four-point subsets of E such that $|C_i \cap C_j| \leq 2$ whenever C_i and C_j are distinct members of $\mathcal{K}$, then the set $\mathcal{C} = \mathcal{K} \cup \{C \subseteq E : |C| = 5, C$ contains no member of $\mathcal{K}\}$ is the collection of circuits of a combinatorial geometry on E. If E contains at least five points, then the combinatorial geometry is non-planar.*

Proof. As it was in proving Proposition 2.3.1, it is routine to verify that the choice of $\mathcal{C}$ satisfies the required conditions of Definition 2:1.1. Any five-point subset of E can contain at most one member of $\mathcal{K}$, ensuring that at least one four-point subset of E does not contain a circuit. From Theorem 2.2.4 we have that E is non-planar. □

Exercise 2.3.3. Let $E = \{1, 2, \ldots, n\}$, and $\mathcal{C}$ consist of all five-point subsets of E. Prove that $\mathcal{C}$ is the set of circuits of a geometry on E. Is it a figure? If so we could call it a figure of n points in general position.

Any example obtained by the method of Proposition 2.3.1 leads to more examples if we successively delete members of $\mathcal{K}$. After each deletion the reduced set is the collection of four-point circuits of a geometry. The following proposition specifies the circuits of this geometry. The proof is similar to that of Proposition 2.3.2 and we omit it.

Proposition 2.3.4. *Let E be any nonempty set and suppose that $\mathcal{K}$ is a collection of four-point subsets of E such that $|C_i \cap C_j| \leq 2$ whenever C_i and C_j are distinct members of $\mathcal{K}$. Then $\mathcal{C} = (\mathcal{K} - \{C_1\}) \cup \{C \subseteq E : |C| = 5, C$ contains no member of $\mathcal{K}\} \cup \{C_1 \cup \{x\} : x \in E - C_1\}$ is the collection of circuits of a combinatorial geometry on E. We say that this geometry is obtained by relaxing the circuit C_1.*

The following family of combinatorial geometries can be obtained in this way. We will see in Chapter 4 that a member of the family is possibly the most intriguing eight-point geometry of all.

Definition 2.3.5. *A* **combinatorial cube** *is a non-planar geometry on $E = \{0, 1, \ldots, 7\}$ that has no three-point circuits, that includes $\{0, 1, 3, 5\}$, $\{0, 2, 3, 6\}$, $\{1, 4, 5, 7\}$, and $\{2, 4, 6, 7\}$ among its circuits, and is such that no pair of its four-point circuits has a three-point intersection.*

Definition 2.3.6. *Let $E = \{0, 1, \ldots, 7\}$ and let $\mathcal{K} = \{\{0, 1, 3, 5\}$, $\{0, 2, 3, 6\}$, $\{1, 4, 5, 7\}$, $\{2, 4, 6, 7\}$, $\{0, 3, 4, 7\}\}$. Then $\mathcal{C} = \mathcal{K} \cup \{C \subseteq E :$*

$|C| = 5, C$ contains no member of $\mathcal{K}\}$ is the collection of circuits of a combinatorial cube on E. We call this geometry the **Vámos cube** *(named after the 20th Century algebraist P. Vámos).*

In a result similar to Lemma 2.3.1, we characterize the geometries obtainable by the methods of Proposition 2.3.3 by an upper bound on the number of points in each line and plane.

Lemma 2.3.2. *A combinatorial geometry may be obtained by the technique of Proposition 2.3.3 if and only if each plane of the geometry contains at most four points and each line of the geometry contains at most two points.*

The proof is similar to the proof of the Lemma 2.3.1, as a point $x \neq a, b, c$ belongs to a plane $a \vee b \vee c$ if and only if $\{a, b, c, x\}$ contains a circuit. We omit the details. We now have a plentiful supply of combinatorial geometries, but are still unsure whether or not they are all figures.

2.4 Sketches of planar geometries and figures

In this section, we return our attention to planar geometries. A drawing of a planar figure is actually a model of the figure. In other words, the planar figure exists if it can be drawn. Therefore if we attempt to draw a planar geometry, and succeed, we know that the geometry is a figure. But even if such an attempt fails it provides a useful visualization of the geometry in the form of a free-hand sketch, as we now see.

Definition 2.4.1. *A* **sketch** *is a collection of points and arcs drawn on the page such that no pair of the points belongs to more than one arc.*

Definition 2.4.2. *Let* **G** *be a planar geometry. A* **sketch** *of* **G** *is a sketch whose points may be labeled by the members of $E(\mathbf{G})$ in such a way that labels belong to a non-trivial line of* **G** *if and only if the points they label are on a common arc.*

We can now visualize any planar geometry from a sketch. We obtain a sketch of the geometry by labeling points of the page by points of the geometry and, for each non-trivial line of the geometry, drawing an arc through points labeled by the members of the line. As in Chapter 1, there is nothing to be gained by drawing an arc through the two points of any trivial line. Sketches of the planar Desargues geometry and planar non-Desargues geometry of Definitions 2.3.3 and 2.3.4, respectively, are in Figure 2.8.

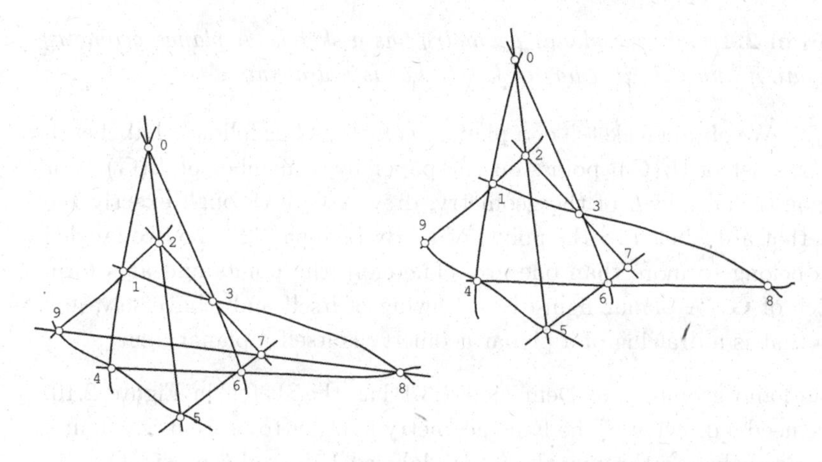

Fig. 2.8 The Desargues geometry and the non-Desargues geometry

The appearance of these sketches encourages us to make the following Definition.

Definition 2.4.3. *A sketch of a planar geometry, in which each of the arcs is a straight line, is a* **drawing** *of the geometry .*

This definition agrees with the original intuitive understanding of drawing that we developed for planar figures in Chapter 1. Each planar figure certainly has a drawing, but not every sketch of a planar figure need be a drawing. The first sketch of the geometry $(E, \mathcal{C})$, where $E = \{1, 2, 3, 4\}$, and $\mathcal{C} = \{\{1, 2, 3\}\}$, in Figure 2.9 is a drawing, but the second is not.

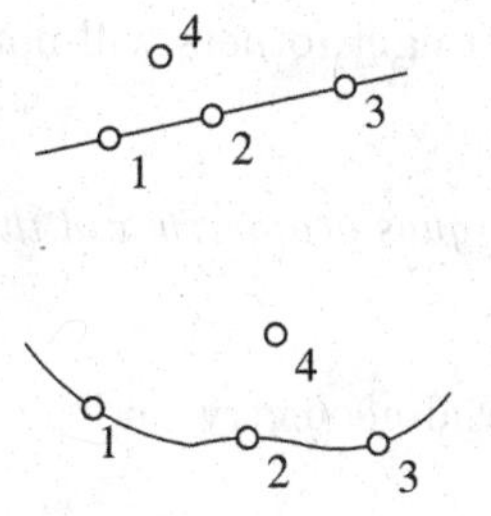

Fig. 2.9 Two sketches of the same geometry

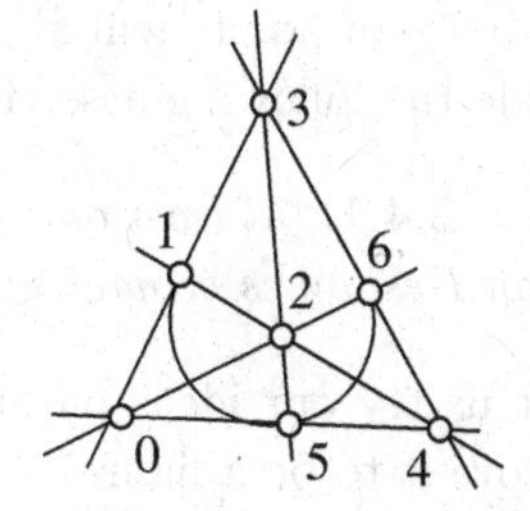

Fig. 2.10 A sketch of the planar Fano geometry

The following theorem sums up the role that sketches and drawings play in our study of planar geometries.

Theorem 2.4.1. *Every planar geometry has a sketch. A planar geometry is a figure if and only if it has a sketch that is a drawing.*

Proof. We obtain a sketch of a planar geometry $\mathbf{G}$ as follows. Label each point of a set of $|E(\mathbf{G})|$ points on the paper by a member of $E(\mathbf{G})$. For each non-trivial line L of the geometry, draw an arc through exactly the points that are labeled by the points of L. By Lemma 2.2.1, no two labeled points belong to more than one arc. Therefore the points and arcs form a sketch of $\mathbf{G}$. A planar figure is a drawing of itself and, conversely, any sketch that is a drawing of a planar geometry is itself a planar figure. □

The Fano geometry of Definition 2.3.1 has the sketch in Figure 2.10. But we need a drawing of the Fano geometry in order to prove it is a figure. If we erased the arc through the points labeled 1, 5, and 6 in Figure 2.10, then we would have a sketch of the non-Fano planar geometry obtained by relaxing the circuit $\{1,5,6\}$ of the Fano geometry. This particular sketch is a drawing as all its arcs are straight. We conclude that the non-Fano planar geometry is a figure.

In every-day life, our freehand drawings tend to be sketches rather than drawings. But we know that, in the case of figures, we could more carefully sketch them - using a straightedge - to give a drawing. We can certainly sketch both the planar Desargues and non-Desargues geometries, but the interesting question is whether or not we can draw them (with each arc a straight line)! When trying to construct a Desargues figure as in Lemma 1.2.2 or Construction 1.2.6, we see that it is only the last line of the construction whose existence is in doubt. We may need to use an arc rather than a line, giving a sketch in place of the hoped-for drawing. Perhaps some choice of points will allow the arc to be straight, others will not. We conclude the following assertion.

Lemma 2.4.1. *At least one of the planar Desargues geometry and the planar non-Desargues geometry is a figure.*

Let us try our ideas on another combinatorial geometry and see if we can prove it to be a figure.

Exercise 2.4.1. Let $E = \{0,1,\ldots,7\}$ and let $\mathcal{K} = \{\{0,1,3\},\{1,2,4\}, \{2,3,5\}, \{3,4,6\},\{4,5,7\}, \{5,6,0\}, \{6,7,1\}, \{7,0,2\}\}$. Verify that $\mathcal{C} =$

$\mathcal{K} \cup \{C \subseteq E : |C| = 4, C$ contains no member of $\mathcal{K}\}$ is the collection of circuits of a planar geometry on the set E. List the non-trivial lines of the geometry. Sketch it, using as many straight arcs as possible.

We have seen above that we are able to sketch any planar geometry. Conversely, we see in the following theorem that we can use a sketch to give us a geometry. In either case the sketch may not be a drawing as the geometry may not be a figure.

Theorem 2.4.2. *Any sketch is a sketch of a planar geometry. In such a sketch, each non-trivial line is the set of labels of points on an arc.*

Proof. We label the points of the sketch by elements of a set E and define $\mathcal{C}$ to be the collection of three-point sets that label points of an arc, together with each four-point subset of E that does not contain any of these three-point sets. It is straightforward to verify that the conditions of Definition 2.1.1 are met by $(E, \mathbf{C})$. □

We now have two ways of testing whether a pair $(E, \mathcal{C})$ is a planar geometry. The first way is to check that each of the four conditions of Definition 2.1.1 is satisfied. The second way is to attempt a sketch of the supposed geometry, use Theorem 2.4.2, and check that a three-point set is a member of $\mathbf{C}$ if and only if its members are on the same arc. In both cases we must also check that each four-point subset of E contains a member of $\mathcal{C}$ so that the geometry will be planar.

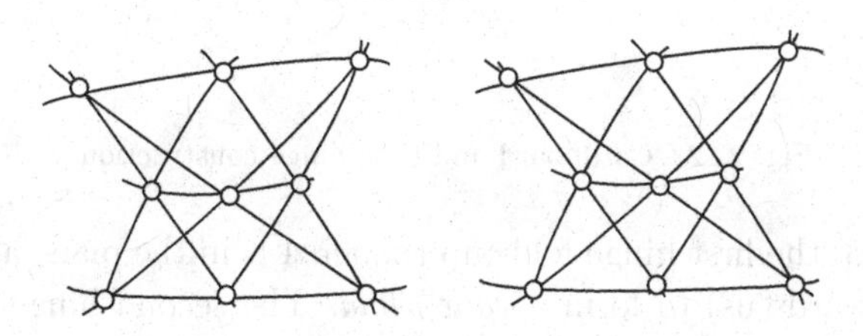

Fig. 2.11 A sketch of a Pappus geometry and a non-Pappus geometry

Exercise 2.4.2. Examine the sketches in Figure 2.11 and use Theorem 2.4.2 to show that each is a sketch of a planar geometry. Label the points of the first sketch in such a way that it is a sketch of a planar geometry having the collection $\{\{0,1,2\}, \{3,5,8\}, \{0,3,4\}, \{2,4,8\}, \{0,5,6\}, \{1,3,7\}, \{4,6,7\}, \{2,5,7\}, \{1,6,8\}\}$ as three-point circuits. Is the second sketch a sketch of a geometry obtained by relaxing a circuit of this geometry?

Definition 2.4.4. *We call the first combinatorial geometry sketched in Figure 2.11 the* **Pappus geometry**. *We call the second combinatorial geometry sketched in Figure 2.11 the* **non-Pappus geometry**.

We conclude from our attempts in Construction 1.2.7 that at least one of the Pappus geometry and the non-Pappus geometry is a figure.

2.5 Models of some combinatorial cubes

At the moment we have no method of determining whether a given non-planar geometry is a figure other than by trying to model it and hoping for a successful completion of the model. In this section, we give the details of a simple method of modeling combinatorial cubes. Cardboard models of many non-planar figures are useful, especially if they are made to fold up for ease of storage. Each piece of cardboard emphasizes the coplanarity of points on it. The procedure shown in Figure 2.12 seems a good way to join pairs of cardboard panels where necessary. It gives quite a good hinge made of adhesive tape which will open out flat and shut fully in one direction.

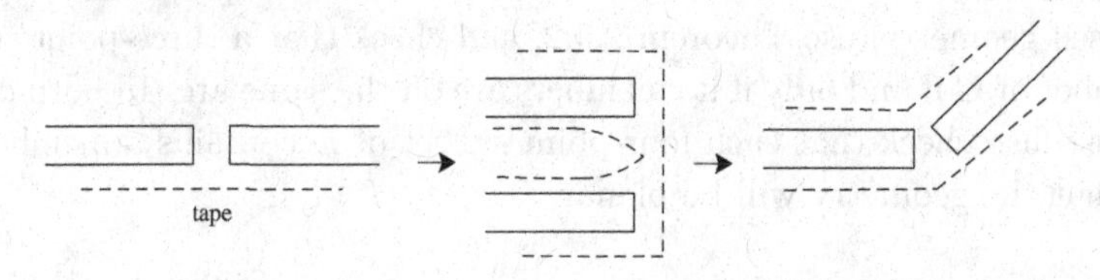

Fig. 2.12 Cardboard and tape-hinge construction

In Figure 2.13, the first hinge folds up (panel A is in the plane and panel B sticks out towards us) to form a *valley fold.* The second hinge folds down (panel A is in the plane and panel B sticks out behind the page) to form a *ridge fold.* If we do not want to emphasize which folding motion is desired to complete the model, we represent the hinge by a thick segment.

The above technique is quite useful for investigating the behaviour of models of various combinatorial cubes, as we see below.

Construction 2.5.1. There is a combinatorial cube on $E = \{0, 1, \ldots, 7\}$ that has only $\{0, 1, 3, 5\}$, $\{0, 2, 3, 6\}$, $\{1, 4, 5, 7\}$, and $\{2, 4, 6, 7\}$ as four-point circuits. It is a relaxation of the Vámos cube. If it is possible to do so, follow the instructions below and make a collapsible model of the cube.

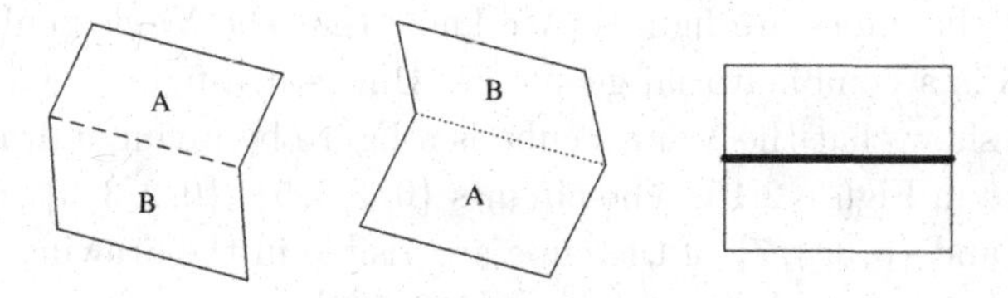

Fig. 2.13 Notation for hinge folding

Draw this model, using the shading techniques suggested in Exercise 1.3.1.

Tape four panels of cardboard together as in Figure 2.14. Make the lefthand and righthand panels a little oversize, fold the model into position and trim the lefthand edge until it butts nicely into the righthand panel. Then trim the righthand edge, leaving tabs, until the two edges butt together nicely. The tabs can then be clipped to the lefthand panel by paper clips. The model easily unfolds flat for storage and carrying.

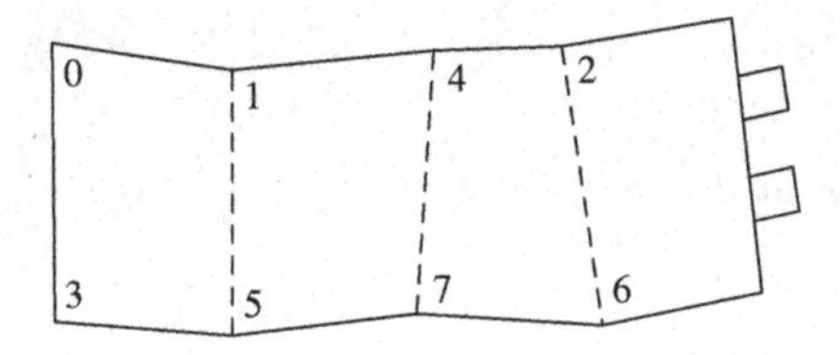

Fig. 2.14 Cardboard model of a combinatorial cube

Construction 2.5.2. If it is possible to do so, follow the instructions below and make a model of the cube on $E = \{0, 1, \ldots, 7\}$ that has only the six sets $\{0, 1, 3, 5\}$, $\{0, 2, 3, 6\}$, $\{1, 4, 5, 7\}$, $\{2, 4, 6, 7\}$, $\{0, 3, 4, 7\}$, and $\{1, 2, 5, 6\}$ as its four-point circuits.

Using four pieces of cardboard, as in the previous model, to represent planes containing $\{0, 1, 3, 5\}$, $\{0, 2, 3, 6\}$, $\{1, 4, 5, 7\}$, and $\{2, 4, 6, 7\}$, tape the joins along $1 \vee 5$, $2 \vee 6$, and $4 \vee 7$ and clip the join along $0 \vee 3$. Do you notice anything interesting about the behavior of this model compared with that of Construction 2.5.1?

The last two constructions have provided models of cubes with exactly four and six, respectively, 4-point circuits. The existence of these models means that the cubes are figures. We know that the Vámos cube of Definition 2.3.6 is a combinatorial geometry. But is it a figure? We may first attempt to show that the Vámos cube is a figure by giving a drawing such as the figure in Figure 2.15. The circuits $\{0,1,3,5\}$, $\{0,2,3,6\}$, $\{1,4,5,7\}$, $\{2,4,6,7\}$, and $\{0,3,4,7\}$ of the cube are visible in the drawing. However, we have to question whether the set $\{1,2,5,6\}$ is necessarily a circuit of the figure? If so, then the figure could not be of the Vámos cube which has only five 4−point circuits. We may second attempt to show that the Vámos cube is a figure by making a model of the Vámos cube, having exactly five 4-point circuits. It seems a simple enough question as to whether a drawing or construction of this cube can be made, but, as we will see in Chapter 4, it is perhaps the most fundamental question of geometry.

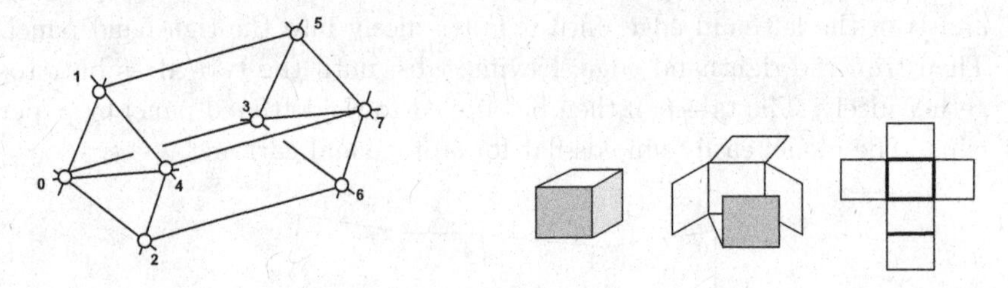

Fig. 2.15 An attempt at drawing the Vámos Cube

Fig. 2.16 A cube unfolded

Exercise 2.5.1. A well-known model of a cube is unfolded in Figure 2.16. Decide which of the six hinged cardboard models in Figure 2.17 would fold up to form a model of a cube.

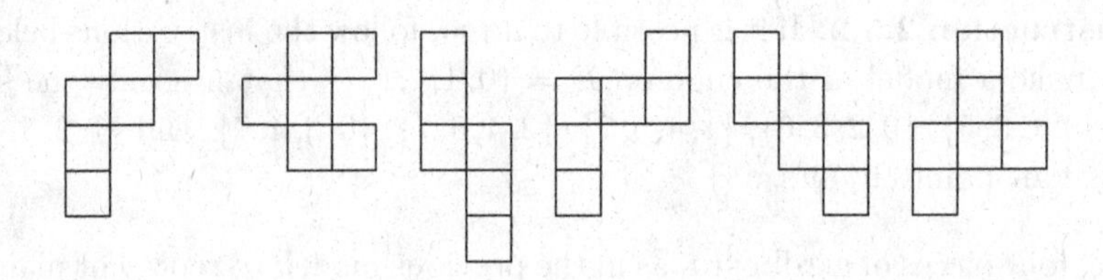

Fig. 2.17 Possible developments of a cube

Exercise 2.5.2. The model illustrated in Figure 2.16 seems much more like a cube as the word is understood in common usage. Is it in some sense a very symmetric model? What is the maximum number of four-point circuits that any combinatorial cube may have?

2.6 Subgeometries of a combinatorial geometry

We have constructed and investigated some geometries in as much detail as possible in an attempt to show that individual examples are figures. In this section we obtain a property of combinatorial geometries that distinguishes them from most other mathematical structures. It goes some way towards explaining why we have had little difficulty in constructing many examples. The lines and planes of a combinatorial geometry $\mathbf{G}$ are distinguished subsets of $E(\mathbf{G})$. But, as we now see, in a very strong sense every subset of $E(\mathbf{G})$ is distinguished by having its own geometric structure.

Definition 2.6.1. *A* **subgeometry** *of a combinatorial geometry* $\mathbf{G}$ *is any ordered pair consisting of a subset* $E' \subseteq E(\mathbf{G})$ *and associated collection* $\mathcal{C}' = \{C : C \in \mathcal{C}(G) \text{ and } C \subseteq E'\}$.

Theorem 2.6.1. *Each subgeometry of a combinatorial geometry is itself a combinatorial geometry.*

Proof. Let the ordered pair $(E', \mathbf{C}')$ be a subgeometry of $\mathbf{G}$. It is a straightforward matter to check that the conditions **C1**, **C2**, and **C4** of Definition 2.1.1 hold for $(E', \mathcal{C}')$. Let us verify that the Elimination Condition **C3** also holds for $(E', \mathbf{C}')$. Suppose that two distinct members C_1 and C_2 of $\mathcal{C}'$ both contain the point x. Then C_1 and C_2 are also in $\mathcal{C}$ and, as the Elimination Condition holds for $\mathbf{G}$, there is some $C_3 \in \mathcal{C}(G)$ satisfying $x \notin C_3 \subseteq C_1 \cup C_2$. But $C_1 \cup C_2 \subseteq E'$, ensuring that C_3 is also in $\mathbf{C}'$ as required. □

The circuits of any subgeometry $(E', \mathbf{C}')$ of the combinatorial geometry $\mathbf{G}$ are determined completely by the ground set E' of the subgeometry. Consequently, in the context of a given geometry $\mathbf{G}$, we often use E' rather than the more cumbersome $(E', \mathbf{C}')$ to stand for the subgeometry and write *the subgeometry* E'.

A nice property of a geometry is that each subset of its points inherits a geometric structure. When points of a drawing of a figure are erased a drawing of a figure remains. This suggests the following result.

Theorem 2.6.2. *Each subgeometry of a combinatorial figure is a combinatorial figure.*

Proof. The circuits of a subgeometry are exactly those of the figure obtained by erasing all points not in the subgeometry. □

Theorem 2.6.1 suggests that there are very many geometries. But we should be cautious. Just because every subset of the point set of a geometry has its own geometric structure does not mean that they are all different. We see this clearly by examining all the four-point subgeometries of a five-point line, for example. Also, our difficulties in determining whether or not drawings, sketches and models are really of "different" combinatorial geometries suggests the need to clarify the meaning of "different".

2.7 Isomorphic combinatorial geometries

In this section, we decide on the meaning of "different" in the context of combinatorial geometries. We then tackle the problem of counting geometries. In Figure 1.5 we noticed that the same figure can be drawn in several ways. It is not always clear whether two sketches represent the same planar geometry or not. We see this difficulty in the following case.

Definition 2.7.1. *An n-gon is* **inscribed** *in another if its vertices belong to successive sides of the second. This second n-gon* **circumscribes** *the first.*

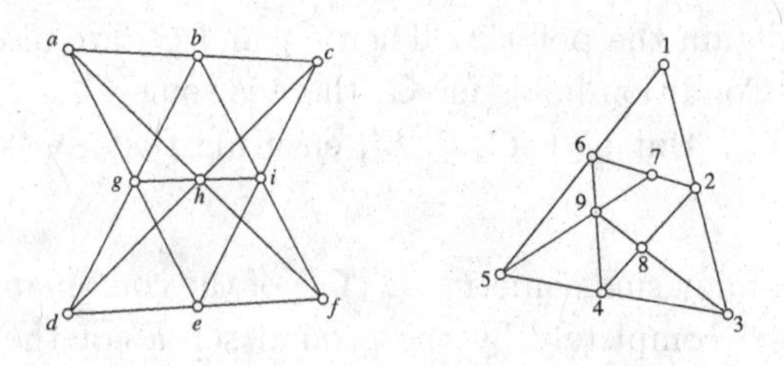

Fig. 2.18 Two figures that share many properties

Example 2.7.1. Show that each each planar geometry [20] sketched in Figure 2.18 consists of a triangle inscribed in a second triangle, which is inscribed in a third, which in turn is inscribed in the first! Show that these two geometries share other properties.

Are the two geometries of Figure 2.18 the same? Experiment may suggest that the first, a Pappus geometry, is a figure as every attempt to draw it seems to work. The second may also be a figure, the sketch given seems to be a drawing although experiment may suggest that it is difficult to make such a drawing.

We certainly know that two geometries are different if, for example, they have a different number of three-point circuits. In fact, any difference defined solely in terms of circuits suffices to distinguish between geometries. On the other hand we know that two drawings are drawings of the same figure if the points of the two can be commonly labeled to give identical lists of circuits. For us, such geometries will be the "same". More precisely:

Definition 2.7.2. *Two combinatorial geometries are* **isomorphic** *if the points of the first can be paired with the points of the second, and paired points given the same label, so that a list of circuits of the first combinatorial geometry is identical to a list of circuits of the second.*

Thus, in Exercise 1.1.1 we are asking: "Which pairs of these figures are isomorphic?" In the following exercise we use Definition 2.7.2 to determine whether geometries are isomorphic.

Exercise 2.7.1. Let $E_1 = \{1, 2, \ldots, 7\}$ and $\mathcal{C}_1 = \{\{5,6,7\}, \{1,2,4,5\}, \{1,3,4,6\}, \{2,3,5,6\}, \{2,3,5,7\}, \{2,3,6,7\}, \{1,2,3,4,7\}, \{1,3,4,5,7\}\}$. Let $E_2 = E_1$ and $\mathcal{K} = \{\{5,6,7\}, \{1,2,4,5\}, \{1,3,4,6\}, \{2,3,5,6\}, \{2,3,5,7\}, \{2,3,6,7\}\}$ and let $\mathcal{C}_2 = \mathcal{K} \cup \{C \subset E : |C| = 5, C \text{ contains no member of } \mathcal{K}\}$. Let $E_3 = \{1, 2, \ldots, 6\}$ and let $\mathcal{C}_3 = \{\{1,2,4,5\},\{1,3,4,6\},\{2,3,5,6\}\}$. Let $E_4 = \{a,b,c,d,e,f,g\}$ and $\mathcal{C}_4 = \{\{e,f,g\}, \{a,b,d,e\}, \{a,c,d,f\}, \{b,c,e,f\}, \{b,c,e,g\}, \{b,c,f,g\}, \{a,b,c,d,g\}, \{a,c,d,e,g\}\}$.

If $\mathbf{G}_i = (E_i, \mathcal{C}_i)$, for $i = 1,2,3,4$, and $\mathbf{G}_5$ and $\mathbf{G}_6$ are the first and second figures, respectively, drawn in Figure 2.19, then determine which pairs of the six geometries are isomorphic.

If a geometry is isomorphic to a figure, then from our point of view they are essentially the same. We blur the distinction between them, using the term "figure" for both. The geometry $(E, \mathcal{C})$, where $E = \{1,2,3,4\}$ and $\mathcal{C} = \{\{1,2,3\}\}$, and Figure 2.20 are isomorphic, we think of each as a figure.

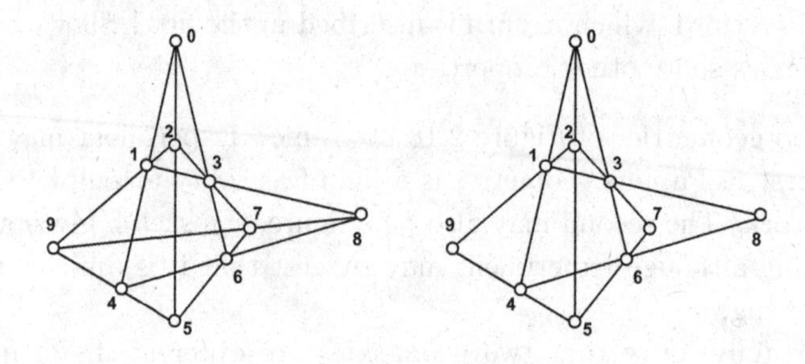

Fig. 2.19 Figures for Exercise 2.7.1

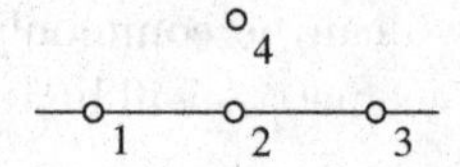

Fig. 2.20 A four-point figure

We look again at the question of whether a planar geometry **G** is a figure in the light of our definition of isomorphism. Pairing each point of a sketch of **G** with its label is certainly a pairing of a set of points of the page with the points of **G**. From Theorem 2.2.2, the pairing is an isomorphism between **G** and the figure of the points of the sketch exactly when the subsets of $E(\mathbf{G})$ that are collinear in **G** are the labels of collinear sets of points on the page. In other words, each arc of the sketch *could* have been drawn straight without re-locating any points - the sketch then being a drawing of **G**.

As we saw in Exercise 2.7.1, deciding whether or not two geometries are isomorphic can be a lengthy business. However, in the case of planar geometries the existence of an isomorphism is sometimes obvious.

Lemma 2.7.1. *Two planar geometries are isomorphic if and only if they have a common sketch.*

Proof. If the two are isomorphic, then each point set can use the same set of labels so that a sketch of one is automatically a sketch of the other. Conversely, any common sketch induces a pairing between labels which is the required isomorphism. □

Exercise 2.7.2. Satisfy yourself that there are only two non-isomorphic combinatorial geometries on a three-element set. One has no circuits and the other has one three-point circuit. We are familiar with both - the first is a triangle and the second is a figure of three collinear points. How many pairwise non-isomorphic four-point geometries are there? Is each a figure? Draw or make a model of each.

Exercise 2.7.3. List all the non-isomorphic combinatorial geometries on the set $E = \{1, 2, \ldots, 5\}$.

It might be interesting to make drawings or models of all those five-point geometries that are figures. We might be able to use tetrahedra and pins, as in Constructions 1.2.1 and 1.2.2, in making the models.

Construction 2.7.1. Show that all five-point (and six-point, if you feel so inclined) combinatorial geometries are combinatorial figures by making a model (a drawing in the case of a planar geometry) of each.

When making tetrahedra to use in this construction we could use the methods of Construction 1.2.1 or Construction 2.5.1. Another technique, which allows complete dismantling of panels for later re-use, is the following:

Construction 2.7.2. Decide on the size of each face of a tetrahedron, for example, and outline it on cardboard. A trouble-free way to choose face sizes is to have all edges the same length, say 6cm. As shown in Figure 2.21, use a hole-punch to make holes at each vertex. Trim as shown, leaving tabs. Crease along the edges of the face, and join adjacent panels by rubber bands.

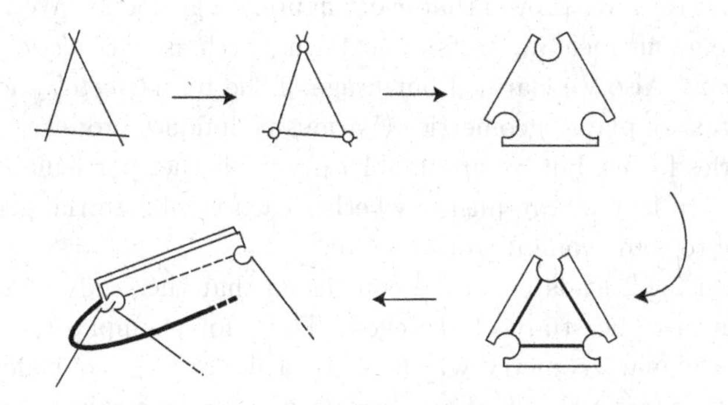

Fig. 2.21 Modeling planes and their intersections

Theorem 2.6.1 and Exercises 2.7.2 and 2.7.3 would possibly lead us to the conclusion that there are many geometries. Here is a list that summa-

Table 2.1 Number of geometries

Set size	point	linear	planar	non-planar	total
1	1	1	1		1
2		1	1		1
3		1	2		2
4		1	3	1	4
5		1	5	3	8
6		1	10	11	21
7		1	24	49	73
8		1	69	617	686
9		1	384	185,981	186,365

rizes what is known of the numbers of pairwise non-isomorphic combinatorial geometries on small sets of points. These numbers were computed for geometries on at most eight elements by Blackburn, Crapo, and Higgs [4], and for geometries on nine elements by Mayhew and Royle [47].

We have now faced with two difficult questions. We have discussed the first, namely: "How can we decide whether or not a given combinatorial geometry is a combinatorial figure?" at some length. For example we asked: "Is a Vámos cube a figure?" In this section, we have met the second: "How can we decide whether or not two combinatorial geometries are isomorphic?" For example, we asked: "Are the two geometries sketched in Figure 2.18 isomorphic?"

Summary of Chapter 2. We gave a simple set of axioms for combinatorial geometries. We proved that every figure is a geometry. We gave exact and unambiguous meaning to familiar terms, such as *line*, *plane*, *collinear*, and *coplanar*. Also we clarified our usage of the terms *sketch* and *drawing* in the context of planar geometries. Various techniques produced examples of geometries for us, but we are unable to say whether particular examples are figures. In fact we are unsure whether each combinatorial geometry is isomorphic to some combinatorial figure.

Theorem 2.4.2 leads us to the conclusion that the study of planar geometries is just the study of sketches. Thus, for example, the question: "Is there a planar geometry which is not a figure?" is equivalent to the question: "Is there a sketch of a planar geometry which cannot be redrawn with all its arcs straight?" We ended the chapter by defining isomorphism and counting small combinatorial geometries.

Chapter 3

PLANAR GEOMETRIES AND PROJECTIVE PLANES

In Chapter 2 we emphasized the role of lines and planes in geometries. We followed a logically sound development that lends weight to our feeling for "straightness" and "flatness" in the case of figures. If we had restricted our investigations to planar geometries, then we could have initially defined them in terms of points and lines - based on our intuitive understanding of points and straight lines. This is the traditional approach to plane geometry. Our approach using the simpler concept of circuit enables us to more easily describe the larger class of combinatorial geometries that interests us. Nevertheless, our use of straight lines in drawing suggests that we examine lines in some detail in this chapter. In particular, we ask: "How can we be certain that two straight lines drawn during a planar construction have a point of intersection?" It is therefore convenient to think of which planar geometries that this requirement leads to in terms of their lines. In Proposition 3.2.1 and Theorem 3.2.1, we list some properties of lines that do not need the notion of circuit for their statement. We prove that these properties suffice to determine whether a collection of subsets of a given set E is the collection of lines of some planar geometry $\mathbf{G}$ on E.

We may continue to prove the existence of some required geometry on the set E, as before, by demonstrating that a collection $\mathcal{C}$ of subsets of E satisfies the requirements of Definition 2.1.1. But instead, where appropriate, we may prove the existence of the geometry by demonstrating that another collection of subsets of E satisfies the requirements of Proposition 3.2.1 or Theorem 3.2.1. Thus the subsets will be the set of lines of a planar geometry on E.

3.1 The intersection of coplanar lines

In this section, we explore the properties of geometries in which each pair of coplanar lines has a point of intersection.

We have used pencil and straightedge to draw many planar figures in constructions - in Construction 1.2.1, we attempted to draw a straight line through an inaccessible point. We did so without thinking too much about the existence of this point, and of other points and straight lines required as part of the construction. More simply, given four points 1, 2, 3, and 4 on a very large page, constructing the intersection of $1 \vee 2$ and $3 \vee 4$ presents no great problem (apart from difficulties of parallelism). This is because our initial set of points sits in a world that seems to have points in abundance. But Lemma 2.2.1 does not guarantee that two lines of a combinatorial geometry will meet in a point, only that they have no more than one common point. The lines may still miss each other.

Perhaps it will clarify our thinking and give us a better understanding of our surroundings if we query our tacit assumptions. If we ask that two lines $1 \vee 2$ and $3 \vee 4$ have a common point, then we are insisting on the existence of a point 5 such that $\{1, 2, 5\}$ and $\{3, 4, 5\}$ are circuits. The Elimination Condition of Definition 2.1.1 applied to the two circuits tells us that $\{1, 2, 3, 4\}$ would then contain a circuit. In other words, two lines can only possibly meet if they are coplanar, not skew. We occasionally see builders putting this result into practice by drawing strings taut between two pairs of points. If the strings just touch it is taken to mean that the four points are coplanar.

Ideally, we would like any two straight lines of our planar constructions to have a common point. This would be guaranteed if our constructions were subgeometries of a planar geometry in which each pair of lines had a common point. Is a combinatorial geometry in which each two coplanar lines have a common point possible? The comment in the paragraph above suggests that we look for geometries satisfying the following condition. An equivalent formulation of this condition was given in Oswald Veblen and John Youngs' excellent book "*Projective Geometry*", which first appeared in 1910 [59].

Definition 3.1.1. **(Veblen-Young Condition)** *If* $\{1, 2, 3, 4\}$ *is a four-point circuit of the combinatorial geometry* $\mathbf{G}$*, then there is a point* 5 *of the geometry so that* $\{1, 2, 5\}$ *and* $\{3, 4, 5\}$ *are circuits.*

We see, by the following lemma, that this is exactly the condition we require of our geometries.

Lemma 3.1.1. *Let* **G** *be a combinatorial geometry. Then the Veblen-Young Condition holds in* **G** *if and only if each two coplanar lines of* **G** *contain a common point.*

Proof. First we suppose that the Veblen-Young Condition holds in **G**. If two lines $A = 1 \vee 2$ and $B = 3 \vee 4$ are coplanar, then Theorem 2.2.4 implies that the set $\{1, 2, 3, 4\}$ contains a circuit. If $\{1, 2, 3, 4\}$ is itself a circuit, then the Veblen-Young Condition guarantees the existence of a point 5 on both $1 \vee 2$ and $3 \vee 4$ as required. If, without loss of generality, $\{1, 2, 3\}$ is a circuit, then 3 is the point common to A and B.

Conversely, suppose that each two coplanar lines of **G** contain a common point. Let $\{1, 2, 3, 4\}$ be a circuit of **G**. Then the lines $1 \vee 2$ and $3 \vee 4$ are distinct and coplanar. Hence, they meet in a fifth point, say 5. Then Theorem 2.2.2 implies that the sets $\{1, 2, 5\}$ and $\{3, 4, 5\}$ are circuits as required by the Veblen-Young Condition. □

In what follows we pursue the attractive consequences of the truth of the Veblen-Young Condition in a planar geometry, and later show that these consequences are available to us in the geometry of our world.

3.2 Projective planes

In this section, we develop the basic properties of those planar geometries for which the Veblen-Young Condition holds. We begin by proving that the property obtained in Corollary 2.2.1 is characteristic of the collection of lines of a planar geometry.

Proposition 3.2.1. *Let E be a nonempty set and* **L** *a collection of subsets of E, each member of* **L** *containing at least two distinct elements of E, so that E and* **L** *satisfy the following condition.*
Condition L1: *Each pair of elements of E is a subset of exactly one member of* **L**.

Suppose that $\mathcal{C} = \{C : |C| = 3$ and C is a subset of a member of **L**$\} \cup \{C : |C| = 4$ *and no three elements of C belong to any one member of $L\}$. Then $\mathcal{C}$ is the set of circuits of a planar geometry on the set E. This geometry is the unique planar geometry on E whose collection of lines is exactly* **L**. *Conversely, the collection of lines of any planar geometry on E satisfies Condition* **L1**.

Proof. The definition of $\mathcal{C}$ is such that its members fulfill Conditions **C1**, **C2**, and **C4** of Definition 2.1.1. Suppose that C_1 and C_2 are distinct three-element members of $\mathcal{C}$ that both contain an element x. If C_1 and C_2 as well both contain an element y distinct from x, then Condition **L1** implies that they are subsets of the same member L of $\mathbf{L}$. Hence $(C_1 \cup C_2) - \{x\}$ is a three-element subset of L. This subset is in $\mathcal{C}$ as required by the Elimination Condition. In all other cases $|(C_1 \cup C_2) - \{x\}| \geq 4$ and $(C_1 \cup C_2) - \{x\}$ contains a member of $\mathcal{C}$, also satisfying the Elimination Condition. Thus $\mathcal{C}$ is the collection of circuits of a geometry on E. The geometry is planar, from Theorem 2.2.4, as each four-point set contains a circuit.

We need to show that the lines of $(E, \mathcal{C})$ are exactly the members of $\mathbf{L}$. We note from Condition **L1** that each pair $\{a, b\}$ of points is contained in exactly one $L \in \mathbf{L}$ and also, by Theorem 2.2.1, in exactly one line, namely $a \vee b$. Also $\{a, b, x\}$ is a circuit if and only if x is in $L - \{a, b\}$. Thus $L = \{x : x \in L\} \cup \{a, b\} = \{x : \{a, b, x\} \text{ is a circuit}\} \cup \{a, b\} = a \vee b$. Any planar geometry with this set of lines has, by Theorem 2.2.2 and Theorem 2.2.4, the same set of circuits. Thus such a geometry is equal to the geometry $(E, \mathcal{C})$.

From Theorem 2.2.1 and Lemma 2.2.1 we have that the lines of any planar geometry on E satisfy Condition **L1**. □

This proposition is a restatement of Theorem 2.4.2, but is not couched in the language of sketches. It cannot offer a characterization of all combinatorial geometries by their lines as, for instance, both the first and the last figures of Figure 1.3 have the same list of lines. However, these geometries are not isomorphic as the former is planar and the latter is not.

We now specialize the characterization of Proposition 3.2.1 to those planar geometries that satisfy an additional requirement.

Definition 3.2.1. *A* **projective plane** *is a planar geometry that satisfies the Veblen-Young Condition and has at least one four-point circuit.*

Proposition 3.2.1, together with the connection between lines and circuits of projective planes derived in Lemma 3.1.1, gives the following useful characterization of projective planes.

Theorem 3.2.1. *Let a set E, and a collection $\mathbf{L}$ of subsets of E, satisfy the following four conditions.*

L1: *Each two elements of E belong to exactly one common member of* $\mathbf{L}$.

L2: *Each two members of* **L** *contain exactly one common element of* E.

L3: *There are four elements of* E*, no three of which belong to any one member of* **L**.

L4: *There are four members of* **L***, no three of which contain any one element of* E.

Suppose that $\mathcal{C} = \{C : |C| = 3$ *and* C *is a subset of a member of* $\mathbf{L}\} \cup \{C : |C| = 4$ *and no three elements of* C *belong to any one member of* $L\}$*. Then* $\mathcal{C}$ *is the set of circuits of a planar geometry on the set* E*. This geometry is a projective plane, and its lines are exactly the members of* **L***. Conversely, the lines of any projective plane on* E *satisfy the four Conditions* **L1** *to* **L4**.

Proof. From Condition **L2**, we have that the empty set is not a member of **L**. If a single-element subset of E were a member of **L**, then Condition **L2** implies that every member of **L** contains this single element. This rules out the possibility that Condition **L4** holds. Therefore, each member of **L** contains at least two distinct elements of E.

We note that Condition **L1** is exactly Condition **L1** of Proposition 3.2.1. Therefore we have that $\mathcal{C}$ is the set of circuits of a unique planar geometry **G** on E.

Each pair of distinct points a and b of **G** belongs to exactly one line of **G** and to exactly one member of **L**. Each line and each member of **L** occur in this way. The defining property above of three-point circuits in $\mathcal{C}$ guarantees that $x \in (a \vee b) - \{a, b\}$ if and only if $\{a, b, x\} \in \mathcal{C}$ if and only if $x \in L$. Thus the unique member L of **L** that contains $\{a, b\}$ is also the line $a \vee b$. Thus the lines of **G** are exactly the members of **L**.

To complete the first half of the proof we need only show that **G** is a projective plane. Condition **L3** ensures the existence of at least one four-point circuit. Suppose that $\{1, 2, 3, 4\}$ is a circuit of **G**. Then, applying Condition **L2** to the lines $1 \vee 2$ and $3 \vee 4$ guarantees the existence of a point 5 belonging to both. Consequently, $\{1, 2, 5\}$ and $\{3, 4, 5\}$ are circuits and the Veblen-Young Condition holds. We now know that the geometry **G** is a projective plane.

To prove the converse, we use Proposition 3.2.1 to verify Condition **L1** and Lemma 3.1.1 to verify Condition **L2** for any projective plane on E. The existence of a four-point circuit guarantees that Condition **L3** is satisfied. If $\{1, 2, 3, 4\}$ is such a circuit, then lines $1 \vee 2$, $2 \vee 3$, $3 \vee 4$ and $1 \vee 4$ satisfy Condition **L4** as, for example, the point $(1 \vee 2) \cap (2 \vee 3) = 2$ is in neither

$3 \vee 4$ nor $1 \vee 4$ as no three members of $\{1, 2, 3, 4\}$ form a circuit. □

Exercise 3.2.1. The circuits of the geometry $(E, \mathcal{C})$ of Exercise 2.3.1 are exactly all the four-point subsets of E. Prove that $(E, \mathcal{C})$ is a planar geometry that fails badly to be a projective plane.

Example 3.2.1. We know from examining the sketch in Figure 2.10 that the sets of numbers in each of the columns of Figure 3.1 are the lines of a particular projective plane, namely the Fano plane, on the seven-point set $\{0, 1, \ldots, 6\}$.

0	1	2	3	4	5	6
1	2	3	4	5	6	0
3	4	5	6	0	1	2

Fig. 3.1 Seven columns of a projective plane

We could use Theorem 3.2.1 directly to verify that Example 3.2.1 does give a projective plane. However, testing Condition **L1** would involve $\binom{7}{2} = 21$ verifications, as would testing Condition **L2**. Likewise, applying Theorem 3.2.1 to the columns of Figure 3.2 would require $\binom{13}{2} + \binom{13}{2} = 156$ checks.

0	1	2	3	4	5	6	7	8	9	10	11	12
1	2	3	4	5	6	7	8	9	10	11	12	0
3	4	5	6	7	8	9	10	11	12	0	1	2
9	10	11	12	0	1	2	3	4	5	6	7	8

Fig. 3.2 Thirteen columns of a projective plane

Exercise 3.2.2. Label the points of Figure 3.3 by 0, 1, through 12 so that the thirteen columns of Figure 3.2 are the lines of the figure. Note that some of the lines of the sketch are curved, but all lines contain four points. Find and sketch a planar Desargues subgeometry in this geometry.

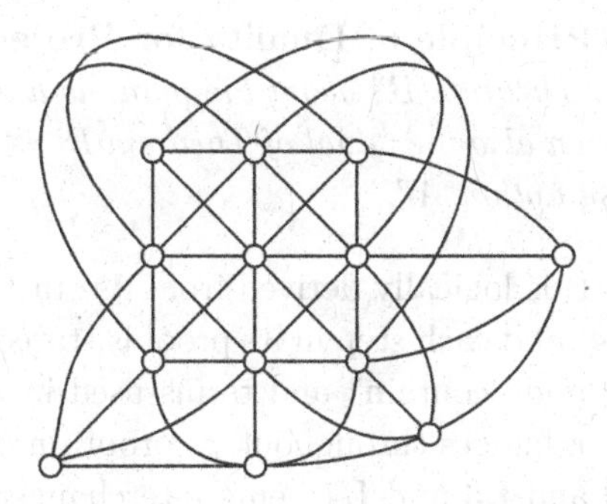

Fig. 3.3 A thirteen point projective plane

Exercise 3.2.3. Check a few of the conditions suggested by Theorem 3.2.1 to indicate whether the set $E = \{0, 1, 2, \ldots, 20\}$ is the set of points and the collection $\mathbf{L}$ of twenty one subsets, $\{\{0+i, 3+i, 5+i, 9+i, 11+i\} : i = 0, 1, 2, \ldots, 20$, addition being modulo $21\}$, of E is the set of lines, respectively, of a projective plane. Do the same with the collection $\{\{0+i, 3+i, 4+i, 9+i, 11+i\} : i = 0, 1, 2, \ldots 20$, addition being modulo $21\}$ in place of $\mathbf{L}$. If either seems to be a projective plane, then look for a familiar combinatorial geometry appearing among its subgeometries.

The symmetry of the characterization of projective planes given in Theorem 3.2.1 is of considerable use. The set of four conditions **L1** to **L4** is unchanged *in toto* by interchanging "point" with "line" and "belong to" with "contain", and this enables us to shorten the process of proving results about these geometries.

Definition 3.2.2. *The* **dual** *of any statement about points and lines of a projective plane is the statement obtained from the original statement by interchanging the words "point" with "line" and "belong to" with "contain".*

For example, the dual of the statement: " ... three points belong to a line" is the statement: " ... three lines contain a point". But we tend to abbreviate commonly used phrases - so " ... three collinear points ..." has its dual statement " ... three concurrent lines ...", "goes through" has dual "lie on", and so on. We need to be aware of dual pairs of abbreviations. Notice that the dual of "n-gon" is again "n-gon", with "side" and "vertex" being dual terms. In particular, "triangle" is a self-dual term. The dual of the dual of any statement is the original statement.

Theorem 3.2.2. (Principle of Duality for Projective Planes) *If, in a projective plane, "Theorem B" about the plane is a logical consequence of "Assumption A", then also the "dual of Theorem B" is a logical consequence of the "dual of Assumption A".*

Proof. Theorem B is logically derived from its starting assumptions and conditions **L1** to **L4**, and each step in its proof is stated in terms of "point", "line", "belong to" and "contain" and terms used in Assumption A. If we make the above interchanges throughout its proof (in particular **L1** and **L2** being interchanged and **L3** and **L4** being interchanged), we obtain a valid argument for the dual theorem from the dual starting assumption. □

The Principle of Duality implies that for every result we prove about a projective plane, we obtain the dual result without any extra work being required. This can be quite helpful, as we see in the following few applications. We begin by revisiting Conjecture 1.2.1.

The Veblen -Young Condition enables us to pursue the possibility of extending the Conjecture 1.2.1 to any combinatorial geometry **G**. Suppose that 123 and 456 are triangles in **G** so that the lines $1 \vee 4$, $2 \vee 5$, and $3 \vee 6$ are concurrent. Then $1 \vee 4$ and $2 \vee 5$ are coplanar, ensuring that $\{1, 2, 4, 5\}$ is a planar set. From Lemma 2.2.2, we have that $1 \vee 2$ and $4 \vee 5$ are coplanar. Further, Lemma 3.1.1 implies that these two lines intersect in a point. Similarly, each other pair of corresponding sides of the two triangles is also coplanar and so intersects in a point. If these three points are collinear, then we have a Desargues subgeometry of **G**; if not, then we have a non-Desargues subgeometry of **G**. Notice that it does not matter whether the triangles 123 and 456 are coplanar. Thus both planar and non-planar subgeometries may arise in this way.

Theorem 3.2.3. *(conjectured Desargues' Theorem) If* 123 *and* 456 *are triangles in a combinatorial geometry* **G** *so that the lines* $1 \vee 4$, $2 \vee 5$ *and* $3 \vee 6$ *are concurrent, then the sides* $1 \vee 2$ *and* $4 \vee 5$ *intersect in a point* 9*, the sides* $2 \vee 3$ *and* $5 \vee 6$ *intersect in a point* 7*, and the sides* $3 \vee 1$ *and* $6 \vee 4$ *intersect in a point* 8*, and the points* 7*,* 8 *and* 9 *are collinear.*

We have not proved this statement, and it is clearly not true for our usual Euclidean geometry. For example, the possibility of the lines $1 \vee 2$ and $4 \vee 5$ being parallel would rule out the existence of the required point 9.

We may use the Principal of Duality in order to obtain information about some consequences of Desargues' Theorem in projective planes.

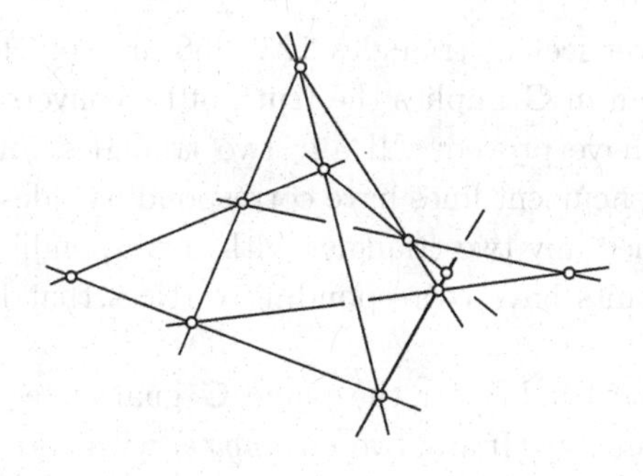

Fig. 3.4 Non-coplanar triangles whose corresponding sides intersect

Theorem 3.2.4. *Let* **G** *be a projective plane. Then Desargues' Theorem is true in* **G** *if and only if the converse of Desargues' Theorem is true in* **G**.

Proof. We may state Desargues' Theorem as: "Any two triangles with corresponding vertices lying on concurrent lines have corresponding sides that intersect in collinear points". The statement of the converse theorem is: "Any two triangles with corresponding sides intersecting in collinear points have corresponding vertices that lie on concurrent lines". We suppose that Desargues' Theorem holds in **G**. Consider two triangles 123, 456 having corresponding sides that intersect in points 7, 8, and 9, respectively, - these three points being collinear.

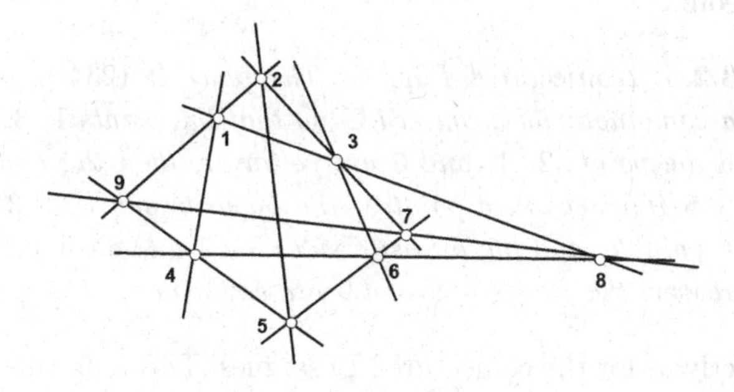

Fig. 3.5 The converse of Desargues' Theorem

We are able to apply Desargues' Theorem to triangles 275 and 184 as the three lines through corresponding vertices are concurrent at 9. Therefore 3, 6, and the point $(1 \vee 4) \cap (2 \vee 5)$ are collinear, that is, the lines through pairs of corresponding vertices of triangles 123, 456 are concurrent. Therefore Desargues' Theorem in **G** implies the truth of its converse in **G**.

Explicitly, we have proved: "If any two triangles with corresponding vertices lying on concurrent lines have corresponding sides that intersect in collinear points, then any two triangles with corresponding sides intersecting in collinear points have corresponding vertices that lie on concurrent lines".

The Principle of Duality for the plane **G** guarantees the truth of the dual statement, namely: "If any two (*triangles*) with corresponding (*sides*) (*intersecting in*) (*collinear*) (*points*) also have corresponding (*vertices*) that (*lie on*) (*concurrent*) (*lines*), then any two (*triangles*) with corresponding (*vertices*) (*lying on*) (*concurrent*) (*lines*) also have corresponding (*sides*) that (*intersect in*) (*collinear*) (*points*)". This just happens to be the required converse of the first implication. □

In the light of the above discussion we may also revisit Conjecture 1.2.2. Again using the Veblen -Young Condition, we pursue the possibility of extending the conjecture to any projective plane **G**. Suppose that 123456 is a planar hexagon in **G** so that the points 1, 3, and 5 are collinear and the points 2, 4, and 6 are collinear. Then each pair of opposite sides is coplanar and, from Lemma 3.1.1, meet in a point. If these points are collinear, then the hexagon has given rise to a Pappus subgeometry of **G**. If these points are not collinear, then the hexagon has given rise to a non-Pappus subgeometry of **G**. This generalized conjecture is historically known as Pappus' Theorem:

Theorem 3.2.5. *(conjectured Pappus' Theorem) If* 123456 *is a planar hexagon in a combinatorial geometry* **G** *so that the points* 1, 3, *and* 5 *are collinear and the points* 2, 4, *and* 6 *are collinear, then the opposite sides* $1 \vee 2$ *and* $4 \vee 5$ *intersect in a point* 9, *the opposite sides* $2 \vee 3$ *and* $5 \vee 6$ *intersect in a point* 7, *and the opposite sides* $3 \vee 4$ *and* $6 \vee 1$ *intersect in a point* 8. *Moreover, the points* 7, 8 *and* 9 *are collinear.*

But, exactly as for the conjectured Desargues' Theorem, this statement may not be true for a geometry **G**. The Principal of Duality gives some information about its consequences in projective planes.

We may rephrase the conjectured theorem as: "Any hexagon whose vertices lie alternately on two lines has the points of intersection of pairs of opposite sides collinear." The dual statement is: "Any hexagon whose sides pass alternately through two points has the lines joining pairs of opposite vertices concurrent."

In Figure 3.6 we have sketched a planar geometry that satisfies the requirements of this dual statement. We remark that we do not know whether this geometry, which we call the "dual Pappus geometry", is a figure. But we next prove that a projective plane for which Pappus' Theorem is valid contains dual Pappus subgeometries but no dual non-Pappus subgeometry.

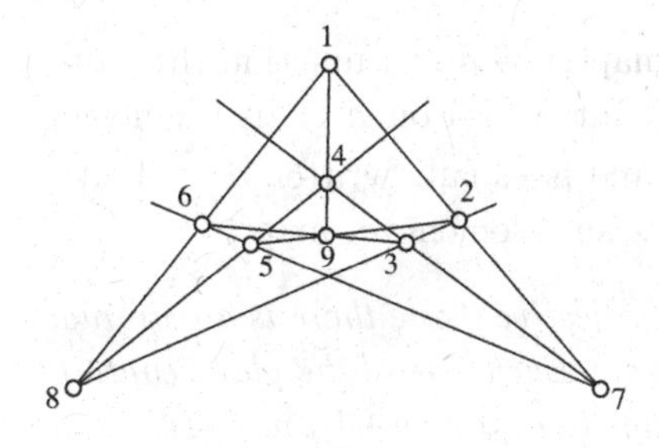

Fig. 3.6 A sketch of a dual Pappus geometry

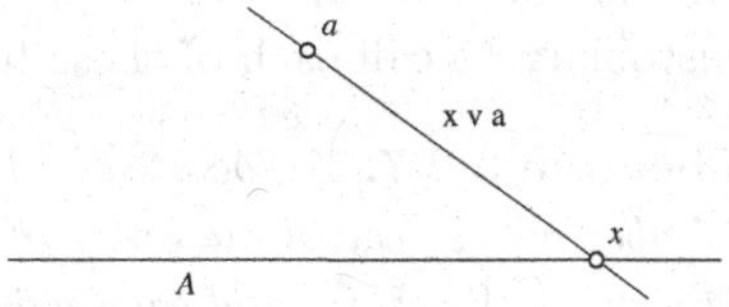

Fig. 3.7 An elementary map

Theorem 3.2.6. *Let* **G** *be a projective plane. Then Pappus' Theorem is true in* **G** *if and only if the following condition is satisfied. If* 123456 *is a hexagon in the plane with lines* $1 \vee 2$, $3 \vee 4$ *and* $5 \vee 6$ *concurrent, and lines* $2 \vee 3$, $4 \vee 5$ *and* $6 \vee 1$ *concurrent, then the lines* $1 \vee 4$, $2 \vee 5$ *and* $3 \vee 6$ *are concurrent.*

Proof. Suppose that Pappus' Theorem is valid in **G**. Consider any hexagon 123456 so that alternate sides are concurrent with the points 7 and 8 respectively as shown in Figure 3.6. Applying Pappus' Theorem to the hexagon 148367 we have that the three lines through pairs of opposite sides of the hexagon 123456 are concurrent as required. The dual of this argument completes the proof of the theorem. □

In any projective plane there are obvious, but surprisingly useful, pairings of lines with points. This self-dual notion, traditionally called an elementary map, leads to a strong limitation on the number of points in any projective plane.

Definition 3.2.3. *In a projective plane let* A *be a line and* a *be a point not in* A*. An* **elementary map** *is a function associating each point* x *of* A *with the line* $x \vee a$ *through* a*. It is usually written* $A \overset{el}{\leftrightarrow} a$.

Lemma 3.2.1. *The elementary map $A \overset{el}{\leftrightarrow} a$ is a bijection between the points on A and the lines through a.*

Proof. Distinct points x and y in A have images $x \vee a$ and $y \vee a$, respectively. If $x \vee a = y \vee a$, then $a \in x \vee y = A$, a contradiction. So the function is one-to-one. If the line X contains a, then it is the image of the point $X \cap A$. This ensures that the function is onto. □

The dual notion of the elementary map $A \overset{el}{\leftrightarrow} a$ is a function that associates each line X through a with the point $X \cap A$ on A. It is the inverse function of the elementary map $A \overset{el}{\leftrightarrow} a$ and is usually written $a \overset{el}{\leftrightarrow} A$. It is customary to call each of these functions an "elementary map".

Theorem 3.2.7. *Associated with each projective plane there is a (cardinal) number $n \geq 2$, called the order of the plane. Each line of the plane contains exactly $n+1$ points, and each point belongs to exactly $n+1$ lines. The plane has exactly $n^2 + n + 1$ points, and exactly $n^2 + n + 1$ lines.*

Proof. Let A, B be any two distinct lines of the projective plane $\mathbf{G}$. The existence of a four-point circuit in $\mathbf{G}$ is guaranteed by Definition 3.2.1. Among the pairwise intersections of the six lines determined by pairs of points of the circuit is a point p on neither A nor B. Then consider the composition $A \overset{el}{\leftrightarrow} p \overset{el}{\leftrightarrow} B$ of a bijection from the points of A to the lines through p and of a bijection from the lines through p to the points on B. This composition is a bijection from the points of A to the points of B. This bijection ensures that A and B contain the same number, say α, of points. Hence each line of $\mathbf{G}$ contains the same number of points.

The dual of this argument guarantees that each two points belong to the same number, say β, of lines. The existence of the elementary map $p \overset{el}{\leftrightarrow} A$ ensures that $\alpha = \beta$.

Let $n = \alpha - 1 = \beta - 1$ and suppose that x is a point of $\mathbf{G}$, distinct from p. Then $p \vee x$ is a line of $\mathbf{G}$. Consequently each point of $\mathbf{G}$ is on a line through p. There are $\beta = n+1$ such lines. Each line contains n points distinct from p, and none of these points is on more than one of the $n+1$ lines. Therefore the plane contains $(n+1)n+1$ points altogether. The dual of this argument completes the proof. □

Exercise 3.2.4. Verify that each of the two projective planes in Example 3.2.1 and Exercise 3.2.2, respectively, has the appropriate number of lines and points.

Example 3.2.2. Inspection of the sketch in Figure 2.10 verifies that the Fano plane satisfies the requirements of Theorem 3.2.1 and is a projective plane. Inspection confirms that each of its lines contains exactly three points, and each of its points belongs to exactly three lines.

Exercise 3.2.5. Write out explicitly the proof that a projective plane of order n has exactly $n^2 + n + 1$ lines. Check that the proof is the dual of part of the proof of Theorem 3.2.7.

As we might expect, because of more restrictive criteria, there are far fewer projective planes than there are geometries. There is a projective plane of order n, for each $n = 2, 3, 4$, and 5, with 7, 13, 21, and 31 points, respectively. Gaston Tarry proved in 1900 that no projective plane of order 6 exists. There is a projective plane of order n, for each $n = 7, 8$, and 9. In 1988 a group at the University of Montreal proved, by eliminating all possibilities with a computer search, that there is no projective plane of order 10. It has been known, for almost one hundred years, that there is a projective plane of each prime power order. The famous 1949 Theorem of R.H. Bruck and H.J. Ryser [5] states that if $n = 4k + 1$ or $n = 4k + 2$, for some positive integer k, and n is not a sum of two squares, then there is no projective plane of order n. This eliminates $n = 14$ as a possibility - but is of no help for $n = 12$. Thus the smallest order for which the question of the existence of a projective plane is unanswered is 12. Such a plane would have 157 points. The existence questions for finite projective planes are more difficult then for infinite projective planes as induction arguments - such as that used in Theorem 3.3.1 - are not available in the finite case.

Problem 3.2.1. *Some people form a lunch club in order to enjoy good food and conversation. They decide that each two members should go to exactly one lunch together; that each two lunches should have some overlap in attendance; and that the President, Secretary, Treasurer, and Auditor should attend the first lunch gathering (which at least a couple of other members were unable to attend). There are fifteen prospective members. Can the Secretary arrange a list of lunches? If not, then what must he do?*

Solution. Suppose that the Secretary has arranged a list of lunches. We write E for the set of members of the club. Then those attending a lunch together form a subset of E, and we denote the collection of these subsets of E by $\mathbf{L}$. We investigate whether or not $\mathbf{L}$ is the collection of lines of a projective plane on E. As two members attend exactly one common lunch together, Condition **L1** of Theorem 3.2.1 holds. But we are not sure about Condition **L2**, as we only are certain that two members of **L** contain at *least* one common member. But suppose that two different members a and b each attend lunches L_1 and L_2. Condition **L1** guarantees $L_1 = L_2$. Therefore Condition **L2** holds.

It seems that the conditions given in Theorem 3.2.1 to define the characteristics of a set of lines of a projective plane are unnecessarily restrictive, as a weaker version of Condition **L2** would suffice to prove the condition in Theorem 3.2.1. We gave it in this stronger form to enable us to more easily see the workings of the Principle of Duality.

In order to be able to apply Theorem 3.2.7 to the Lunch Club we must show Conditions **L3** and **L4** hold. We know that at least two members, say a and b, are unable to attend the first lunch. The only lunch that a and b both attend has one member of the first lunch also present. Without loss of generality, we may suppose that it is neither the Secretary nor Treasurer. We have that the set {Secretary, Treasurer, a, b} satisfies Condition **L3** as neither a nor b can attend the only lunch that the Secretary and Treasurer both attend. The four lunches attended by both a and b, by both b and the Secretary, by both the Secretary and the Treasurer, and by both the Treasurer and b, respectively, satisfy Condition **L4**.

Therefore for the Secretary to succeed in his task, the list of attendees at each lunch would need to be the lines of a projective plane. One, at least, of these lines would contain at least four points. Therefore the plane would have order at least $4-1=3$ and at least thirteen members would be required. The Secretary should either discourage two of those interested, or he should search for prospective members, needing six more than the original fifteen. Actually finding a projective plane of order 3 or 4 in order to arrange the lunch attendances, may be a little difficult. Perhaps Exercise 3.2.3 or the list of columns of Figure 3.2 would be helpful for the Secretary.

The following deceptively simple result [63] concerning the gregarious nature of our species hinges on deducing that a collection of subsets of people at a party satisfies Conditions **L1** and **L2**, but not Condition **L3**, of Theorem 3.2.1. The simplest known proof requires the use of eigenvalues of a matrix.

Theorem 3.2.8. *In a party of n people, suppose that every pair of people has exactly one common friend. Then there is a person at the party who is a friend of everyone else.*

3.3 Subgeometries of a projective plane

But back to our original reason for looking at projective planes; we wondered if the straight lines of planar constructions can be expected to intersect? This leads us to ask the following question: Is it possible for our planar figures to be subgeometries of a projective plane? Yes, in fact as we show in this section, any planar geometry is a subgeometry of a projective plane.

Theorem 3.3.1. *Each planar geometry is a subgeometry of some projective plane.*

Proof. We prove this by constructing a suitable projective plane point-by-point inductively. Suppose that $\mathbf{G}$ is a projective plane. If $E(\mathbf{G})$ does not include a four-point circuit, then we define $E_1 = E(\mathbf{G}) \dot{\cup} \{\{1,2,3,4\}\}$, and $\mathcal{C}_1 = \mathbf{C}(\mathbf{G}) \cup \{$ all four-point subsets of E_1 not containing a three-point member of $\mathcal{C}\}$. Otherwise put $(E_1, \mathcal{C}_1) = \mathbf{G}$ and suppose that, for the natural number n, a planar geometry $(E_n, \mathcal{C}_n)$ is defined. We list the pairs of lines that have an empty intersection in this geometry, and put $E_{n+1} = E_n \dot{\cup} x_n$, where x_n is a new point (or symbol) added to each of the first pair L_1, L_2 of lines that have an empty intersection in $(E_n, \mathbf{C_n})$. Also, for each $a \notin L_1 \cup L_2$, we add $\{a, x_n\}$ to the collection of lines of E_n. Then Proposition 3.2.1 ensures that $L_1 \cup \{x_n\}$, $L_2 \cup \{x_n\}$, the sets $\{a, x_n\}$, together with all lines of $(E_n, \mathcal{C}_n)$ excluding L_1, L_2 are the lines of a planar geometry $(E_{n+1}, \mathcal{C}_{n+1})$. It may be helpful to think of a sketch of $(E_{n+1}, \mathcal{C}_{n+1})$ being obtained from a sketch of $(E_n, \mathcal{C}_n)$ by adding a common point representing x to both an arc representing L_1 and to an arc representing L_2. We have inductively defined a geometry $(E_n, \mathcal{C}_n)$ for each natural number n, with $E_1 \subseteq \ldots \subseteq E_n \subseteq \ldots$.

Let E be the set $\bigcup_{n=1}^{\infty} E_n = E_1 \dot{\cup} \{x_1, x_2, \ldots\}$ and let $\mathbf{L}$ be the collection of those subsets of E that each consist of a line of E_n, for some natural number n, together with all points added to it in each E_m, $m \geq n$. Then E and $\mathbf{L}$ satisfy the four conditions of Theorem 3.2.1 - and so define a projective plane that has $\mathbf{G}$ as a subgeometry. □

This proof is an example of an induction argument. Things to be done are listed, the list is attended to one item at a time and, even though the list is allowed to grow, each item listed is attended to.

Definition 3.3.1. *The projective geometry of the above theorem is called the* **free completion** *of the planar geometry* $\mathbf{G}$ *and is written* $F(\mathbf{G})$.

Exercise 3.3.1. Prove that $F(\mathbf{G}) = \mathbf{G}$ if and only if $\mathbf{G}$ is a projective plane.

How well-behaved is the free completion of $\mathbf{G}$? In order to find out we extend Definition 1.2.2 to include geometries.

Definition 3.3.2. *A combinatorial geometry is* **closed** *if each of its points belongs to at least three of its non-trivial lines.*

Lemma 3.3.1. *Let* $\mathbf{G}$ *be a planar geometry. Then each closed subgeometry of the free completion* $F(\mathbf{G})$ *of* $\mathbf{G}$ *is also a subgeometry of* $\mathbf{G}$ *itself.*

Proof. We argue as in Lemma 1.2.1. Suppose that X is a closed subgeometry of $F(\mathbf{G})$, and suppose further that x is the last point of X that is added during the construction of $F(\mathbf{G})$. Suppose the point x were in, say, $E_n - E_{n-1}$. Then X would also be a subgeometry of $(E_n, \mathcal{C}_n)$. Distinct lines of X would belong to distinct lines of $(E_n, \mathcal{C}_n)$. But x would belong to only two lines of $(E_n, \mathcal{C}_n)$, contradicting the requirement that it is in at least three lines of the closed subgeometry X of $(E_n, \mathbf{C}_n)$. Thus $X \subseteq E_1 = E(\mathbf{G})$. □

The last result implies that anything of interest in the free completion of $\mathbf{G}$ is already contained in $\mathbf{G}$ itself. For example, no Desargues geometry would appear as a subgeometry during the construction of $F(\mathbf{G})$ unless it were already in $\mathbf{G}$. This causes Desargues' Theorem to fail very badly for free completions generally. In particular, any attempt to add to a planar figure in order to construct a Desargues figure would be doomed to failure if the projective plane in which it was attempted were a free completion of the initial figure.

But we notice that a planar geometry may be a subgeometry of different projective planes. For example, any four-point circuit C can be thought of as a subgeometry of the Fano plane, or of the free completion $F(C)$. If $\mathbf{G}$ is not itself a projective plane, but does contain a four-point circuit, then it can be shown that the free completion $F(\mathbf{G})$ contains infinitely many points. R. P. Dilworth and others conjecture that, if $E(\mathbf{G})$ is a finite set, it might be possible to find a finite projective plane containing $\mathbf{G}$ as a subgeometry. It is not known if this can always be done.

Our motivation for studying projective planes is to enable us to be confident that the lines of our world have a common point whenever possible - to enable us to carry out constructions of planar figures, in particular, using pencil and straightedge. We established, in Theorem 3.3.1, that it is certainly possible to consider any planar figure as a subgeometry of a projective plane. But will it be sensible to do so? How do the properties of projective planes obtained as a consequence of Theorem 3.2.1 agree with our experience of the world around us? We wear a Euclidean straight-jacket of parallel lines and it will not be easy to cast it off, just as it would be difficult for fish in a pond to imagine other surroundings. Our minds are too "immersed" in their Euclidean pond. But we will try.

3.4 An extended Euclidean plane

In this section, we efficiently embed each planar figure as a subgeometry of a nicely behaved projective plane. Until now we have agreed that we see and understand our shared world in similar but unspecified ways. Each of us has his or her own meaning for the words which describe the varied facets of this world. We need few of these words to talk of the things which have interested us in our discussion so far of figures. *Point*, *straight line*, *plane*, *collinear*, and *coplanar* virtually exhaust the list of these words.

Some of our confidence that we do share an understanding is based on our long exposure at school and college to a common grounding of Euclidean geometry. This is a particular, and very appropriate, mathematical model of our physical surroundings. We are taught how to think of points, of lines, of parallel pairs of straight lines, of planes, of distance, and so on.

We first make sure that we are in agreement about our intuitive understanding of the set of all points of a plane of Euclidean geometry. If you are happy with the notion of parallel straight lines, then you and I can proceed, for the moment, with our exploration of the world in terms of the combinatorial geometries of Definition 2.1.1. In later chapters we will make use of other properties of a Euclidean plane, but for the moment this will suffice.

If, on the other hand, you are uncertain of your understanding of the concept of a Euclidean plane, then we can make it more precise by introducing Cartesian coordinates . We define a Cartesian point to be an ordered pair (x, y) of real numbers. We define the Cartesian plane to be the set of all such pairs, together with the associated properties induced by the known behaviour of the real number system. Further, we define a Cartesian line to be any set of Cartesian points of the form $\{(x, y) : y = mx + c$, for some pair m, c of real numbers $\}$, or $\{(a, y) : y$ any real number, for some real number

$a\}$. We use the name $y = mx + c$ for the first set and the name $y = a$ for the second set. We define two lines $y = m_1x + c_1$ and $y = m_2x + c_2$ as parallel if $m_1 = m_2$. Any two lines $x = a_1$ and $x = a_2$ are also defined to be parallel. The simple arithmetic of simultaneous equations tells us that pairs of parallel lines are exactly those pairs of lines which have no common point - each pair of non-parallel lines sharing a point. It is straightforward to show that the Cartesian plane satisfies the requirements of Proposition 3.2.1. Therefore the Cartesian plane is a planar geometry.

It is common practice to take the Cartesian points as labels of Euclidean points, allocated by drawing axes in the usual way as in Figure 3.8.

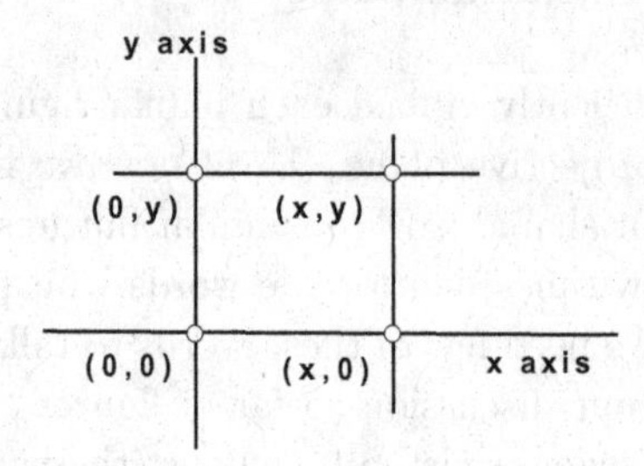

Fig. 3.8 Cartesian coordinate axes

It is our understanding, based on much experience, that this pairing of Euclidean points with their Cartesian labels, usually called "Cartesian coordinates", is an isomorphism between a Euclidean plane and the Cartesian plane. If this is the case, then the two are essentially "the same". This gives us the opportunity to bring the forces of algebra to bear on our geometric problems, and conversely gives us the opportunity to gain insight into algebraic questions via geometry.

So strong is our belief in this ismorphism that we blur the distinction between the two, talking of "the Euclidean point (x, y)" rather than "the Euclidean point with label (x, y)". The computer drawing program that you used in Construction 1.2.6, and elsewhere, treats each point as a pair of real numbers. The program uses the arithmetic of the real number system to obtain its lines and points, even though the appearance of its screen does not give this impression.

The question of the appropriateness of Euclidean geometry as a description of our world is thus bound up with the corresponding question of the suitability of the real number system for our arithmetic. Translating our geometric questions into algebraic terms is no magic solution. This translation gives equally difficult questions about the structure of the real numbers. But it is another useful tool in our kit-bag.

We turn now to the construction of a projective plane that contains all the figures in any given Euclidean plane. We denote the Euclidean plane by E^2. Is it a projective plane? Certainly, Conditions **L1**, **L3**, and **L4** of Theorem 3.2.1 hold for the collection of lines, and Condition **L2** almost holds. Condition **L2** fails only for pairs of parallel straight lines. Perhaps we could enlarge E^2 a little so that the enlarged geometry satisfies Condition **L2** and is a projective plane. If we add a new element to the set of points of E^2, and add the element also to each of two parallel straight lines, then the two lines would then satisfy Condition **L2**.

Definition 3.4.1. *Let L be a straight line of E^2. Consider all straight lines of E^2 parallel to L. We choose a symbol, and call it the* **direction**, *or* **ideal point**, *or* **point at infinity**, *of each of these straight lines.*

We add this symbol to the points of E^2, and we repeat the process, defining a new symbol for each family of parallel straight lines (see Figure 3.9). Let E be the set consisting of all the Euclidean points of E^2 together with all the directions defined in this way.

We next enlarge each straight line L, and each straight line parallel to L, by the inclusion of their direction. This symbol is not added to any other line.

Definition 3.4.2. *Let L be a straight line of E^2. The set obtained by enlarging the set L of Euclidean points by the inclusion of the direction of L is an* **extended Euclidean line**. *We call it the extension of L.*

Each extended Euclidean line is a subset of E, extensions of two non-parallel straight lines intersect in a Euclidean point, and extensions of two parallel straight lines intersect in a direction. Therefore the set E together with the collection of extended Euclidean lines satisfies Condition **L2** of Theorem 3.2.1.

Conditions **L3** and **L4** are also satisfied, and condition **L1** almost holds. There is clearly exactly one extended Euclidean line containing any two Euclidean points a and b. Likewise, if a is a Euclidean point and b is a direction, then there is a unique extended Euclidean line through a in the direction b. But there is no extended Euclidean line containing two ideal points. We can remedy this defect by adding one more line. This line consists exactly of the set of all directions to our proposed collection of lines.

Definition 3.4.3. *The* **ideal line,** *or* **line at infinity,** *of E^2 is the set of all directions of E^2.*

Now we check that these are indeed the points and lines of a projective plane.

Proposition 3.4.1. *Let E^2 be any Euclidean plane. Let E be the set consisting of the Euclidean points and the directions of E^2. If $\mathbf{L}$ is the collection of subsets of E consisting of all extended Euclidean lines of E^2 and the ideal line of E^2, then $\mathbf{L}$ is the collection of lines of a projective plane on E.*

Proof. The unique extended Euclidean line that contains two distinct Euclidean points a and b is the extension of the straight line that contains a and b. The ideal line contains neither a nor b. So a and b belong to exactly one common member of $\mathbf{L}$. The unique extended Euclidean line containing a and a direction c is the extension of the unique straight line through a in the direction c. Again, the ideal line does not contain a. So a and c belong to exactly one common member of $\mathbf{L}$. Finally, we note that no extended Euclidean line contains two directions, but that the ideal line does. We have shown that the collection $\mathbf{L}$ satisfies Condition **L1** of Theorem 3.2.1.

Two non-parallel straight lines have extensions whose intersection is exactly the common Euclidean point of the straight lines. The extensions of two parallel straight lines each contain a common direction, but the extensions have no other element in common. Any extended Euclidean line shares one direction with the ideal line. Consequently, $\mathbf{L}$ satisfies Condition **L2** of Theorem 3.2.1.

Any four-point circuit $\{1, 2, 3, 4\}$ of E^2 satisfies Condition **L3**. The circuit determines the Euclidean lines $1 \vee 2$, $2 \vee 3$, $3 \vee 4$ and $1 \vee 4$ whose extensions satisfy Condition **L4**. □

Definition 3.4.4. *We call the projective plane of Proposition 3.4.1 the* **extended Euclidean plane** *obtained from the Euclidean plane E^2 and denote it by EE^2.*

Lemma 3.4.1. *The set of circuits of EE^2 consists of each collinear three-point subset of Euclidean points, each three-point set consisting of two Euclidean points and the direction of the straight line through the two, each* *three-point set of directions, and each four-point set not containing any of these three-point sets.*

Proof. The list of circuits comes directly from Theorem 3.2.1. □

Theorem 3.4.1. *The Euclidean plane E^2 is a subgeometry of the extended Euclidean plane EE^2.*

Proof. The circuits of the subgeometry on the Euclidean points of E^2 specified by Lemma 3.4.1 are those that we agreed on in Definition 1.3.1 as the circuits of the euclidean plane E^2. Therefore the two geometries are one. □

Now we return to our discussion at the beginning of this chapter, asking ourselves whether a Euclidean plane, or an extended Euclidean plane, is the more appropriate "surrounding" for our paper - that is for the drawings of Chapter 1.

The choice is between two planar geometries. The first choice is a Euclidean plane E^2. The advantages of choosing E^2 are our familiarity with it and the simplicity of drawing using just a pencil and a straightedge (that is, an unmarked ruler). Against these advantages we have the disadvantage of two straight lines in our attempted drawings not necessarily containing a common point. The second choice is an extended Euclidean plane EE^2. The advantage of choosing it is the certainty that any point of intersection called for in a construction will exist. The choice is between *affine geometry* and *projective geometry* (where the special nature of parallel lines is overcome). We are choosing the "pond" in which we work and think.

For the moment we will select EE^2 as the natural home of planar figures. We will look back from time to time to see if this choice is paying off. So from now on we extend the meaning of *planar figure* given intuitively in Chapter 1 to include all subgeometries of EE^2 - not just those contained in E^2. In other words, we allow drawings that may include directions.

Definition 3.4.5. *A* **planar figure** *is a subgeometry of some extended Euclidean plane.*

It is true, although we will not prove it here, that any finite planar figure in this extended sense can be drawn entirely in E^2, that is, it is also a figure in the original sense.

We call any set of extended Euclidean lines of EE^2 that have a common direction *parallel*. In the context of an extended Euclidean plane their common direction is a point of the geometry. Thus extended Euclidean lines are parallel if and only if they are extensions of parallel straight lines. Thus we may talk of "parallel lines" without confusion. The notation $A||B$ continues to mean that the lines A and B are parallel. As is customary, we indicate this in drawings by drawing the same number of arrow-heads on each of A and B. Thus in Figure 3.9 the first two lines are parallel, as are the last two.

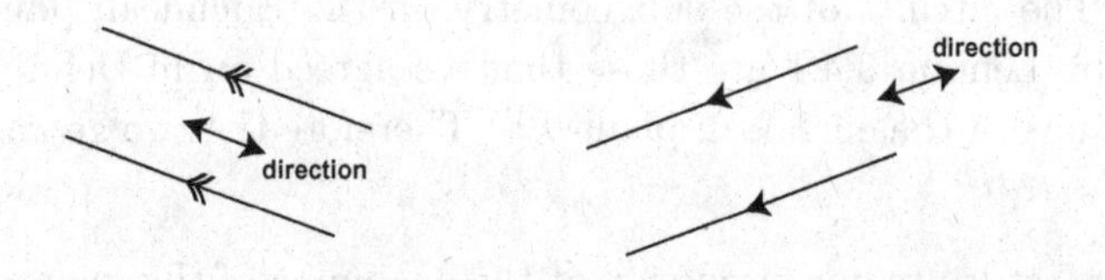

Fig. 3.9 A notation for directions and parallel lines

We represent the direction of a line by a double-headed arrow, and the ideal line by the symbol L_∞. Thus in Figure 3.9 the first direction belongs to each of the first pair of lines, and the second direction belongs to the second pair. We use these notations in a drawing of any figure that includes ideal elements. For example, five of the figures of Figure 3.10 contain directions.

Example 3.4.1. Figure 3.10 contains three 3-point circuits, three triangles and two pentagons.

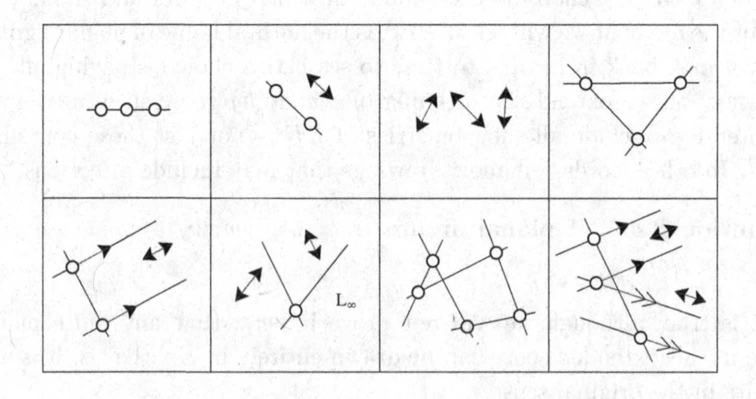

Fig. 3.10 Examples of figures in an extended Euclidean plane

Construction 3.4.1. Draw four examples of 4-gons in EE^2 with, respectively, no-, one-, two adjacent-, and two opposite-, vertices ideal.

We want to take advantage of guaranteed points of intersection of pairs of lines. We can now ask, with assurance, for the intersection of any pair of lines of a planar figure. But how do we carry out constructions quickly and reliably?

3.5 Pencil and roller constructions

What recipes or constructions can we carry out - and what tools do we need to carry them out - in an extended Euclidean plane? In this section, we answer these questions. A hint can be obtained from the following result, traditionally known as "Playfair's Axiom". The axiom was introduced in an 1860 textbook on geometry authored by the English teacher John Playfair (1748-1819). This textbook is significant for including a translation of the first six books of Euclid's "Elements" into the English language .

Lemma 3.5.1. (Playfair's Axiom) *Through any given Euclidean point p there is a line parallel to a given line L.*

Proof. In the language of extended Euclidean planes we are claiming that there is exactly one line through two given points. This is exactly Condition **L1** of Theorem 3.2.1 for the two points p and the direction of L □

This suggests that the appropriate tools for constructions in EE^2 are pencil and *roller.* A roller was an indispensable instrument of book-keepers in the days before mechanical setting-out of accounts books was possible. It is a smooth cylinder that enables us to draw a line through two Euclidean points (using it exactly as a straightedge, or unmarked ruler), or to draw a line through a Euclidean point and a direction (placing it in the right direction and rolling it until it is on the point and then drawing the line). It also enables us to tell if two lines are parallel, by placing it on one and rolling it to the other. If they are not parallel, then, by using the roller as a straightedge, we can locate the intersection point. In other words, a roller and a pencil enable us to do exactly what we know is possible in EE^2 - drawing the lines guaranteed to exist by Condition **L1**, and locating the points guaranteed to exist by Condition **L2** of Theorem 3.2.1. Summing up, we have the following theorem:

Theorem 3.5.1. *The constructions that are always possible in any projective plane are a line through two points and a point of intersection of two lines. In an extended Euclidean plane these are exactly the constructions possible with pencil and roller.*

We practice with our new drawing implements in the following constructions.

Construction 3.5.1. Draw, using a roller and pencil, the four 4-gons of Construction 3.4.1.

Construction 3.5.2. Assuming Desargues' Theorem to be true for any extended Euclidean plane, devise a construction, using roller and pencil, for a Desargues figure that contains exactly one ideal point. Draw the figure.

Construction 3.5.3. Assuming Desargues' Theorem to be true for any extended Euclidean plane, devise a construction for a Desargues figure that contains exactly three ideal points. Draw the figure.

Construction 3.5.4. Let us assume Desargues' Theorem to be true. Let L and M be two lines which do not intersect on the page, and p be a point neither in L nor in M. Mark a point, 0, on the page. Draw lines A, B, and C through the point 0. Label the intersection of A and L by 1, the intersection of A and M by 4, the intersection of C and L by 3, and the intersection of C and M by 6. Draw $p \vee 1$ and $p \vee 4$. Then label $(p \vee 1) \cap B$ by 2 and $(p \vee 4) \cap B$ by 5. Draw lines $2 \vee 3$ and $5 \vee 6$. Label the point $(2 \vee 3) \cap (5 \vee 6)$ by b. Draw the line $p \vee b$. Verify that the line $p \vee b$ obtained by the construction is concurrent with L and M.

Notice that formally this process is just that required in solving Problem 1.2.1. But it broadens Problem 1.2.1 as we are now able to use points in EE^2 - not merely points in E^2. We need not worry whether or not lines are parallel, for example, if A and L are parallel, then 1 is a direction. Thus we draw $p \vee 1$ with roller and pencil by placing the roller along L, rolling it to p, and drawing $p \vee 1$.

Our use of extended Euclidean planes enables us to see that many apparently different figures are really examples of the same figure. For example, the above construction covers different possibilities. The given lines L and M could be parallel as in Figure 3.11, or the given point p could be a direction as in Figure 3.12.

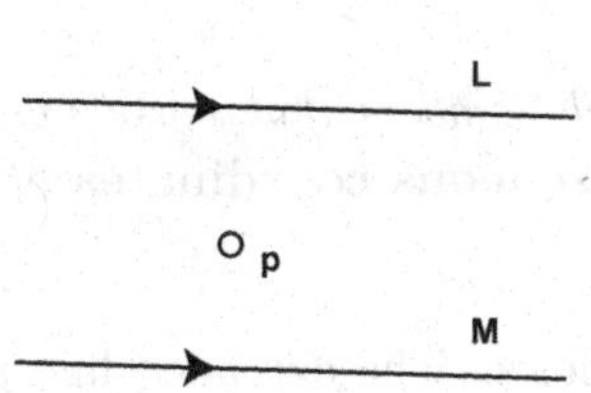

Fig. 3.11 Parallel lines and a Euclidean point

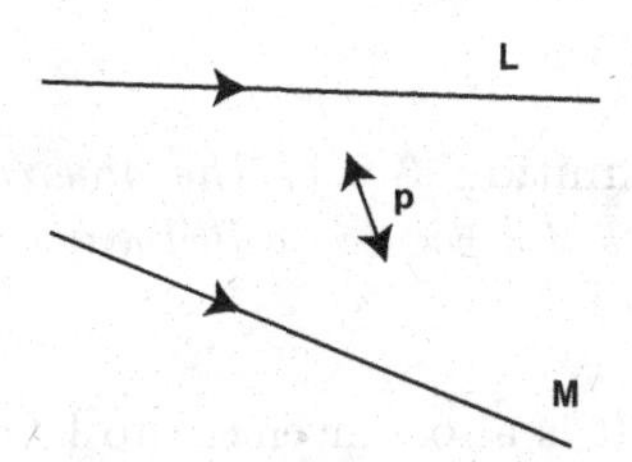

Fig. 3.12 Non-parallel lines and a direction

In each case Construction 3.5.4 gives the required line through the point p.

3.6 Coordinatizing an extended Euclidean plane

In this section, we begin with the usual Cartesian coordinates of each Euclidean point of an extended Euclidean plane. Using them we attach labels to all the points of the extended Euclidean plane. The extension enables us to transform questions about geometry into questions about arithmetic and *vice versa*.

We attach labels to the points as follows. Each label is an ordered triple of real numbers. The point having Cartesian coordinates (x, y) is labeled $(1, x, y)$. The direction of the line $y = mx + c$ is labeled $(0, 1, m)$. The direction of the line $x = a$ is labeled $(0, 0, 1)$. We go further, giving more than one label to each point by allocating the label (k, kx, ky), for each non-zero k, to the point (x, y), and the label $(0, k, km)$, for each non-zero k, to the direction of the line $y = mx + c$, and the label $(0, 0, k)$, for each non-zero k, to the direction of the line $x = a$. We sum up this process in the following lemma.

Lemma 3.6.1. *Each ordered triple of real numbers, not all of which are zero, is a label of exactly one point of a given extended Euclidean plane. Every point of the extended Euclidean plane is labeled. Two ordered triples* (x_0, x_1, x_2) *and* (x'_0, x'_1, x'_2) *label the same point exactly when* $kx_0 = x'_0$, $kx_1 = x'_1$ *and* $kx_2 = x'_2$, *for some non-zero real number* k.

Proof. If $x_0 \neq 0$, then the triple (x_0, x_1, x_2) labels the Euclidean point $(x_1/x_0, x_2/x_0)$. If $x_0 = 0$, and $x_1 \neq 0$, then (x_0, x_1, x_2) labels the direction of $y = (x_2/x_1)x$. If both $x_0 = x_1 = 0$, then (x_0, x_1, x_2) labels the direction of $x = 0$. □

Definition 3.6.1. *The ordered triple of real numbers that make up any label of a point is called a set of* **real homogeneous coordinates** *of the point.*

It is also convenient to label lines, as follows. The extended line $y = mx + c$ is labeled by any ordered triple $[kc, km, -k]$, for any non-zero real number k. The extended line $x = a$ is given the label $[ka, -k, 0]$, for any non-zero real number k. The ideal line is given the label $[k, 0, 0]$, for each non-zero real k. We use square brackets in order to distinguish labels of lines from labels of points.

Lemma 3.6.2. *Each ordered triple of real numbers, not all of which are zero, is a label of exactly one line of a given extended Euclidean plane. Each line of the extended Euclidean plane is labeled. Two ordered triples* $[x_0, x_1, x_2]$ *and* $[x'_0, x'_1, x'_2]$ *label the same line exactly when* $kx_0 = x'_0, kx_1 = x'_1$ *and* $kx_2 = x'_2$, *for some non-zero real number* k.

Proof. The proof is analogous to that of the previous lemma. □

The justification for this seemingly *ad hoc* allocation of labels is the simplicity of the associated arithmetic of an inner product space.

Let us write R^3 for the three-dimensional vector space on the set $\{(x_0, x_1, x_2) : \text{each } x_i \text{ is a real number}\}$, with respect to componentwise addition, componentwise multiplication by scalars, and a componentwise inner product.

Thus: $(x_0, x_1, x_2) + (x'_0, x'_1, x'_2) = (x_0 + x'_0, x_1 + x'_1, x_2 + x'_2)$,
$a(x_0, x_1, x_2) = (ax_0, ax_1, ax_2)$, and
$(x_0, x_1, x_2).(x'_0, x'_1, x'_2) = x_0x'_0 + x_1x'_1 + x_2x'_2$, for each real number a, and for each pair of members (x_0, x_1, x_2) and (x'_0, x'_1, x'_2) of R^3.

We now regard each two-dimensional subspace of R^3 as the set of its one-dimensional subspaces, giving the following geometry.

Theorem 3.6.1. *If* E *is the set of one-dimensional subspaces of* R^3, *and* $\mathbf{L}$ *is the set of two-dimensional subspaces of* R^3, *then* E *and* $\mathbf{L}$ *are the sets of points and lines, respectively, of a projective plane. We call this geometry the real homogeneous plane.*

Proof. The sets E and $\mathbf{L}$ satisfy Conditions **L1** to **L4** of Theorem 3.2.1, ensuring the existence of the required projective plane. □

Proposition 3.6.1. *Each extended Euclidean plane is isomorphic to the real homogeneous plane.*

Proof. Let p be a given point of an extended Euclidean plane. There is a unique one-dimensional subspace of R^3 that contains any label of p. Pair p with this subspace. To show that this pairing is the required isomorphism we need to establish that it induces a pairing of circuits as required by Definition 2.7.2. This is done in a case-by-case investigation of the circuits of the extended Euclidean plane listed in Lemma 3.4.1. We require a set of labels of a circuit of n points to span an $(n-1)$-dimensional subspace of R^3, for $n = 3$, and for $n = 4$, and conversely. We will not do this. □

As a consequence of this isomorphism we identify the two projective planes, giving the following simple algebraic tests in any extended Euclidean plane.

Proposition 3.6.2. *In an extended Euclidean plane a point* (x_0, x_1, x_2) *belongs to the line* $[x'_0, x'_1, x'_2]$ *if, and only if,* $x_0x'_0 + x_1x'_1 + x_2x'_2 = 0$.

Three points (x_0, x_1, x_2), (x'_0, x'_1, x'_2), *and* (x''_0, x''_1, x''_2) *are collinear if and only if the determinant* (†) *is zero.*

$$(\dagger) \qquad \begin{vmatrix} x_0 & x_1 & x_2 \\ x'_0 & x'_1 & x'_2 \\ x''_0 & x''_1 & x''_2 \end{vmatrix}$$

Three lines $[x_0, x_1, x_2]$, $[x'_0, x'_1, x'_2]$, *and* $[x''_0, x''_1, x''_2]$ *are concurrent if and only if determinant* (†) *is zero.*

Proof. The inner product of (x_0, x_1, x_2) and (x'_0, x'_1, x'_2) is zero if and only if (x_0, x_1, x_2) is orthogonal to (x'_0, x'_1, x'_2). From the definition of $[x'_0, x'_1, x'_2]$ we have that this is equivalent to (x_0, x_1, x_2) belonging to the two-dimensional orthogonal subspace that is the line labeled $[x'_0, x'_1, x'_2]$. The second test is a restatement of the requirement that three collinear points belong to a common two-dimensional subspace of R^3. The last test follows from the second after noting that $[x_0, x_1, x_2]$, $[x'_0, x'_1, x'_2]$, and $[x''_0, x''_1, x''_2]$ are concurrent if and only if a non-zero member of R^3 is orthogonal to each of (x_0, x_1, x_2), (x'_0, x'_1, x'_2), and (x''_0, x''_1, x''_2). This requires the three vectors to belong to a two-dimensional subspace of R^3. □

Proposition 3.6.2 reduces questions about an extended Euclidean plane to questions in the arithmetic of a real inner-product space - a highly satisfactory state of affairs. It gives us two possible methods of investigating any problem we meet.

Example 3.6.1. The point that has Cartesian coordinates $(5,8)$ has homogeneous coordinates $(1,5,8)$. The line through $(0,2)$ with slope 3 has homogeneous coordinates $[2,3,-1]$. The inner product $1\times 2+5\times 3+8\times(-1)=-9\neq 0$. Therefore the point $(5,8)$ is not on the line.

The new techniques we have developed above enable us to investigate Desargues' Theorem successfully in any extended Euclidean plane.

Lemma 3.6.3. *Let three distinct points of the real homogeneous plane be collinear. Then their labels (a_0,a_1,a_2), (b_0,b_1,b_2), and (c_0,c_1,c_2) can be chosen so that $a_i=b_i+c_i$, for each $i\in\{0,1,2\}$.*

Proof. The three points are collinear. Consequently their labels are linearly dependent. The result follows on division by the non-zero coefficient of the label of the point labeled by (a_0,a_1,a_2). □

Theorem 3.6.2. *Desargues' Theorem holds in the real homogeneous projective plane.*

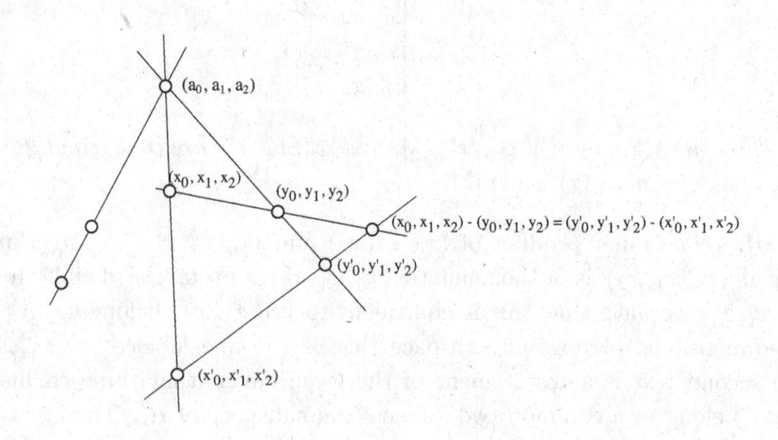

Fig. 3.13 Proving Desargues' Theorem

Proof. Consider any pair of triangles, whose pairs of corresponding vertices belong to concurrent lines as partially shown in the sketch in Figure 3.13 Using Lemma 3.6.3, we choose the labels so that

$(a_0, a_1, a_2) = (x_0, x_1, x_2) + (x'_0, x'_1, x'_2) = (y_0, y_1, y_2) + (y'_0, y'_1, y'_2) = (z_0, z_1, z_2) + (z'_0, z'_1, z'_2)$. Therefore $(y_0, y_1, y_2) - (z_0, z_1, z_2) = (z'_0, z'_1, z'_2) - (y'_0, y'_1, y'_2)$, $(z_0, z_1, z_2) - (x_0, x_1, x_2) = (x'_0, x'_1, x'_2) - (z'_0, z'_1, z'_2)$, and $(x_0, x_1, x_2) - (y_0, y_1, y_2) = (y'_0, y'_1, y'_2) - (x'_0, x'_1, x'_2)$. Each side of these equations is a name for the intersection of two corresponding sides of one of the two given triangles, as each is a linear combination of points in a side and therefore itself in the side. The sum of, say, the lefthand sides of the three equations is zero, that is the three intersections belong to a two-dimensional subspace - they are collinear in the plane. □

Exercise 3.6.1. Prove that Pappus' Theorem holds in the real homogeneous plane.

We have made some progress. If we agree that the real homogeneous projective plane is isomorphic to any extended Euclidean plane, then we have shown that Desargues' Theorem holds for our planar figures. If not, then we must wait a little longer. The question of the truth of Desargues' Theorem for non-planar figures remmains unanswered. Let us press on and answer another question that has been bothering us.

3.7 Sylvester's Theorem and the Fano plane

In this section, we finally lay the question of whether each combinatorial geometry is a combinatorial figure to rest. We have been a little vague about whether certain geometries are figures. We are still undecided about the Fano plane, for example. However, in 1893 the Irish mathematician J.J. Sylvester [65, 11], [13] asked if there are, in any finite non-linear but planar collection E of Euclidean points, two points that are not collinear with any other point of E. In 1944 T. Gallai proved that the answer is in the affirmative. Any drawing of a finite projective plane of order n would have $n + 1$ points on each straight line. Consequently, no sketch of any finite projective plane is a drawing of the plane. Sadly, no finite projective plane is a figure. So our attempts to draw, rather than sketch, a Fano plane will always fail.

We prove a version of Sylvester's Theorem that suffices for our purposes.

Theorem 3.7.1. *No finite projective plane is a figure.*

Proof. Suppose to the contrary that some finite projective plane has a drawing. Consider the following four Euclidean points of the drawing: the farthest to the right of the page (labeled 1), the farthest to the left (labeled 2), the highest on the page (labeled 3), and the lowest on the page (labeled 4). From Lemma 3.1.1 we know that $1 \vee 4$ and $2 \vee 3$ intersect at a point 5 of the projective plane. By our choice of 1, 2, 3, and 4 we deduce that 5 is an ideal point as shown in Figure 3.14.

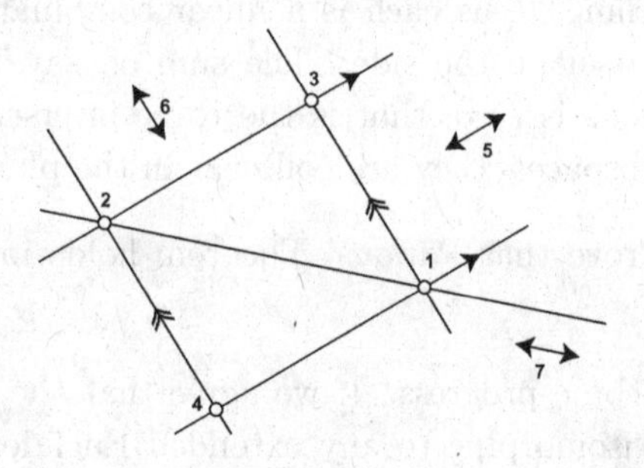

Fig. 3.14 An attempt to draw a Fano plane

In this way we show also that $1 \vee 3$ and $2 \vee 4$ intersect in an ideal point 6. The ideal line $5 \vee 6$ belongs to the drawing and intersects $1 \vee 2$ in an ideal point 7. The line $3 \vee 7$ intersects $1 \vee 4$ in a Euclidean point. This contradicts the choice of the point 1. So there is no such drawing. of any finite projective plane in an extended Euclidean plane. □

Corollary 3.7.1. *A Fano plane is not a figure.*

The broader implication of Theorem 3.7.1 is important enough for us to state separately.

Theorem 3.7.2. *Not every geometry is a figure.*

Summary of Chapter 3. We have examined, in some detail, projective planes. These are planar geometries in which each two lines intersect non-emptily. We showed that every planar geometry is a subgeometry of a projective plane. In the case of planar figures we did this quite efficiently in an extended Euclidean plane. This guarantees our ability to construct figures by the use of roller and pencil. Our introduction of homogeneous coordinates gives us another method of proof, and establishes a link between geometric truth and algebraic truth. We also established that geometries other than figures do exist.

Chapter 4

NON-PLANAR GEOMETRIES AND PROJECTIVE SPACES

We decided in Chapter 3 that an extended Euclidean plane is a convenient and realistic surrounding for the planar figures we draw on paper. Although our world is not confined to the page, the methods of Chapter 3 guide us in this chapter to an appropriate setting for non-planar figures.

4.1 Projective spaces

In this section, we formulate some desirable properties of non-planar geometries. We then investigate in more detail those geometries that possess these properties. When constructing figures, we want straight lines to intersect whenever reasonable, that is, in the case that the straight lines are coplanar. This we know to be equivalent to the truth of the Veblen-Young Condition. But experience suggests that a straight line and a plane may also share a common point. As is the case for two coplanar straight lines, only the possibility of parallelism seems to stand in the way of constructing such a common point.

The next condition on the circuits of a combinatorial geometry **G** ensures that each line and plane of the geometry have nonempty intersection.

Definition 4.1.1. (Strong Veblen-Young condition) *If* $\{1', 2', 3', 4'\}$ *is a four-point circuit in the combinatorial geometry* **G**, *then there is a point* $5'$ *of the geometry so that* $\{1', 2', 5'\}$ *and* $\{3', 4', 5'\}$ *are circuits. If* $\{1, 2, 3, 4, 5\}$ *is a five-point circuit in* **G**, *then there is a point* 6 *of the geometry so that* $\{1, 2, 6\}$ *and* $\{3, 4, 5, 6\}$ *are circuits.*

We next prove that this is exactly the condition we require.

Lemma 4.1.1. *Let* **G** *be a combinatorial geometry. Then the Strong*

Veblen-Young Condition holds in **G** *if and only if both*

(i) each two coplanar lines of **G** *contain a common point, and*

(ii) each line and plane of **G** *contain a common point.*

Proof. First suppose that conditions (i) and (ii) hold in the geometry **G**. Further suppose that $\{1', 2', 3', 4'\}$ is a four-point circuit of **G**, and that $\{1, 2, 3, 4, 5\}$ is a five-point circuit of **G**. Then, from Lemma 3.1.1, we have that there is a point $5'$ so that $\{1', 2', 5'\}$ and $\{3', 4', 5'\}$ are circuits.

It follows from *(ii)* that the line $1 \vee 2$ intersects the plane $3 \vee 4 \vee 5$ in some point 6. If 6 and 1 were the same point, then $\{1, 3, 4, 5\}$ would contain a circuit. If 6 and 3 were the same point, then $\{1, 2, 3\}$ would be a circuit. Therefore, the point $6 \notin \{1, 2, 3, 4, 5\}$. Thus the point 6 is in the line $1 \vee 2$ and $\{1, 2, 6\}$ is a circuit. Also, 6 is in the plane $3 \vee 4 \vee 5$ and, from Definition 2.2.3, there is a circuit C with $6 \in C \subseteq \{3, 4, 5, 6\}$. If the point 3 were not in C, then $\{4, 5, 6\}$ would be a circuit. Hence the Elimination Condition would ensure that $(\{1, 2, 6\} \cup \{4, 5, 6\}) - \{6\} = \{1, 2, 4, 5\}$ contained a circuit. But $\{1, 2, 3, 4, 5\}$ is itself a circuit. Hence, the only remaining possibility is that $\{3, 4, 5, 6\}$ is a circuit. Thus we have shown that conditions (i) and (ii) together imply the Strong Veblen-Young Condition.

Conversely, suppose that the Strong Veblen-Young Condition holds in **G**, and that L is a line, and P is a plane, of **G**. We may write $L = 1 \vee 2$ and $P = 3 \vee 4 \vee 5$ for points 1, 2, 3, 4 and 5. There is nothing to prove if two of the points are equal. If the set $\{1, 2, 3, 4, 5\}$ is a circuit, then the existence of $6 \in P \cap L$ is assured by our assumption. If the set $\{1, 2, 3, 4, 5\}$ is not a circuit, then, by Condition **C2** of Definition 2.1.1, it properly contains a circuit. If $\{1, 2, 3\}$ is a circuit, then the point 3 is in $P \cap L$. If $\{1, 3, 4\}$ is a circuit, then the point 1 is in $P \cap L$. The set $\{3, 4, 5\}$ is not a circuit as the three points 3, 4 and 5 were chosen to be non-collinear. If $\{1, 3, 4, 5\}$ is a circuit, then the point 1 is in $P \cap L$. If $\{1, 2, 4, 5\}$ is a circuit, then the first part of the Strong Veblen-Young Condition enables us to use Lemma 3.1.1 to prove the existence of a common point of the lines $1 \vee 2 = L$ and $4 \vee 5$. Noting that $4 \vee 5 \subseteq P$, we have shown that P and L have a common point. Hence the Strong Veblen-Young Condition implies (i) and (ii). □

We cannot get a stronger result than that given in Lemma 4.1.1. This is because Lemma 2.2.5 ensures that a plane and any line not in the plane intersect in at most one point. We now examine combinatorial geometries for which the Strong Veblen-Young Condition is valid

Definition 4.1.2. *A* **projective space** *is a geometry that satisfies the*

Strong Veblen-Young Condition and has at least one five-point circuit.

As we would hope and expect, projective planes and projective spaces are closely related.

Lemma 4.1.2. *Each plane of a projective space is a projective plane and each line of a projective space is non-trivial.*

Proof. Let P be a plane of a projective space $\mathbf{G}$. Then $E(\mathbf{G})$ contains some five-point circuit $\{1, 2, 3, 4, 5\}$. Without loss of generality, the point 1 is not in the plane P. From Lemma 4.1.1 we have that $(1 \vee i) \cap P$ is a point, say i', for each $i = 2, 3, 4, 5$. The four points $2'$, $3'$, $4'$, and $5'$ are coplanar. If, for example, $\{2', 3', 4'\}$ were a circuit, then $\{1, 2, 3, 4\} \subseteq 1 \vee 2' \vee 3'$, ensuring that $\{1, 2, 3, 4\}$ would contain a circuit. But this is not the case. Similar arguments show that no three-point subset of $\{2', 3', 4', 5'\}$ is a circuit. Thus $\{2', 3', 4', 5'\}$ is a circuit in P. This, together with the first part of the Strong Veblen-Young condition, guarantees that P satisfies Definition 3.2.1 and hence is a projective plane. Lemma 2.2.3 ensures that each line of $\mathbf{G}$ belongs to a projective plane and is consequently non-trivial. □

In Chapter 3, following Proposition 3.2.1, we referred to the first and the last figures of Figure 1.3. The two geometries are on the same set of points and have the same list of lines. But one is planar and the other is not. This example establishes that the collection of lines alone cannot uniquely specify a combinatorial geometry, unless the geometry is known to be planar. We also need information about the collection of planes of the geometry. We now collect together some properties of the lines and planes of a combinatorial geometry with a view to finding a list of properties that will characterize two collections of subsets of a set E as the lines and planes, respectively, of a projective space on E.

Proposition 4.1.1. *In any non-planar geometry, (i) each two points belong to exactly one line, (ii) each two planes contain at most one line, (iii) if two points belong to a plane, the line containing both is also in the plane, (iv) if two planes contain some point, any line contained in both also contains the point, (v) each line, and point not belonging to it, belong to exactly one plane, (vi) each line and plane not containing it contain at most one common point, and (vii) there are four points not belonging to any one plane.*

Proof. These results follow, in order, from Theorem 2.2.1, Lemma 2.2.4,

Lemma 2.2.2, Lemma 2.2.3, Lemma 2.2.5, Theorem 2.2.4 and the existence of a five-point circuit in the geometry. □

In a result analogous to Theorem 3.2.1, we specialize these requirements and thereby characterize a projective space in terms of properties of the collection of its lines and the collection of its planes.

Theorem 4.1.1. *Let a set E, a collection* **L** *of subsets of E, and another collection* **P** *of subsets of E, satisfy the following eight conditions.*
Condition L1*: Each two elements of E belong to exactly one common member of* **L***.*
Condition P1*: Each two members of* **P** *contain exactly one common member of* **L***.*
Condition P2*: If two elements of E belong to a member of* **P***, then the unique member of* **L** *containing both is a subset of this member of* **P***.*
Condition P3*: If two members of* **P** *contain some element of E, then the unique member of* **L** *contained in both also contains the element of E.*
Condition P4*: Each member of* **L***, and element of E not belonging to it, are contained in exactly one member of* **P***.*
Condition P5*: Each member of* **L** *and member of* **P** *not containing it contain exactly one common element of E.*
Condition P6*: The set E contains a subset of five elements, no four of them belonging to any one member of* **P***.*
Condition P7*: The set* **P** *contains a subset of five elements, no four of them containing any one member of E.*

Suppose that $\mathcal{C} = \{C : |C| = 3$ and C is a subset of a member of **L**$\} \cup \{C : |C| = 4$, *C is contained in a member of* **P***, and no three elements of C belong to any one member of $L\} \cup \{C : |C| = 5$ and no four elements of C belong to any one member of $P\}$. Then $\mathcal{C}$ is the set of circuits of a non-planar geometry on the set E. This geometry is a projective space, its lines are exactly the members of* **L***, and its planes are exactly the members of* **P***.*

Conversely, the set **L** *of lines and the set* **P** *of planes of any projective space satisfy the eight conditions* **L1***, and* **P1** *to* **P7***.*

Proof. We first suppose that the eight conditions hold, and we prove that $(E, \mathcal{C})$ is a combinatorial geometry. Clearly the requirements **C1**, **C2**, and **C4** of Definition 2.1.1 hold for the collection $\mathcal{C}$. Exactly as in the proof of Proposition 3.2.1, we see that two three-point circuits that share two common elements satisfy the Elimination Condition. If two distinct sets

C_1 and C_2 are both members of $\mathcal{C}$ and $|C_1 \cup C_2| = 4$, then this is the only possibility. Suppose that $a \in C_1 \cap C_2$.

If $|C_1 \cup C_2| = 5$, then the possibilities are; first $|C_1| = |C_2| = 3$ and $|C_1 \cap C_2| = 1$; second $|C_1| = 3$, $|C_2| = 4$ and $|C_1 \cap C_2| = 2$; and third $|C_1| = 4 = |C_2|$ and $|C_1 \cap C_2| = 3$.

In the first case we may write $C_1 = \{a, b, c\}$ and $C_2 = \{a, d, e\}$. On the one hand, if C_1 and C_2 are both subsets of the same member L of $\mathbf{L}$, then $\{b, c, d\} \subseteq L$, also ensuring that $(C_1 \cup C_2) - \{a\} \supseteq \{b, c, d\} \in \mathcal{C}$. If on the other hand, $C_1 \subseteq L_1$ and $C_2 \subseteq L_2 \neq L_1$, then there is a member P of $\mathbf{P}$ that contains both L_1 and d, and therefore contains a and d, and therefore L_2. Thus $\{b, c, d, e\}$ is in P, and therefore contains a member of $\mathcal{C}$.

In the second case we may write $C_1 = \{a, b, c\} \subseteq L \in \mathbf{L}$, $C_2 = \{a, b, d, e\} \subseteq P \in \mathbf{P}$. As $a, b \in P$ and $L \subseteq P$, then $\{b, c, d, e\}$ is a subset of P and consequently contains a member of $\mathcal{C}$. In the third case we may write $C_1 = \{a, b, c, d\}$, $C_2 = \{a, b, c, e\}$ and there is one member L of $\mathbf{L}$ containing both a and b and one member P of $\mathbf{P}$ containing both L and c, and thus containing $\{a, b, c\}$. Hence d and e are also both in P, and $\{b, c, d, e\}$ contains a member of $\mathcal{C}$.

If $|C_1 \cup C_2| \geq 6$, then $(C_1 \cup C_2) - \{a\}$ contains a member of $\mathcal{C}$. Thus $\mathcal{C}$ satisfies the Elimination Condition and is the collection of circuits of a unique geometry $\mathbf{G}$ on E.

From Condition **P5** we have that the empty set is not a member of $\mathbf{L}$. If a single-element subset of E were a member of $\mathbf{L}$, then, from Condition **P5**, every member of $\mathbf{P}$ would contain this single element. This rules out the possibility of Condition **P7** holding. Therefore each member of $\mathbf{L}$ contains at least two distinct elements of E. Each pair of distinct points a and b of $\mathbf{G}$ therefore belongs to exactly one line of $\mathbf{G}$ and to exactly one member of $\mathbf{L}$. Each line and each member of $\mathbf{L}$ occurs in this way. The defining property above of three-point circuits in $\mathcal{C}$ guarantees that $x \in (a \vee b) - \{a, b\}$ if and only if $\{a, b, x\} \in \mathcal{C}$ if and only if $x \in L - \{a, b\}$. Thus the unique member L of $\mathbf{L}$ that contains $\{a, b\}$ is also the line $a \vee b$ and so the lines of $\mathbf{G}$ are exactly the members of $\mathbf{L}$. Similarly the member of $\mathbf{P}$ containing three non-collinear points a, b and c is the plane $a \vee b \vee c$, and the planes of $\mathbf{G}$ are exactly the members of $\mathbf{P}$. From **P6** we have that the geometry is non-planar, and has a five-point circuit.

Let L_1 and L_2 be two distinct lines contained in some plane P. For any point a not contained in the plane P, the planes $P_1 = a \vee L_1$ and $P_2 = a \vee L_2$ intersect in a line L. This line L intersects the plane P in a point b, and b is in each of the planes P_1, P_2, and P. As $L_1 = P \cap P_1$, we know that $b \in L_1$. Similarly $b \in L_2$ and so L_1 and L_2 have a common point. This, together with **P5**, enables us to use Lemma 4.1.1 to prove the Strong Veblen-Young Condition. Thus **G** is a projective space.

Conversely, in any projective space with the set **L** of lines and the set **P** of planes, the condition **L1** follows from Theorem 2.2.1, the condition **P2** is a consequence of Lemma 2.2.2, the conditions **P4**, **P5**, and **P6** follow from Lemma 2.2.3, Lemma 4.1.1, and Definition 4.1.2 respectively. Suppose that P and $Q = x \vee y \vee z$ are any two planes of the projective space. Their intersection is at most a line and, without loss of generality, $x \notin P$, and so $x \vee y$ and $x \vee z$ intersect P in two points, from **P5**. It follows from Lemma 2.2.2 that $P \cap Q$ is exactly a line, ensuring the truth of **P1**. The last requirement, **P7**, follows by choosing five of the planes defined by each three-point subset of a five-point circuit of the projective space. □

We can consequently be certain that, if the planes and lines of a combinatorial geometry satisfy the eight conditions **L1**, and **P1** to **P7**, then the geometry is a projective space and, from Lemma 4.1.1, each two coplanar lines of the geometry meet in a point and each line and plane of the geometry meet in a point.

Exercise 4.1.1. The circuits of the geometry $(E, \mathcal{C})$ of Exercise 2.3.3 are exactly all the five-point subsets of E. Prove that $(E, \mathcal{C})$ fails badly to be a projective space.

Exercise 4.1.2. Consider the claim that the following columns of numbers are the planes of a projective space on the fifteen point set $E = \{0, 1, \ldots, 14\}$.

0	1	2	3	4	5	6	7	8	9	10	11	12	13	14
1	2	3	4	5	6	7	8	9	10	11	12	13	14	0
2	3	4	5	6	7	8	9	10	11	12	13	14	0	1
4	5	6	7	8	9	10	11	12	13	14	0	1	2	3
5	6	7	8	9	10	11	12	13	14	0	1	2	3	4
8	9	10	11	12	13	14	0	1	2	3	4	5	6	7
10	11	12	13	14	0	1	2	3	4	5	6	7	8	9

Assuming the claim to be correct, and using Condition **P1** of Theorem 4.1.1, list all the lines of the projective space. List those lines that are in the plane consisting of the points of the first column. Do you recognize this planar subgeometry of the projective space?

The numbers of planes and points in the above example, together with the nature of the eight characterizing conditions of Theorem 4.1.1, suggest the possibility of some form of Principle of Duality for projective spaces. This is indeed the case and, as for projective planes, we can use it in proofs. The following changes leave the eight conditions, **L1** and **P1** to **P7**, *in toto* unchanged.

Definition 4.1.3. *The* **dual** *of any statement about points, lines and planes of a projective space is the statement obtained from the original statement by interchanging the words "point" with "plane" and "belong to" with "contain".*

Theorem 4.1.2. (Principle of Duality for Projective Spaces) *If, in a projective space, "theorem B" about the space is a logical consequence of "assumption A", then also "the dual of theorem B" is a logical consequence of "the dual of assumption A".*

Proof. Theorem B is logically derived from its starting assumptions and conditions **L1** and **P1** to **P7** and each step in its proof is stated in terms of "point", "line", "plane", "belong to" and "contain" and terms used in assumption A. If we make the above interchanges throughout its proof (in particular **L1** and **P1** being interchanged, **P2** and **P3** being interchanged, **P4** and **P5** being interchanged and **P6** and **P7** being interchanged), then we obtain a valid argument for the dual theorem from the dual starting assumption. □

The following proof illustrates the use of the Principle of Duality for Projective Spaces.

Lemma 4.1.3. *If* **G** *is a projective space, then the intersection of each three planes of* **G** *is either a point or a line.*

Proof. We know from Definition 2.2.3 and Theorems 2.2.3 and 2.2.1 that, in a combinatorial geometry, and particularly in a projective space, three points belong to a unique plane unless they are collinear. If the three points are collinear, then they belong to a unique line. Using the Principle of Duality for Projective Spaces: three (*planes*) (*contain*) a unique (*point*) unless they (*contain a common line*), in which case they (*contain only this line*). □

In Chapter 3 we used the fact that each two lines have an intersection point to pair the points on a line with the lines through a point, using the elementary map of Definition 3.2.3. Bearing in mind the properties listed in Theorem 4.1.1, and the Principle of Duality for Projective Spaces, we could obtain results analogous to Lemma 3.2.1 and Theorem 3.2.7.

Exercise 4.1.3. Make a guess at a bijection that could play a role in a projective space analogous to that of an elementary map in a projective plane. Make a guess at statements of a lemma and a theorem analogous to Lemma 3.2.1 and Theorem 3.2.7, respectively. Check that your guesses are consistent with the results of Exercise 4.1.2.

4.2 Subgeometries of a projective space

In this section, we prove that projective spaces are, in one crucial respect, well-behaved combinatorial geometries. The following proposition is closely related to Desargues' Theorem and leads us to the conclusion that not every geometry is a subgeometry of a projective space. This is the price we must pay for the advantage of other desirable properties of projective spaces.

Proposition 4.2.1. *Let* **G** *be a projective space, and suppose* $\{a, b, c, d\}$ *and* $\{a, b, e, f\}$ *are circuits in distinct planes* P_1, P_2, *respectively, of* **G**. *The lines* $a \vee b$, $c \vee d$ *and* $e \vee f$ *intersect in one common point if and only if* $\{c, d, e, f\}$ *is a circuit contained in a plane* P_3 *distinct from* P_1 *and* P_2.

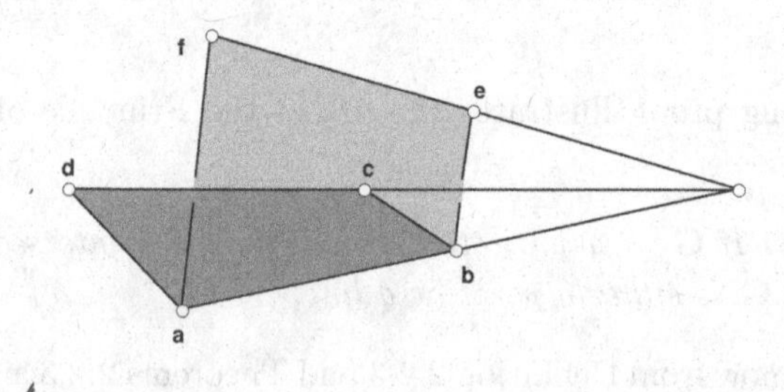

Fig. 4.1 The existence of a third 4-point circuit

Proof. Suppose that $\{c, d, e, f\}$ is a subset of the plane P_3. Then, from Lemma 4.1.3, the intersection $P_1 \cap P_2 \cap P_3$ is a line or a point. If it were a line, then it would be the line $a \vee b$, and the set $\{a, b, c, d, e, f\}$ would be planar, which is not the case. The intersection is therefore a point, and is in the common line of each pair of P_1, P_2, P_3 as required.

Conversely, suppose that the lines $a \vee b$, $c \vee d$ and $e \vee f$ intersect in one common point. Lemma 4.1.1 applied to the lines $c \vee d$ and $e \vee f$ proves that the set $\{c, d, e, f\}$ is planar. If $\{c, d, e\}$ were a circuit then $e \in c \vee d \subset P_1$ and so $P_2 = a \vee b \vee e = P_1$, which is not the case. Similarly no three-point subset of $\{c, d, e, f\}$ is a circuit. Therefore $\{c, d, e, f\}$ is a circuit contained in some plane P_3. If P_1 and P_3 were the same plane, then the set $\{a, b, c, d, e, f\}$ would be planar; a contradiction. Similarly $P_2 \neq P_3$. □

In Theorem 3.3.1 we proved that any planar geometry sits inside a projective plane, although the projective plane itself may not be very "nice". We now see how different the non-planar situation is.

Theorem 4.2.1. *A Vámos cube is not a subgeometry of any projective space.*

Proof. We recall that the Vámos cube of Definition 2.3.6 is the combinatorial geometry $\mathbf{G}$, where $E(\mathbf{G}) = \{0, 1, \ldots, 7\}$, $\mathcal{K} = \{\{0, 1, 3, 5\}, \{0, 2, 3, 6\}, \{1, 4, 5, 7\}, \{2, 4, 6, 7\}, \{0, 3, 4, 7\}\}$, and $\mathcal{C}(G) = \mathcal{K} \cup \{C \subseteq E : |C| = 5, C$ contains no member of $\mathcal{K}\}$. Let us suppose the cube were a subgeometry of some projective space. Applying Proposition 4.2.1 to circuits $\{0, 1, 3, 5\}$, $\{1, 4, 5, 7\}$ and $\{0, 3, 4, 7\}$ we have that the lines $0 \vee 3$, $1 \vee 5$, and $4 \vee 7$ would be concurrent. Thus the point $(0 \vee 3) \cap (4 \vee 7)$ would be on the line $1 \vee 5$. Similarly, by applying the Proposition to circuits $\{2, 4, 6, 7\}$, $\{0, 2, 3, 6\}$ and $\{0, 3, 4, 7\}$ we would have the point $(0 \vee 3) \cap (4 \vee 7)$ on the line $2 \vee 6$. Hence the lines $1 \vee 5$ and $2 \vee 6$ would contain a common point (see Figure 4.2) and, using the Elimination Condition **C3** of Definition 2.1.1, we would have that $\{1, 2, 5, 6\}$ contains a circuit. But this is not so. So our assumption has led to a contradiction, and the Vámos cube cannot be a subgeometry of any projective space. □

4.3 Desargues' Theorem and projective spaces

We were able to prove in Theorem 3.3.1 that each planar geometry is a subgeometry of some projective plane. An unfortunate consequence of this is the less than ideal behaviour of the projective planes themselves. For example, our comments in the paragraph following Lemma 3.3.1 emphasized the fact that Desargues' Theorem fails in some projective planes. Certainly

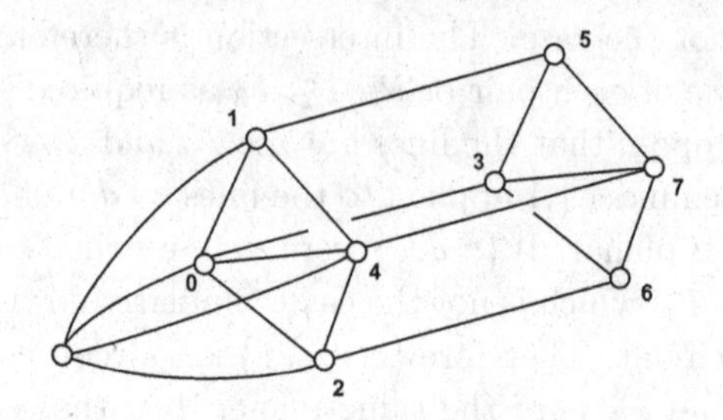

Fig. 4.2 A Vámos Cube Extension

Theorem 4.2.1 ensures that no spatial analogue of Theorem 3.3.1 is possible. But the very result, Lemma 4.2.1, which is the key to proving that a Vámos cube is not a subgeometry of any projective space is also a key to proving the following welcome results.

Proposition 4.3.1. *Let* 123 *and* 456 *be non-coplanar triangles in a projective space* **G**. *Then the three lines* $1 \vee 4$, $2 \vee 5$ *and* $3 \vee 6$ *are concurrent if and only if the three points* $(1 \vee 2) \cap (4 \vee 5)$, $(2 \vee 3) \cap (5 \vee 6)$, *and* $(1 \vee 3) \cap (4 \vee 6)$ *are collinear.*

Proof. Suppose that the lines $1 \vee 4$, $2 \vee 5$, and $3 \vee 6$ are concurrent. From Lemma 4.1.1, the sets $\{1, 2, 4, 5\}$, $\{2, 3, 5, 6\}$, and $\{1, 3, 4, 6\}$ are each planar and from Lemma 4.1.2 each of the intersections $(1 \vee 2) \cap (4 \vee 5)$, $(2 \vee 3) \cap (5 \vee 6)$ and $(1 \vee 3) \cap (4 \vee 6)$ is a point. Each of these three points is in both of the distinct planes $1 \vee 2 \vee 3$ and $4 \vee 5 \vee 6$, and from Lemma 2.2.4 the three points are collinear.

Conversely, suppose that $(1 \vee 2) \cap (4 \vee 5)$, $(2 \vee 3) \cap (5 \vee 6)$, and $(1 \vee 3) \cap (4 \vee 6)$ are collinear points. Lemma 4.1.1 ensures that $\{1, 2, 4, 5\}$, $\{2, 3, 5, 6\}$, and $\{1, 3, 4, 6\}$ are each planar sets and, as $1 \vee 2 \vee 3 \neq 4 \vee 5 \vee 6$, they are in distinct planes. If each of the sets $\{1, 2, 4, 5\}$, $\{2, 3, 5, 6\}$, and $\{1, 3, 4, 6\}$ is a circuit, then we use Proposition 4.2.1 to prove the lines $1 \vee 4$, $2 \vee 5$, and $3 \vee 6$ concurrent. If they are not all circuits an examination of the possibilities leads to the same conclusion in each case. Figure 4.3 illustrates one such possibility. □

Theorem 4.3.1. *Desargues' Theorem and its converse are true in each projective space. That is, the lines through each pair of corresponding vertices of two triangles are concurrent if and only if the points of intersection of each pair of corresponding sides of the triangles are collinear.*

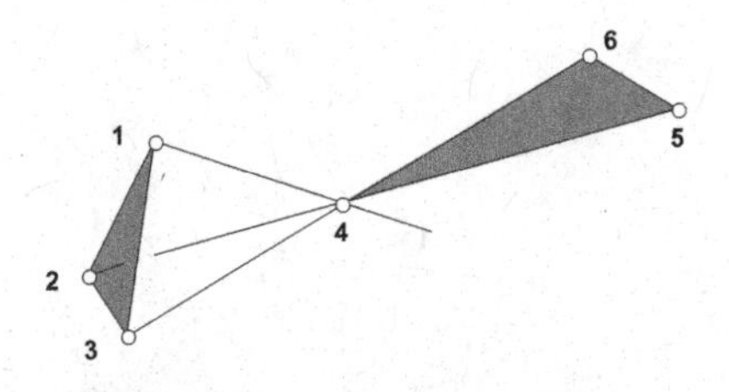

Fig. 4.3 Neither $\{1, 2, 4, 5\}$ nor $\{1, 3, 4, 6\}$ is a circuit

Proof. Proposition 4.3.1 provides the result for any non-coplanar pair of triangles. Now suppose that $1'23$ and $4'56$ are a coplanar pair of triangles in the projective space. Suppose that the lines $1' \vee 4'$, $2 \vee 5$ and $3 \vee 6$ intersect at the point 0. Our proof relies on the fact that the triangles lie in a non-planar geometry, and we construct a non-planar Desargues geometry containing triangles 123, 456 whose "shadows" are $1'23$ and $4'56$ respectively.

We begin by choosing a point $p \notin 1' \vee 2 \vee 3$. We have from Lemma 4.1.2 that there is a point 1, distinct from $1'$ and p, on the line $p \vee 1'$. As $0 \vee 4'$ meets $p \vee 1$ in a point, the points 0, 1, p, and $4'$ are coplanar. From Lemma 4.1.1 we have that the point $4 = (0 \vee 1) \cap (p \vee 4')$ exists. Therefore the three lines $1 \vee 4$, $2 \vee 5$ and $3 \vee 6$ intersect at the point 0. We apply Proposition 4.3.1 to the non-coplanar pair of triangles 123 and 456, deducing that the points $(1 \vee 2) \cap (4 \vee 5) = 9$, $(2 \vee 3) \cap (5 \vee 6) = 8$, $(1 \vee 3) \cap (4 \vee 6) = 7$ exist and are collinear.

As the line $p \vee 1' = p \vee 1$ meets the line $2 \vee 9 = 1 \vee 2$ in a point, the points $1'$, 2, p and 9 are coplanar and the line $1' \vee 2$ intersects the line $p \vee 9$ in the unique point of $p \vee 9$ that is in the plane $1' \vee 2 \vee 3$. We call this point $9'$. From a similar argument we have that $4' \vee 5$ meets $p \vee 9$ at the point $9'$. Similarly $1' \vee 3$ and $4' \vee 6$ meet at the point $8'$ of intersection of $p \vee 8$ and $1' \vee 2 \vee 3$. It is immediate that $2 \vee 3$ and $5 \vee 6$ meet at the intersection $7'$ of $p \vee 7$ and $1' \vee 2 \vee 3$. As 7, 8 and 9 are collinear, all three points $7'$,$8'$, $9'$ are in both planes $p \vee 7 \vee 8$ and $1' \vee 2 \vee 3$, and so are collinear. Therefore Desargues' Theorem is valid for the projective space.

Conversely, let $1'23$ and $4'56$ be a coplanar pair of triangles and suppose that the points $(1' \vee 2) \cap (4' \vee 5)$, $(2 \vee 3) \cap (5 \vee 6)$ and $(1' \vee 3) \cap (4' \vee 6)$ are collinear. Lemma 4.1.2 guarantees that the plane containing the two triangles is a projective plane. From Theorem 3.2.4 we have that the lines $1' \vee 4'$, $2 \vee 5$ and $3 \vee 6$ are concurrent. □

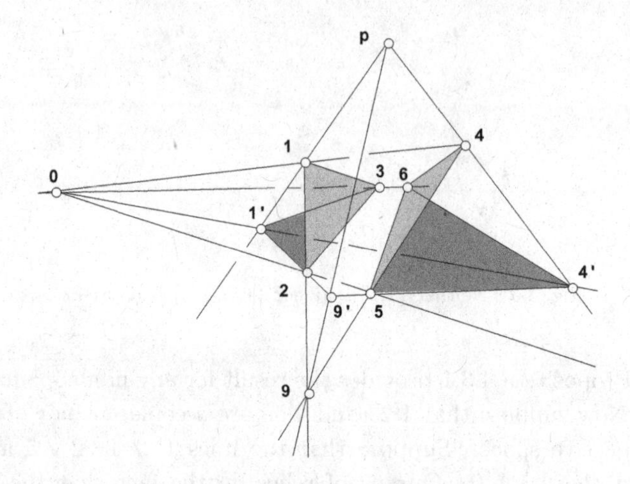

Fig. 4.4 Two non-coplanar triangles 123 and 456

This result makes the idea of investigating the possibility of embedding combinatorial figures in a projective space very attractive. Our original motivation for studying projective spaces was to see if all the combinatorial figures of our usual experience could be thought of as subgeometries of some convenient projective space - enabling us to extend these figures by construction as easily as possible. We now see that this would have the added advantage of Desargues Theorem being valid for figures.

A critical difference to the case of planar figures exists. In Theorem 3.3.1 we proved that each planar geometry lies in some projective plane, but Theorem 4.2.1 indicated that not every non-planar geometry is a subgeometry of a projective space. Does our experience of the world around us suggest that figures may be thought of in the context of a projective space?

4.4 Extended Euclidean space

In this section, we construct a projective space that contains all figures. We use a method analogous to that which proved successful in Chapter 3. Let us write E^3 for usual Euclidean space. If it were a projective space, from Lemma 4.1.1 we know that coplanar straight lines would intersect - in fact Lemma 4.1.2 insists that each plane would be a projective plane. But our experience in constructing EE^2 suggests that we may need to add new symbols to E^3 in order to give a geometry in which Condition **L1** of

Theorem 3.2.1 holds.

Definition 4.4.1. *Let L be a straight line of E^3. Consider all straight lines of E^3 parallel to L. We choose a symbol, and call it the* **direction,** *or* **ideal point,** *or* **point at infinity,** *of each of these straight lines.*

We add this symbol to the points of E^3, and we repeat the process, defining a new symbol for each family of parallel straight lines. Let E be the set consisting of all the Euclidean points of E^3 and all the directions defined in this way. We next enlarge each straight line L, and each straight line parallel to L, by the inclusion of their direction. This symbol is not added to any other straight line.

Definition 4.4.2. *Let L be a straight line. The set obtained by enlarging the set L of Euclidean points by the inclusion of the direction of L is an* **extended Euclidean line.** *We call it the extension of L.*

If a straight line is in a Euclidean plane we add its direction to the points of that plane, giving an enlarged Euclidean plane exactly as in Definition 3.4.4.

Definition 4.4.3. *Let P be a Euclidean plane of E^3. The set obtained by enlarging the set P of Euclidean points by the inclusion of the direction of each straight line L in P is an* **extended Euclidean plane.** *We call it the* **extension** *of P.*

As in the construction of an extended Euclidean plane in Chapter 3, the collection of extended Euclidean lines contained in an extended Euclidean plane almost satisfies the requirements of Theorem 3.2.1. Only Condition **L1** fails for pairs of directions. We remedy this defect as before.

Definition 4.4.4. *Let P be a Euclidean plane. The set of all directions in the extension of P is an* **ideal line.**

We observe that two extended Euclidean planes share the same ideal line if and only if they are extensions of parallel Euclidean planes. Each extended Euclidean line, each ideal line, and each extended Euclidean plane is a subset of the set E. This set E of Euclidean points and directions, with its associated collection of extended Euclidean lines, ideal lines and collection of extended Euclidean planes almost satisfies the requirements of Theorem 4.1.1.

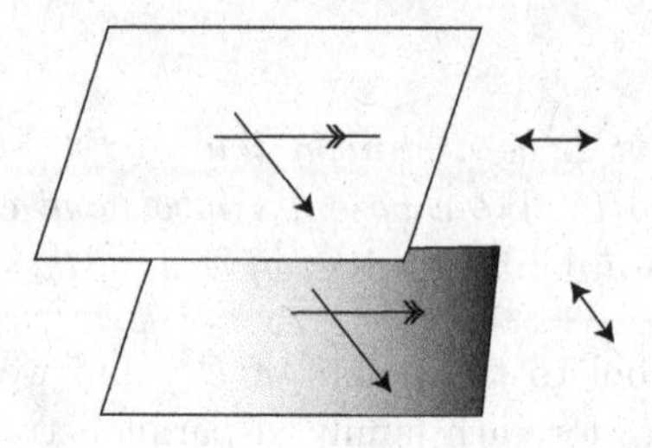

Fig. 4.5 Two parallel planes and shared directions

Only condition **P5** fails. Each extended Euclidean line and each Euclidean point not in it are contained in exactly one extended Euclidean plane. Each extended Euclidean line and each direction not in it are contained in exactly one extended Euclidean plane, and each ideal line and Euclidean point, are contained in exactly one extended Euclidean plane. But **P5** fails for an ideal line and a direction not in the ideal line. To try to remedy the situation we add one more set, consisting of all directions, to the proposed collection of planes.

Definition 4.4.5. *The* **ideal plane**, *or* **plane at infinity**, *of* E^3 *is the set of all directions.*

Now we check that we have defined a projective space.

Proposition 4.4.1. *Let* E^3 *be Euclidean space. Let* E *be the set consisting of all Euclidean points and directions. Let* **L** *be the collection of subsets of* E *consisting of all extended Euclidean lines, and all ideal lines. Let* **P** *be the collection of subsets of* E *consisting of all extended Euclidean planes, and the ideal plane. Then* **L** *and* **P** *are, respectively, the collections of lines and planes of a projective space on* E.

Proof. We prove that the eight requirements of Theorem 4.1.1 are satisfied. The set **L** was chosen so that Condition **L1** holds. Two non-parallel extended Euclidean planes intersect only in the extension of their common straight line, two parallel extended Euclidean planes intersect in the ideal line of their common directions, and an extended Euclidean plane and the ideal plane also intersect in the ideal line of directions of the extended Euclidean plane. Consequently Condition **P1** holds.

The set **P** was chosen so that Condition **P2** holds. The argument we have just given also proves that **P3** is valid.

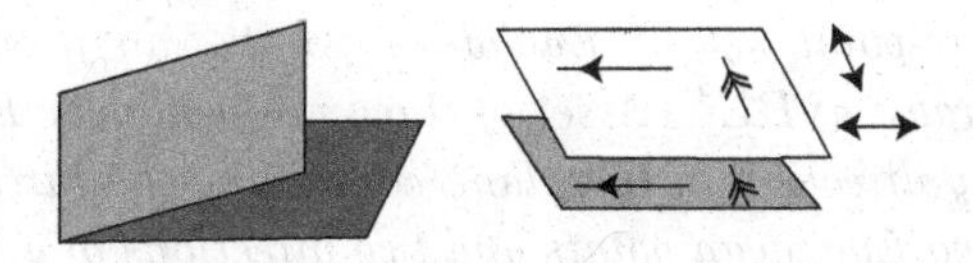

Fig. 4.6 Two pairs of planes

An extended Euclidean line and a Euclidean point not in it belong to exactly one extended Euclidean plane, and not to the ideal plane. An ideal line and a Euclidean point belong to the one member of the parallel family of extended Euclidean planes sharing this ideal line of direction which contains the Euclidean point. An extended Euclidean line, and a direction not in it, belong to the extended Euclidean plane containing the line and the different direction given. An ideal line and another direction belong only to the ideal plane. Thus Condition **P4** is satisfied.

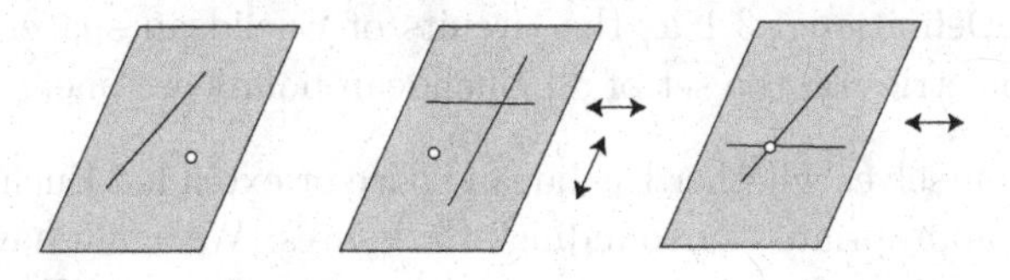

Fig. 4.7 The union of a line and a point is a planar set

An extended Euclidean plane and extended Euclidean line not in the plane share either a Euclidean point or a direction. An extended Euclidean plane and an ideal line not in the plane share the direction common to the plane and to the extension of each member of the parallel family of Euclidean planes that contain the ideal line. The ideal plane and an extended Euclidean line share the direction of the line. Consequently Condition **P5** holds. The existence of a five-point circuit in Euclidean space ensures the truth of the last two conditions **P6** and **P7**. □

Definition 4.4.6. *We call the projective space of Proposition 4.4.1* **extended Euclidean space** *and denote it by EE^3.*

Lemma 4.4.1. *A linear three-point set of Euclidean points is a circuit of EE^3. A set of two Euclidean points and the direction of the line through the two is a circuit of EE^3. A set of three directions in any extended Euclidean plane is a circuit of EE^3.*

A planar four-point set of Euclidean points, no three of which are collinear, is a circuit of EE^3. A set of three non-collinear Euclidean points together with any direction of the plane containing the three is a circuit of EE^3. A set of two Euclidean points and two directions of a Euclidean plane through the two is a circuit of EE^3. A set of four directions, no three being directions in a common Euclidean plane, is a circuit of EE^3.

A five-point set that does not contain any of the three-point or four-point sets specified above is a circuit of EE^3

Proof. This result follows directly from Theorem 4.1.1. □

Theorem 4.4.1. *Euclidean space is a subgeometry of extended Euclidean space.*

Proof. The circuits of the subgeometry on the set of Euclidean points of EE^3 are specified by Lemma 4.4.1. They are those chosen on intuitive grounds in Definition 1.3.1 as the circuits of Euclidean space. Therefore the two geometries on the set of all Euclidean points are one. □

We ask ourselves whether Euclidean space or extended Euclidean space is the more appropriate "surrounding" for figures. We again have a choice, as we had for planar figures, but this time between E^3 and EE^3 rather than between E^2 and EE^2. For the moment we choose the conceptual simplicity of EE^3, that is, of projective geometry. We extend the meaning of figure to include all subgeometries of EE^3, not just those of E^3.

Definition 4.4.7. *A* **combinatorial figure** *is a subgeometry of extended Euclidean space.*

In other words, the geometric structure associated, by Definition 2.6.1, with any subset of Euclidean and ideal points is a figure. This extension agrees with, and includes, the enlarged meaning for planar figures given in Definition 3.4.5.

It is true, although we will not prove it, that any finite combinatorial figure in the above extended sense can be modeled entirely in E^3, that is, it is also a figure in the original sense of Chapter 1.

At last we are able to definitely decide whether a particular non-planar geometry is a figure.

Theorem 4.4.2. *The Vámos cube is not a combinatorial figure.*

Proof. Any figure is a subgeometry of extended Euclidean space. Theorem 4.2.1 does not allow a Vámos cube to be a subgeometry of any projective space, in particular of extended Euclidean space. □

There is an immediate practical consequence of Theorem 4.4.2. In Construction 2.5.2 we modeled a combinatorial cube on the set $E = \{0, 1, \ldots, 7\}$ that has six four-point circuits. The four 4-point circuits $\{0, 1, 3, 5\}$, $\{0, 2, 3, 6\}$, $\{1, 4, 5, 7\}$, and $\{2, 4, 6, 7\}$ were each modeled by a cardboard panel. The set $\{0, 3, 4, 7\}$ is one of the remaining two four-point circuits. In order to make an accurate model the points 0, 3, 4, and 7 of the model should be coplanar. We found it difficult to trim the pieces of cardboard so that they fitted well when assembled.

Proposition 4.2.1 suggests the following method of achieving this. We cut two pieces of cardboard, one modeling the circuit $\{1, 4, 5, 7\}$, and the other modeling the circuit $\{0, 1, 3, 5\}$. Then we hinge them with sticky tape. Then the following consequence of Proposition 4.2.1 gives a way of trimming the panels.

Lemma 4.4.2. *Let two panels with vertex sets* $\{0, 1, 3, 5\}$ *and* $\{1, 4, 5, 7\}$*, respectively, share a hinge. Then the vertices* 0*,* 3*,* 4*, and* 7 *are coplanar in one non-collapsed position of the hinged pair if and only if they are coplanar in all positions of the hinge.*

Proof. By applying Proposition 4.2.1 to a figure modeled by the hinged pair in one position, we have that the points are coplanar if and only if the lines $1 \vee 5$, $0 \vee 3$, and $4 \vee 7$ are concurrent. This point of concurrence is on the line of the hinge and remains un-moved as the hinge opens and shuts. Therefore the points 0, 3, 4, and 7 remain coplanar in any position of the hinge. □

Thus we may trim as shown in Figure 4.8. No matter how much the hinge is opened in its final assembly position, the edge modeling $0 \vee 3$ and the edge modeling $4 \vee 7$ are coplanar as required to model a four-point circuit $\{0, 3, 4, 7\}$. Similarly, we make another hinged pair of cardboard panels to model the circuits $\{2, 4, 6, 7\}$ and $\{0, 2, 3, 6\}$, and then trim them so that the edge modeling $0 \vee 3$ and the edge modeling $4 \vee 7$ are also coplanar in any position of the hinge. To complete the model we need only fit the two hinged pairs together.

Suppose we cut and assemble the four cardboard pieces so that the four points 0, 3, 4, and 7 are coplanar, as required to model a four-point circuit. Then Theorem 4.2.1 implies that the vertices 1, 2, 5, and 6 will be coplanar

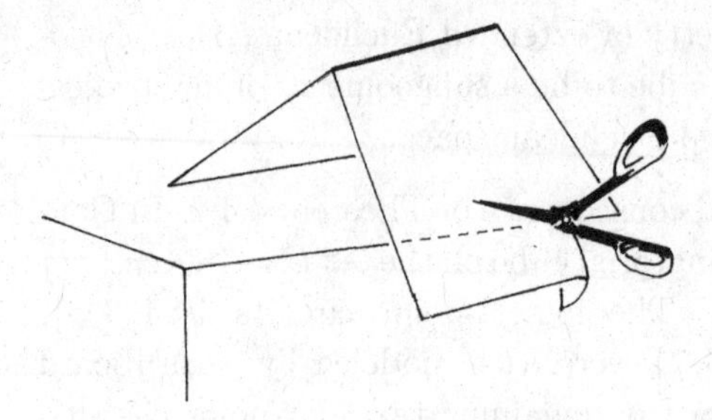

Fig. 4.8 Trimming a model of a combinatorial cube

as required. For if this were not so we would have modeled the Vámos cube. We have just shown that this is impossible in Theorem 4.2.1.

But we may even go further to simplify the construction. From Proposition 4.2.1 we have that in a successful model all four lines $0 \vee 3$, $1 \vee 5$, $4 \vee 7$, and $2 \vee 6$ are concurrent. This suggests that a template similar to that in Figure 4.9 might be useful in order to eliminate any trimming required at the "ends of the box". Three hinges may then be taped together, and we can be confident that the model will fold into position without error.

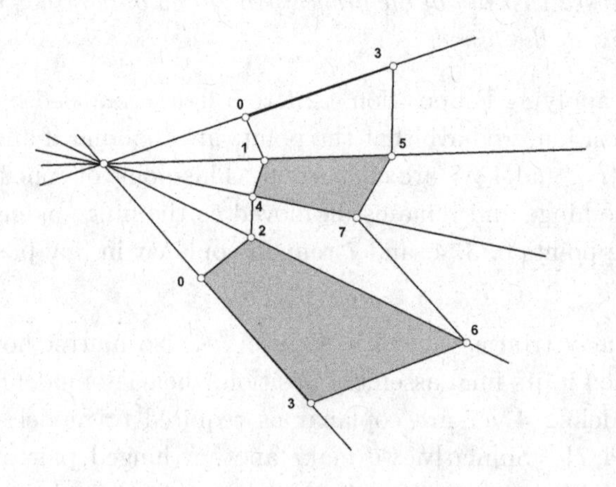

Fig. 4.9 Designing a model of a cube

We call any extended Euclidean lines of EE^3 that have a common direction "*parallel*". In the context of extended Euclidean space their common direction is a point of our geometry. Thus extended Euclidean lines are parallel if and only if they are extensions of parallel straight lines. Thus we may talk of "parallel lines" without confusion in the context of both Euclidean geometry and extended Euclidean geometry. The notation $A||B$ continues to mean that the lines A and B are parallel. As before, we indicate this in drawings by drawing the same number of arrow-heads on each of A and B. A double-headed arrow remains our notation for a direction. These conventions agree with, and extend, those we agreed on in Chapter 3. When drawing non-planar figures it is often the convention to draw a 4-gon that has opposite sides parallel in order to represent the plane containing it. Thus two planes meeting in an extended Euclidean line or an ideal line are often shown as the first or second figure, respectively, in Figure 4.10.

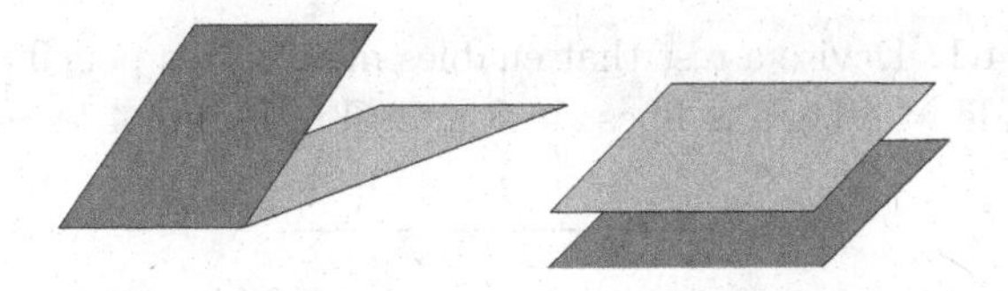

Fig. 4.10 A convention for representing planes

We have used this convention already in, for example, Figures 1.5 and 2.19, and in Proposition 4.4.1.

Construction 4.4.1. Suppose that all the vertices of a non-planar 4-gon in extended Euclidean space are Euclidean points. Draw, and make a model of, the 4-gon. Do the same for three other non-planar 4-gons containing, respectively, one, two adjacent, and two opposite, ideal vertices.

Construction 4.4.2. Draw a combinatorial cube whose vertices are Euclidean. Draw a combinatorial cube, one of whose vertices is ideal. Draw a combinatorial cube four of whose vertices are ideal. Can you think of a satisfactory way of making a model of the Euclidean parts of the last two cubes?

Theorem 4.4.3. *Desargues' Theorem is true in extended Euclidean space, and in each extended Euclidean plane.*

Proof. Theorem 4.3.1 ensures that Desargues Theorem holds for the projective space EE^3. Let two triangles 123 and 456 belong to an extended Euclidean plane EE^2. Each pair of corresponding sides of the two triangles intersects in a point of the projective plane EE^2. Suppose that the lines $1 \vee 4$, $2 \vee 5$, and $3 \vee 6$ are concurrent. As the two triangles belong to the projective space EE^3, from Theorem 4.3.1 we have that the three points of intersection are collinear. □

Thus the assumptions underlying Constructions 1.2.6, 3.5.2, 3.5.3 and 3.5.4, and Problem 1.2.1 are valid and our world is *Desarguesian*, the proof for non-coplanar triangles being much easier than the proof for planar triangles. The planar case proof is only possible because the plane of the two triangles is in a projective space.

There could be no guarantee that a planar universe would be desarguesian. What sort of non-desarguesian geometry, and consequent physical structures, could exist in such a universe is a matter open to conjecture.

Exercise 4.4.1. Devise a test that enables me to use a pencil and roller in order to decide whether the lines A, B and C in Figure 4.11 concurrent.

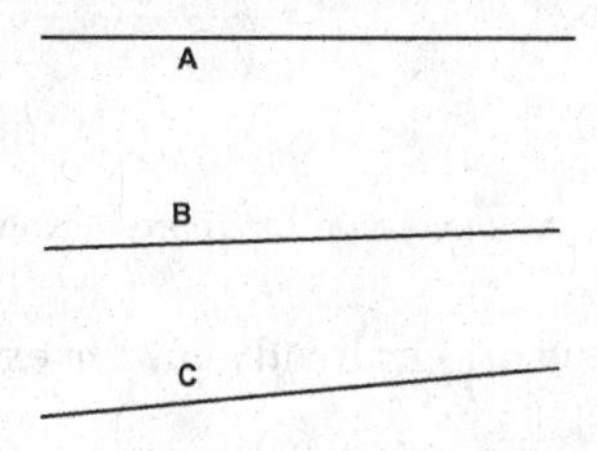

Fig. 4.11 Three lines on the page

Exercise 4.4.2. Condition **P7** of Theorem 4.1.1 ensures that extended Euclidean space contains five planes, no four of which contain a common point. Describe a combinatorial figure whose points are the points of intersection of each three of these planes.

Construction 4.4.3. Make a model of a non-planar Desargues figure.

We can now summarize the position regarding Desargues geometries.

Theorem 4.4.4. *A planar Desargues geometry is a combinatorial figure, a planar non-Desargues geometry is not a combinatorial figure. A non-planar Desargues geometry is a combinatorial figure. There is no non-planar non-Desargues geometry.*

Proof. The first and second claim follow from Lemma 1.2.2 and Theorem 4.3.1. Similarly, Theorem 4.3.1 ensures that an attempt to construct a non-planar Desargues figure will succeed. Suppose that 123 and 456 are two triangles, not in the same plane, and that the lines $1 \vee 4$, $2 \vee 5$, and $3 \vee 6$ are concurrent. Suppose further that $1 \vee 2$ and $4 \vee 5$ meet in the point 9. Suppose that the other two pairs of corresponding sides of the triangles meet in 8 and 7, respectively. Each of the points 7, 8, and 9 is in the plane $1\vee2\vee3$, by Lemma 2.2.2. Similarly each point is in $4\vee5\vee6$. If the points 7, 8, and 9 were non-collinear, then we would have $1\vee2\vee3 = 7\vee8\vee9 = 4\vee5\vee6$. But this is not possible as triangles 123 and 456 are not coplanar. Thus the assumption of the existence of a non-planar non-Desargues geometry leads to a contradiction, and so no such geometry can exist. □

Construction 4.4.4. Let X be a combinatorial figure isomorphic to that drawn in Figure 4.12. Suppose further that the points labeled 1 and 2 in X are each ideal. Deduce a consequent property of the line through the points of X that are labeled 4, and 6 and the line through the points of X that are labeled 5 and 7. Draw the figure X.

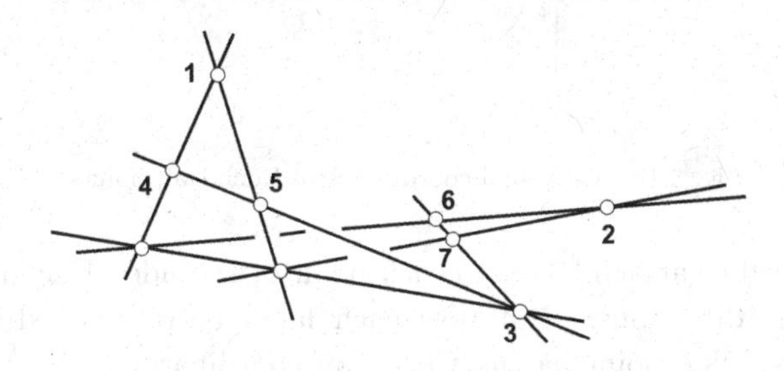

Fig. 4.12 A figure of Euclidean points

4.5 Coordinatizing extended Euclidean space

In this section, in a treatment analogous to that given in Section 3.6, we attach labels to the points of extended Euclidean space. As well, we assign labels to the planes of extended Euclidean space. This coordinatization enables us to identify extended Euclidean space with an isomorphic geometry defined on the set of one-dimensional subspaces of an inner product vector space, transforming questions about geometry into questions about algebra and *vice versa.* The rules, given in Theorem 4.5.1, for investigating problems of extended Euclidean space are easy to use in practice and certainly justify their place in any continuing examination of Euclidean geometry. In particular they provide us with a convenient introduction to perpendicularity.

We begin by recalling the usual Cartesian coordinates of Euclidean space. It is the practice to label each Euclidean point by an ordered triple, (x, y, z), of real numbers, as in Figure 4.13. So common is this practice that we blur the distinction between label and point, talking of "the Euclidean point (x, y, z)" rather than "the Euclidean point with label (x, y, z)".

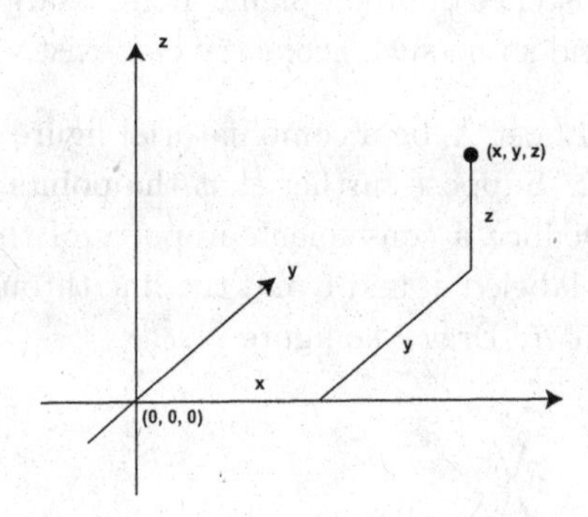

Fig. 4.13 Cartesian coordinates of Euclidean points

We proceed to attach labels to each point of extended Euclidean space, including the directions. But now each label is an ordered 4-tuple of real numbers. The point having Cartesian coordinates (x, y, z) is labeled $(1, x, y, z)$. Our labeling of the directions of an extended Euclidean plane in Section 3.6 was equivalent to labeling the direction of the line $(0, 0) \vee (a, b)$ by $(0, a, b)$. We use the spatial analogue of this in labeling the direction of the line through the points with Cartesian coordinates $(0, 0, 0)$ and (a, b, c), where not all of a, b, and c are zero. This direction is labeled by $(0, a, b, c)$. We go further, giving more than one label to each point of extended Euclidean space by also allocating the label (kx_0, kx_1, kx_2, kx_3), for each non-zero real number k, to the point with label (x_0, x_1, x_2, x_3). We sum up this process in the following lemma:

Lemma 4.5.1. *Each ordered 4-tuple of real numbers, not all of which are zero, is a label of exactly one point of extended Euclidean space. Each point of extended Euclidean space is labeled. Two ordered 4-tuples (x_0, x_1, x_2, x_3) and (x'_0, x'_1, x'_2, x'_3) label the same point exactly when $kx_0 = x'_0$, $kx_1 = x'_1$, $kx_2 = x'_2$ and $kx_3 = x'_3$ for some non-zero real number k.*

Proof. For $x_0 \neq 0$, the 4-tuple (x_0, x_1, x_2, x_3) labels the Euclidean point that has Cartesian coordinates $(x_1/x_0, x_2/x_0, x_3/x_0)$. The 4-tuple $(0, x_1, x_2, x_3)$ labels the direction of the line that contains the origin and the Euclidean point having Cartesian coordinates (x_1, x_2, x_3). Thus each ordered 4-tuple labels one point of extended Euclidean space.

Each direction belongs to a unique extended Euclidean line through the origin, ensuring that we have labeled each point of extended Euclidean space.

Two labels with non-zero first coordinates label the same Euclidean point if and only if they are multiples of one another. Two Euclidean points that are collinear with the origin have Cartesian coordinates that are multiples of one another, ensuring a well-defined labeling of each direction by 4-tuples that are multiples of one another. □

Definition 4.5.1. *The ordered 4-tuple of real numbers that makes up any label of a point is called a set of* **real homogeneous coordinates** *of the point.*

It is also convenient to label planes. The extended Euclidean plane whose Euclidean points satisfy the equation $ax + by + cz = d$ is labeled by the ordered 4-tuple $[ka, kb, kc, -kd]$, for each non-zero real number k. The ideal plane is given the label $[0, 0, 0, k]$, for each non-zero real k. We use square brackets in order to distinguish labels of planes from labels of points.

Definition 4.5.2. *The ordered 4-tuple of real numbers that makes up any label of a plane is called a set of* **real homogeneous coordinates** *of the plane.*

Lemma 4.5.2. *Each ordered 4-tuple of real numbers, not all of which are zero, is a label of exactly one plane of extended Euclidean space. Each plane of extended Euclidean space is labeled. Two ordered 4-tuples* $[x_0, x_1, x_2, x_3]$ *and* $[x'_0, x'_1, x'_2 x'_3]$ *label the same plane exactly if* $kx_0 = x'_0, kx_1 = x'_1$, $kx_2 = x'_2$ *and* $kx_3 = x'_3$ *for some non-zero real number* k.

Proof. The proof is similar to that of Lemma 4.5.1. □

As in Section 3.6, the justification for this seemingly *ad hoc* allocation of labels is the simplicity of the associated arithmetic of an inner product space. We now give the basic rules for applying this arithmetic to geometric purposes.

Let us write R^4 for the four-dimensional vector space on the set $\{(x_0, x_1, x_2, x_3) : \text{each } x_i \text{ is a real number}\}$, with respect to componentwise addition, componentwise multiplication by scalars, and a componentwise inner product. Thus we have:

$(x_0, x_1, x_2, x_3) + (x'_0, x'_1, x'_2, x'_3) = (x_0 + x'_0, x_1 + x'_1, x_2 + x'_2, x_3 + x'_3)$,
$a(x_0, x_1, x_2, x_3) = (ax_0, ax_1, ax_2, ax_3)$, and
$(x_0, x_1, x_2, x_3).(x'_0, x'_1, x'_2, x'_3) = x_0x'_0 + x_1x'_1 + x_2x'_2 + x_3x'_3$, for each real number a and each pair of members (x_0, x_1, x_2, x_3) and (x'_0, x'_1, x'_2, x'_3) of R^4.

We now regard each subspace of R^3 as the set of its one-dimensional subspaces, giving the following geometry on the set of one-dimensional subspaces of R^4.

Theorem 4.5.1. *Let E be the set of one-dimensional subspaces of R^4. Let* **L** *be the set of two-dimensional subspaces of R^4, and let* **P** *be the set of three-dimensional subspaces of R^4. Then E,* **L** *and* **P** *are the sets of points, lines and planes, respectively, of a projective space. We call this geometry real homogeneous space.*

Proof. The sets E, **L** and **P** satisfy Conditions **L1** and **P1** to **P7** of Theorem 4.1.1, ensuring the existence of the required projective space. □

Proposition 4.5.1. *Extended Euclidean space is isomorphic to real homogeneous space.*

Proof. We start by pairing each point of extended Euclidean space with the unique one-dimensional subspace of R^4 that contains any one of the labels of the point. To show that this pairing is the required isomorphism, we need to establish that it induces a pairing of circuits as required by Definition 2.7.2. This is done in a case-by-case investigation of the circuits of extended Euclidean space listed in Lemma 4.4.1. We require a set of labels of a circuit of n points to span an $(n-1)$-dimensional subspace of R^4, for each $n = 3, 4$, and 5 and conversely. We will not do this. □

The isomorphism can also be demonstrated rigorously in the following way. Extra conditions are added to the requirements of Euclidean space. These requirements, which are intuitively quite acceptable, typically involve the ideas of pairs of points "separating" other pairs of points, and of "distance". Extended Euclidean space is defined as before and it is then shown that, to within isomorphism, that there is only one geometry satisfying these conditions. Real homogeneous space is then shown, using properties of the real number system, to satisfy the conditions. Thus real homogeneous space and extended Euclidean space are isomorphic. We will not carry out this lengthy process.

As a consequence of this isomorphism we identify the two projective spaces, giving the following simple algebraic tests in extended Euclidean space.

Proposition 4.5.2. *In extended Euclidean space a point* (x_0, x_1, x_2, x_3) *belongs to the plane* $[x'_0, x'_1, x'_2, x'_3]$ *if, and only if,* $x_0x'_0 + x_1x'_1 + x_2x'_2 + x_3x'_3 = 0$.

Four points (x_0, x_1, x_2, x_3), (x'_0, x'_1, x'_2, x'_3), $(x''_0, x''_1, x''_2, x''_3)$ *and* $(x'''_0, x'''_1, x'''_2, x'''_3)$ *are coplanar if and only if the determinant* (†) *is zero.*

$$(\dagger) \qquad \begin{vmatrix} x_0 & x_1 & x_2 & x_3 \\ x'_0 & x'_1 & x'_2 & x'_3 \\ x''_0 & x''_1 & x''_2 & x''_3 \\ x'''_0 & x'''_1 & x'''_2 & x'''_3 \end{vmatrix}$$

Four planes $[x_0, x_1, x_2, x_3]$, $[x'_0, x'_1, x'_2, x'_3]$, $[x''_0, x''_1, x''_2, x''_3]$ *and* $[x'''_0, x'''_1, x'''_2, x'''_3]$ *are concurrent if and only if determinant* (†) *is zero.*

Proof. The inner product of (x_0, x_1, x_2, x_3) and (x'_0, x'_1, x'_2, x'_3) is zero if and only if (x_0, x_1, x_2, x_3) is orthogonal to (x'_0, x'_1, x'_2, x'_3). The definition of $[x'_0, x'_1, x'_2, x'_3]$ implies that this is equivalent to (x_0, x_1, x_2, x_3) belonging to the three-dimensional orthogonal subspace that is the line labeled $[x'_0, x'_1, x'_2, x'_3]$.

The second test is a restatement of the requirement that coplanar points belong to a common three-dimensional subspace of R^4.

The last test follows from the second after noting that $[x_0, x_1, x_2, x_3]$, $[x'_0, x'_1, x'_2, x'_3]$, $[x''_0, x''_1, x''_2, x''_3]$, and $[x'''_0, x'''_1, x'''_2, x'''_3]$ are concurrent if and only if a non-zero member of R^4 is orthogonal to each of (x_0, x_1, x_2, x_3), (x'_0, x'_1, x'_2, x'_3), $(x''_0, x''_1, x''_2, x''_3)$, and $(x'''_0, x'''_1, x'''_2, x'''_3)$. This requires the four vectors to belong to a three-dimensional subspace of R^4. □

Proposition 4.5.2 reduces questions about extended Euclidean space to questions about the arithmetic of a real inner product space. It gives us two possible methods of investigating any problem we meet. For example, we now have two methods of testing whether two lines are parallel. The first method is by the procedures of Section 3.5, lying a roller along one line and rolling it to the other line. The second method is the algebraic technique of checking that the two lines share a point labeled by a 4-tuple that has a zero first coordinate.

Example 4.5.1. Let us test in two ways whether or not the lines $(1,0,0,0) \vee (1,2,1,0)$ and $(1,0,1,0) \vee (1,4,3,0)$ are parallel.

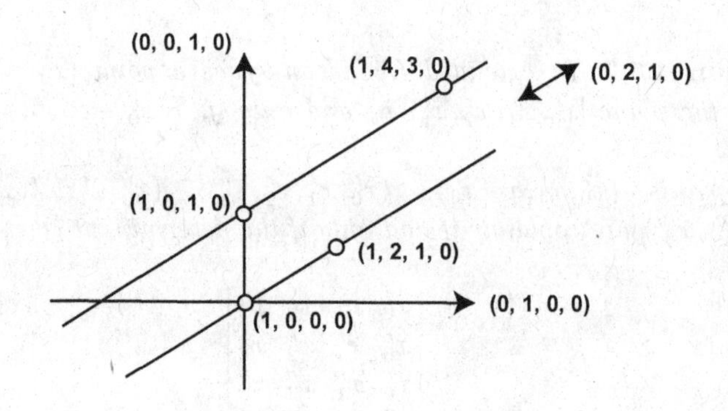

Fig. 4.14 Two parallel lines

We can draw the two straight lines and use a roller to check that they are parallel. We see that the direction of the first line is any linear combination of $(1,0,0,0)$ and $(1,2,1,0)$ having 0 for its first coordinate. The 4-tuple $(1,0,0,0)-(1,2,1,0)=(0,-2,-1,0)$ is the required direction. Similarly, $(1,0,1,0)-(1,4,3,0)=(0,-4,-2,0)=2(0,-2,-1,0)$, the same direction, belongs to the second line.

In Theorem 3.6.2 we gave an algebraic proof of Desargues' Theorem for two coplanar triangles. An algebraic proof, without the restriction of planarity, is straightforward and we give it as an example of the power of algebraic methods. We need the following useful result.

Lemma 4.5.3. *Let three distinct points of extended Euclidean space be collinear. Then they have labels* (a_0,a_1,a_2,a_3), (b_0,b_1,b_2,b_3), *and* (c_0,c_1,c_2,c_3), *respectively, so that* $a_i=b_i+c_i$, *for each* $i=0,1,2,3$.

Proof. Labels of the three collinear points belong to a two-dimensional subspace of R^4. Consequently their labels are linearly dependent. Thus $\alpha(a_0,a_1,a_2,a_3)+\beta(b_0,b_1,b_2,b_3)+\gamma(c_0,c_1,c_2,c_3)=(0,0,0,0)$, for some real α, β and γ. As labels of distinct points are not multiples of one another, none of α, β and γ is zero. The result follows from $(\alpha a_0,\alpha a_1,\alpha a_2,\alpha a_3)=(-\beta b_0,-\beta b_1,-\beta b_2,-\beta b_3)+(-\gamma_0,-\gamma c_1,-\gamma c_2,-\gamma c_3)$. $\square$

Theorem 4.5.2. *Desargues' Theorem is true in extended Euclidean space.*

Proof. Let $(x_0, x_1, x_2, x_3)(y_0, y_1, y_2, y_3)(z_0, z_1, z_2, z_3)$ and $(x'_0, x'_1, x'_2, x'_3)(y'_0, y'_1, y'_2, y'_3)(z'_0, z'_1, z'_2, z'_3)$ be two triangles in real homogeneous space so that the lines through corresponding vertices are concurrent, as in Figure 4.15

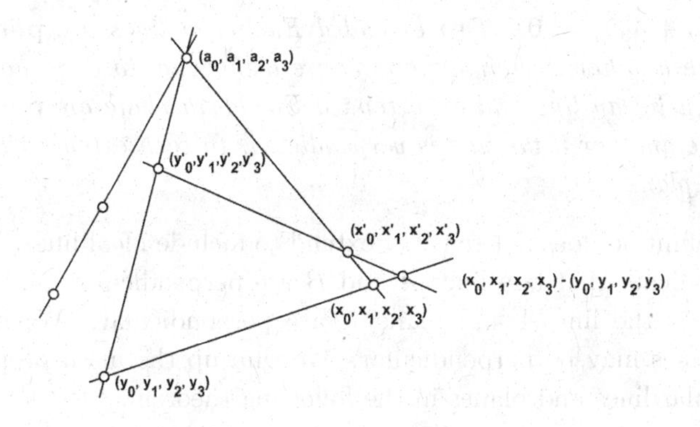

Fig. 4.15 Proving Desargues' Theorem

Using Lemma 4.5.3 we may choose labels so that $(a_0, a_1, a_2, a_3) = (x_0, x_1, x_2, x_3) + (x'_0, x'_1, x'_2, x'_3) = (y_0, y_1, y_2, y_3) + (y'_0, y'_1, y'_2, y'_3) = (z_0, z_1, z_2, z_3) + (z'_0, z'_1, z'_2, z'_3)$. Therefore $(y_0, y_1, y_2, y_3) - (z_0, z_1, z_2, z_3) = (z'_0, z'_1, z'_2, z'_3) - (y'_0, y'_1, y'_2, y'_3)$, $(z_0, z_1, z_2, z_3) - (x_0, x_1, x_2, x_3) = (x'_0, x'_1, x'_2, x'_3) - (z'_0, z'_1, z'_2, z'_3)$, and $(x_0, x_1, x_2, x_3) - (y_0, y_1, y_2, y_3) = (y'_0, y'_1, y'_2, y'_3) - (x'_0, x'_1, x'_2, x'_3)$. Each side of these equations is a name for the intersection of two corresponding sides of the two given triangles, as each is a linear combination of points in a side and therefore itself in the side. The sum of, say, the lefthand side of each equation is zero, that is the three intersections belong to a two-dimensional subspace - they are collinear. □

We note that this proof works equally well, regardless of whether the two triangles are coplanar or not.

4.6 Perpendicular lines and planes

In this section, we use pairs of orthogonal directions in the inner product space R^4 in order to define and develop some properties of perpendicular extended Euclidean lines and planes. We conclude by giving, in Theorem 4.8.1, conditions for the existence of a special tetrahedron. This result is invaluable in our investigation, in Chapter 8, of the distortion that occurs in perspective drawings of rectangular boxes .

Definition 4.6.1. *Two directions* (a_0, a_1, a_2, a_3) *and* (b_0, b_1, b_2, b_3) *are* **orthogonal** *to one another if their inner product is zero, that is, if* $a_1b_1 + a_2b_2 + a_3b_3 = 0$. *Two extended Euclidean lines are* **perpendicular** *to one another if their directions are orthogonal to one another. An extended Euclidean line and an extended Euclidean plane are* **perpendicular** *to one another if the line is perpendicular to each extended Euclidean line in the plane.*

This definition does not sensibly extend to include ideal lines. We write $A \perp B$ to indicate that two lines A and B are perpendicular, and $A \perp P$ to indicate that the line A and plane P are perpendicular. We note that two skew lines may be perpendicular. We sum up the main properties of perpendicular lines and planes in the following theorem:

Theorem 4.6.1. *(i) Let A and B be parallel lines and A and C be perpendicular lines. Then B and C are perpendicular to one another.*

(ii) Let A and B be coplanar, but not parallel, lines and let the line C be perpendicular to each of A and B. Then C is perpendicular to the plane that contains both A and B.

(iii) Let p be a Euclidean point and A be an extended Euclidean line, $p \notin A$. Then there is a unique extended Euclidean line through p, intersecting A in a Euclidean point, and perpendicular to A.

(iv) Let p be a Euclidean point and P be an extended Euclidean plane. Then there is a unique line through p and perpendicular to P. This line intersects P in a Euclidean point called the foot of the perpendicular from p to P.

(v) Let a be a Euclidean point and A be an extended Euclidean line. Then there is a unique plane through a and perpendicular to A. This plane intersects A in a Euclidean point.

Proof. To prove (i) we need only observe that A and B share a common direction, and this direction is orthogonal to the direction of C. In order to prove (ii) we let a, b, and c be the directions of A, B, and C, respectively. Any direction d in the plane P that contains $A \cup B$ is collinear with a and b. Consequently any label of d is a linear combination of a label of a and a label of b. This ensures that d is also orthogonal to c, as required. The line A contains a direction a. Within the 3-dimensional subspace $A \vee p$ of R^4, the two-dimensional subspace $a^\perp$ that is orthogonal to the one-dimensional subspace of labels of the direction a is a line and contains a direction $d \neq a$. Then the line $p \vee d$ meets A in a point (as both are two-dimensional subspaces of a three-dimensional vector space). This point is not d and therefore is a Euclidean point. The directions a and d are orthogonal giving $(p \vee d) \perp A$ as required in (iii). The proof of (iv) is similar to that of (iii). We select the direction d in the 1-dimensional subspace orthogonal to the 3-dimensional subspace P of R^4. The required line is $p \vee d$. We prove (v) by choosing the 3-dimensional subspace of R^4 orthogonal to the direction of A. This subspace does not contain the direction of A. Thus it is a plane that meets the line A in a Euclidean point. □

We use the notation of Figure 4.16 to indicate that two intersecting lines, or a line and a plane, are perpendicular to one another.

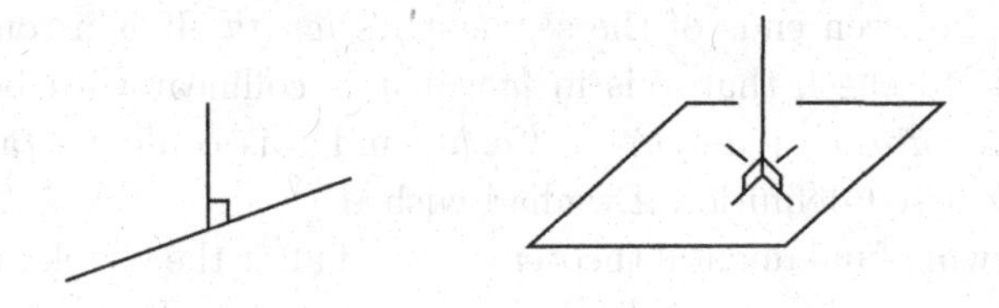

Fig. 4.16 Perpendicular lines and planes

We note in passing that part (ii) of Theorem 4.6.1 is exactly the information used by a builder, when erecting a post of a wall, to test by set-square that the post is perpendicular to the floor. This is the reason for the usual notation, shown in Figure 4.16. Just as we now have two methods of testing lines for parallelism, we also have two methods of testing for perpendicularity. We may either prove the inner product of the directions of the two lines in question to be zero, or we may use a set-square in the traditional way as would a carpenter.

4.7 Pythagoras' Theorem, length and angle

Our understanding of the idea of the distance between a pair of Euclidean points as an allocation of a number to a pair of points is strengthened by a lifetime of using ruler and tape measure. This process does not extend sensibly to include directions, but we can phrase our definition of distance between any two Euclidean points in terms of homogeneous coordinates. Thus the *distance* between two Euclidean points $(1, a_1, a_2, a_3)$ and $(1, b_1, b_2, b_3)$ is the number $\sqrt{(a_1 - b_1)^2 + (a_2 - b_2)^2 + (a_3 - b_3)^2}$. This definition allows us two techniques for measuring, namely an algebraic calculation and the usual use of a marked straightedge, or ruler. Writing "$d(a, b)$" for the distance between points a and b, we can prove the following three basic properties of distance directly from the definition:

Theorem 4.7.1. *Let a, b and c be three Euclidean points. Then*
(i) $d(a, b) \geq 0$, *with equality holding only if* $a = b$,
(ii) $d(a, b) = d(b, a)$, *and*
(iii) (Triangle Inequality) $d(a, b) + d(b, c) \geq d(a, c)$.

The *segment* consisting of all the Euclidean points between the Euclidean points a and b is familiar to anyone who draws part of a Euclidean line with a straightedge. We write $[ab]$ to stand for the segment and call the distance between ends of the segment its *length*. It is a routine piece of arithmetic to check that x is in $[ab]$ if it is collinear with both a and b and satisfies $d(a, x) + d(x, b) = d(a, b)$, and this could be taken as an algebraically based definition if we had wished.

The following fundamental theorem, named after the Greek mathematician Pythagoras, links perpendicularity and distance. Its truth justifies a method of constructing the large wooden set-squares that are often used to guarantee that boxing for large rectangular cement slabs is in shape, although the agreement of the algebraic test of perpendicularity with practical physical tests relies on an appropriate physical choice of axes for allocating Cartesian coordinates to Euclidean points.

Theorem 4.7.2. (Pythagoras' Theorem) *Let a, b, and c be distinct Euclidean points. Then* $(a \vee b) \perp (b \vee c)$ *if and only if* $(d(a, b))^2 + (d(b, c))^2 = (d(c, a))^2$.

Proof. If $a = (1, a_1, a_2, a_3)$, $b = (1, b_1, b_2, b_3)$, and $c = (1, c_1, c_2, c_3)$, then the direction of $a \vee b$ is $(0, b_1 - a_1, b_2 - a_2, b_3 - a_3)$, and the direction of $b \vee c$ is $(0, c_1 - b_1, c_2 - b_2, c_3 - b_3)$. Thus $(a \vee b) \perp (b \vee c)$ if and only if $\sum_{i=1}^{3}(b_i - a_i)(c_i - b_i) = 0$. This can be rewritten as the condition $\sum_{i=1}^{3}(b_i - a_i)^2 + (c_i - b_i)^2 - (a_i - c_i)^2 = 0$. □

It is interesting to note that the test for perpendicular lines provided by Pythagorus' Theorem is independent of the particular choice of a in the line $a \vee b$ and c in the line $b \vee c$.

Let a, b, and c be three distinct Euclidean points. The set $[ab] \cup [bc]$ is an *angle*. We denote this angle by $a\hat{b}c$ or $c\hat{b}a$ and speak of "*the angle abc*" or "*the angle cba*". The point b is the *apex* of the angle. The angle $a\hat{b}c$ is an *acute* angle if $(d(a,b))^2 + (d(b,c))^2 < d(a,c))^2$, the angle $a\hat{b}c$ is a *right* angle if $(d(a,b))^2 + (d(b,c))^2 = d(a,c))^2$, and the angle $a\hat{b}c$ is an *obtuse* angle if $(d(a,b))^2 + (d(b,c))^2 > d(a,c))^2$. It follows that $a\hat{b}c$ is a right angle if and only if the lines $a \vee b$ and $b \vee c$ are perpendicular. If the vertices of the triangle abc are Euclidean points, then the angles $[ab] \cup [bc]$, $[bc] \cup [ca]$ and $[ca] \cup [ab]$ are the *angles of the triangle*. The triangle is *acute-angled* if all three angles are acute. The following two results are immediate consequences of Pythagoras' Theorem.

Lemma 4.7.1. *Let a, b, and c be three non-collinear Euclidean points. If one of the angles of the triangle abc is a right angle, then the other two angles are acute. If b is between a and the Euclidean point a', and one of $a\hat{b}c$ and $a'\hat{b}c$ is acute, then the other is obtuse.*

We could phrase a definition of the *measure* of the angle, in terms of coordinates, but it is not a simple process. Where needed we will accept the traditional use of a protractor as a practical means of measuring, or assigning a number to, an angle. The most common units of measurement for angles are *degrees* and *radians*, a protractor being usually graduated to measure in degrees. Our definition, based on Pythagoras' Theorem, enables us to conveniently distinguish between acute angles, right angles and obtuse angles. By protractor, the measure of an acute angle is less than 90°, and a right angle measures exactly 90°. For example, in Figure 4.17, $a_1\hat{b}c$ is acute, $a_2\hat{b}c$ is a right angle, and $a_3\hat{b}c$ is obtuse.

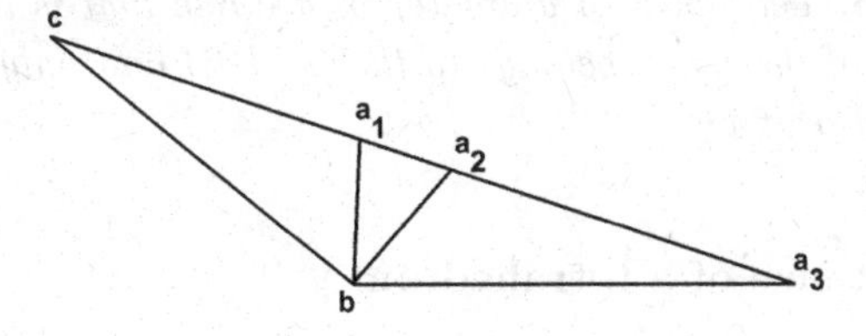

Fig. 4.17 Measuring angles

We are often concerned with modeling triangles. If the vertices of a triangle abc are Euclidean points, then we can model the triangle by cutting out a piece of cardboard bounded by the segments $[ab]$, $[bc]$, and $[ca]$. These segments are often called the *sides* of the triangle. We speak of the lengths of these segments as "the lengths of the sides of the triangle".

Simple experimenting, by drawing intersecting arcs with a compass, makes the following converse of Condition (iii) of Theorem 4.7.1 plausible:

Theorem 4.7.3. *There is a triangle with the Euclidean vertices a, b, and c if and only if $d(a,b) \leq d(b,c) + d(c,a)$, $d(b,c) \leq d(c,a) + d(a,b)$, and $d(c,a) \leq d(a,b) + d(b,c)$.*

Theorem 4.7.4. *Two triangles abc and $a'b'c'$ are* similar *if, for some real number k, $d(a,b) = kd(a',b')$, $d(b,c) = kd(b',c')$ and $d(c,a) = kd(c',a')$.*

Note that each two triangles in Figure 4.18 are similar.

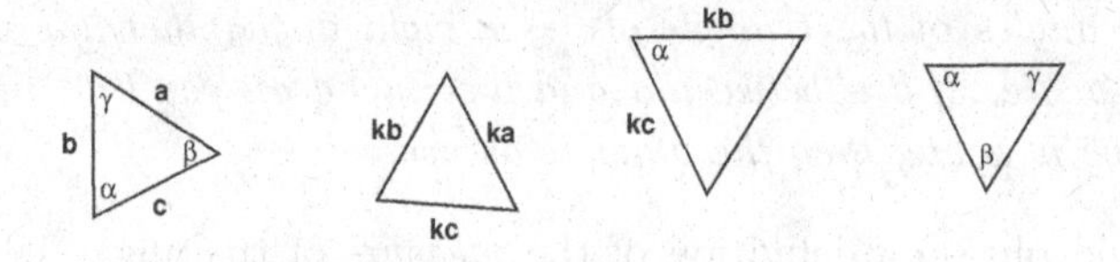

Fig. 4.18 Four similar triangles

Suppose that c is a Euclidean point in the plane P, and that r is any positive real number. The *circle* in P with *center* c and *radius* $r \geq 0$ is the planar set of points $\{p : d(c,p) = r\}$. If points a and b are collinear with c and belong to the circle, then $[ab]$ is called a *diameter* of the circle. The following result follows from the definitions of perpendicularity and distance and Pythagoras' Theorem.

Theorem 4.7.5. *Let $[ab]$ be a diameter of a circle that is in the plane P. Then a point x of the plane belongs to the circle if and only if the triangle axb is right-angled at x.*

4.8 The existence of a tetrahedron

In this section, we prove that the existence of a right-angled tetrahedron depends on the existence of a certain acute-angled triangle. This result helps our analysis in Chapter 8 of distortion in perspective drawings .

Definition 4.8.1. *We say that a tetrahedron $pv_1v_2v_3$ is* **right-angled** *at p if each two of the lines $p \vee v_1$, $p \vee v_2$, and $p \vee v_3$ are perpendicular.*

As the following theorem is neither commonly available nor intuitively obvious we give a detailed proof.

Theorem 4.8.1. *Let v_1, v_2 and v_3 be three non-collinear Euclidean points. Then a Euclidean point p exists so that the tetrahedron $pv_1v_2v_3$ is right-angled at p if and only if the triangle $v_1v_2v_3$ is acute-angled.*

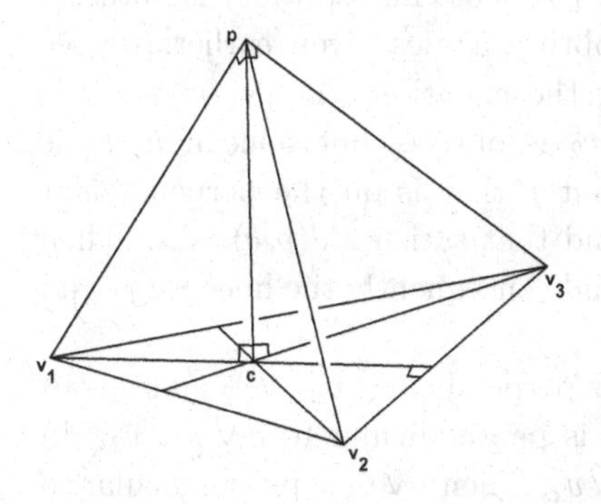

Fig. 4.19 A tetrahedron $pv_1v_2v_3$, right-angled at p

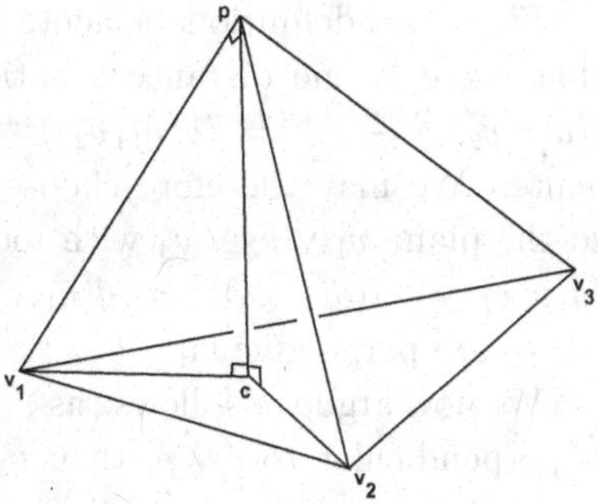

Fig. 4.20 Right-angles at p

Proof. First we suppose that a point p exists so that the tetrahedron $pv_1v_2v_3$ is right-angled at p. We repeatedly use Theorem 4.6.1. Let c be the foot of the perpendicular from p to the plane $v_1 \vee v_2 \vee v_3$. As $p \vee v_1$ is perpendicular to $p \vee v_2$ and to $p \vee v_3$, then it is perpendicular to every extended Euclidean line in $p \vee v_2 \vee v_3$. In particular it is perpendicular to the line $v_2 \vee v_3$. As $p \vee c$ is perpendicular to each extended Euclidean line in the plane $v_1 \vee v_2 \vee v_3$, it is perpendicular to the line $v_2 \vee v_3$. Thus $v_2 \vee v_3$ is perpendicular to both $p \vee v_1$ and to $p \vee c$, and hence to each extended Euclidean line in the plane $p \vee c \vee v_1$. In particular $v_2 \vee v_3$ is perpendicular to the line $v_1 \vee c$. In a similar way we prove that the line $v_2 \vee c$ is perpendicular to the line $v_3 \vee v_1$, and the line $v_3 \vee c$ is perpendicular to the line $v_1 \vee v_2$. The lines $v_1 \vee c$, $v_2 \vee c$, and $v_3 \vee c$ are called the *altitudes* of the triangle $v_1v_2v_3$ and their point of concurrence is the *orthocenter* of the triangle.

From an application of Pythagoras' Theorem to each of the right-angled triangles v_1pv_3 and v_2pv_3 we have that $d(v_1, v_3) > d(p, v_1)$ and $d(v_2, v_3) > d(p, v_2)$. Combining these facts with an application of Pythagoras' Theorem to the triangle v_1pv_2 gives that $(d(v_1, v_3))^2 + (d(v_2, v_3))^2 > (d(v_1, v_2))^2$ and the angle $v_1\hat{v}_3v_2$ is therefore acute. Similarly, each angle $v_2\hat{v}_1v_3$ and $v_3\hat{v}_2v_1$ is acute and the triangle $v_1v_2v_3$ is acute-angled as required.

Conversely, we suppose that the triangle $v_1v_2v_3$ is acute angled. Then let u_k be the intersection of the line $v_i \vee v_j$ with the perpendicular from the point v_k, for $\{i,j,k\} = \{1,2,3\}$. If u_k is not in the segment $[v_iv_j]$, but, without loss of generality, has v_j between it and v_i, then $v_k\hat{v}_ju_k$ is acute, and $v_k\hat{v}_jv_i$ is obtuse. This is a contradiction to our assumption that the triangle $v_1v_2v_3$ is acute-angled. Therefore each u_k is in the segment $[v_iv_j]$. Let $c = (v_1 \vee u_1) \cap (v_2 \vee u_2)$. Then $v_2\hat{c}u_1$ is acute and so $v_2\hat{c}v_1$ is obtuse.

From the definition of acute and obtuse angles given earlier, we see that there is one distance x satisfying the equation $((d(c,v_1))^2 + x^2) + ((d(c,v_2))^2 + x^2) = (d(v_1,v_2))^2$ if $v_1\hat{c}v_2$ is obtuse, but none if $v_1\hat{c}v_2$ is acute. We may therefore choose a point p that is on the perpendicular to the plane $v_1 \vee v_2 \vee v_3$ with foot c and that satisfies $d(p,c) = x$. Then $(d(p,v_1))^2 + ((d(p,v_2))^2 = (d(v_1,v_2))^2$ and consequently the lines $p \vee v_1$ and $p \vee v_2$ are perpendicular.

We now argue as follows: as $v_1 \vee v_3$ is perpendicular to $c \vee v_2$ and $v_1 \vee v_3$ is perpendicular to $c \vee p$, then $v_1 \vee v_3$ is perpendicular to $c \vee p \vee v_2$. In particular, $v_1 \vee v_3$ is perpendicular to $p \vee v_2$. Then $p \vee v_2$ is perpendicular to $p \vee v_1 \vee v_3$ so $p \vee v_2$ is perpendicular to $p \vee v_3$. Similarly $p \vee v_1$ is perpendicular to $p \vee v_3$. Thus the tetrahedron $pv_1v_2v_3$ is right-angled at p. □

We note that there are exactly two suitable tetrahedra, one obtained by choosing the point p above the plane $v_1 \vee v_2 \vee v_3$ and the other by choosing p the same distance from c, but below the plane.

Summary of Chapter 4. We have examined projective spaces in some detail. These are non-planar geometries in which each pair of coplanar lines have a non-empty intersection, and in which each plane and line have a non-empty intersection. Not every geometry is a subgeometry of a projective space - contrasting with the case of planar geometries and projective planes. However, Desargues' Theorem is valid in every projective space.

Combinatorial figures can be thought of as subgeometries of a convenient projective space, namely extended Euclidean space. Consequently, we now know Desargues' Theorem to be true in the Euclidean geometry that describes our world.

We laid a foundation for future algebraic investigations into the geometry of extended Euclidean space by allocating homogeneous coordinates to each point and plane, providing an alternative technique for obtaining geometric results, and establishing a link between geometric truth and algebraic truth.

Chapter 5

PERSPECTIVE DRAWINGS

Drawing, in the usual artistic sense, is an attempt to convey information about a non-planar figure to a viewer via a planar figure. Many different methods of depicting subjects have evolved over centuries of artistic endeavour. For example, in some cultures, the more important of two individuals is drawn larger - regardless of the individuals place in the scene. In pre-Renaissance Russian icons, distant subjects were often drawn larger than the same-sized subjects closer to the artist.

It was only during the Renaissance that a clearly understood basis for a more realistic treatment of the relationship between subject and image evolved. This understanding, using scientific method and mathematics, was one of the great achievements of the Renaissance. It began with, sometimes incorrect, complicated rules for drawing very special subjects. The painting "*The Last Supper*" (circa 1308) by di Buoninsegna, shown in Figure 5.1, is a not wholly successful product of these beginnings. The understanding developed throughout the 15th and 16th centuries in the work of Alberti, Brunelleschi, da Vinci and many others [44], and began the modern theory of perspective drawing. It culminated with the development of projective geometry by Desargues in his systematic study of perspective drawing in the 17th century.

We use our knowledge of extended Euclidean space to begin such a study. This space allows us to define and obtain basic properties of perspective drawings and learn how to construct simple examples. We simplify and unify many, apparently *ad hoc*, standard drawing techniques by deriving them in the context of extended Euclidean space. But we should remember that art is not just the realistic recording of scenes on canvas. Post-Renaissance artists use skills that are not scientifically analyzed to give a message to a viewer's emotions - just as pre-Renaissance artists did.

Fig. 5.1 The Last Supper

5.1 The definition and basic properties of a perspective drawing

We may think of an artist's canvas as a transparent screen through which the artist views a figure with one eye and traces what is seen. Accordingly, we define the process of perspective drawing as a function that associates each point of one combinatorial figure, the subject of the drawing, with a point of another combinatorial figure, the artist's drawing. We can think of perspective rendition as the process of drawing and we can think of a perspective drawing as the result of this process.

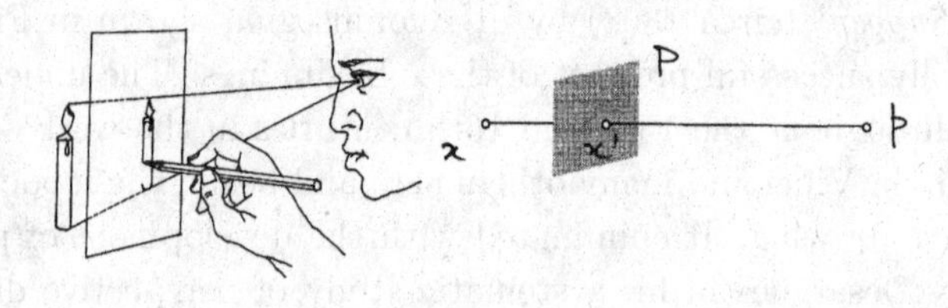

Fig. 5.2 Drawing a point in practice and in theory

Definition 5.1.1. *Let E be a combinatorial figure, p be a point that is not in E, and P be an extended Euclidean plane not containing p. The function that associates each point x of E with the point $x' = (x \vee p) \cap P$ of P is a* **perspective rendition** *of the figure E. We say that x' is the image of x, for each x in E, and call the combinatorial figure $E' = \{x' | x \in E\}$ a perspective drawing of E, drawn from the viewpoint p in the plane P.*

Example 5.1.1. An image in a camera obscura and a photograph taken by a pinhole camera are examples of perspective drawings.

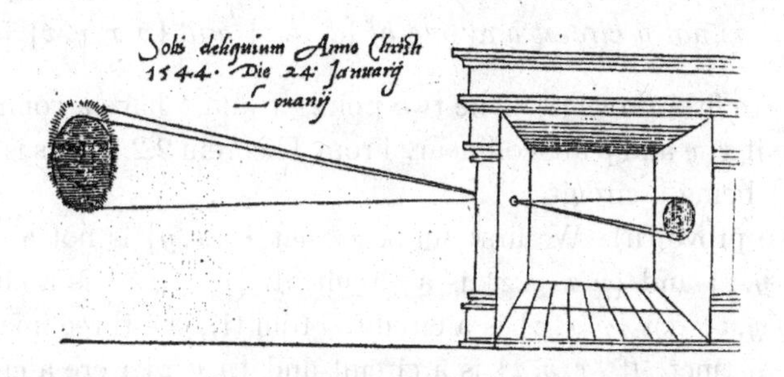

Fig. 5.3 A 16th century illustration of a *camera obscura*

Each perspective drawing is a planar figure. A useful feature of this form of art is exactly this property, perspective drawings being quite easy to carry and stack. However this same feature guarantees that, in general, a perspective drawing is not isomorphic to the subject drawn. It is the "squashing" of images, as in two of the four perspective renditions shown in Figure 5.4, that may prevent a perspective drawing of a figure being isomorphic to the figure. We now specify the conditions under which this "squashing" occurs.

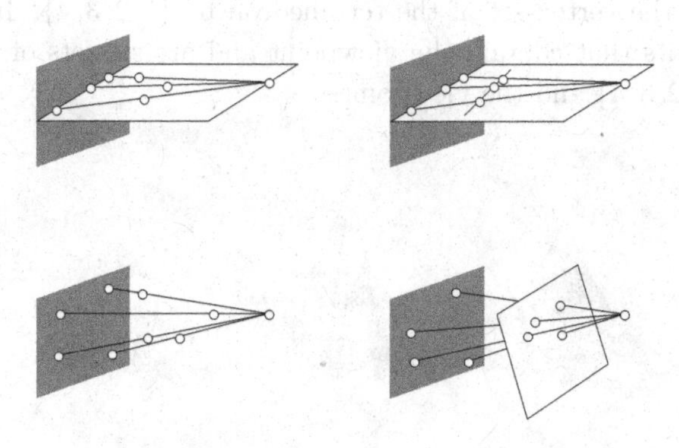

Fig. 5.4 Circuits in perspective drawings

Proposition 5.1.1. *Let E' be a perspective drawing of a combinatorial figure E, drawn from a viewpoint p. Suppose that x, y and z are three* *points of E. Then (i) The two points x and y have a common image if and only if the set $\{p, x, y\}$ is a circuit; and*

(ii) The set of images of x, y and z is a circuit if and only if both the set $\{p, x, y\}$ is not a circuit and one of $\{x, y, z\}$ and $\{p, x, y, z\}$ is a circuit.

Proof. We first prove (i). The two points x and y have a common image if and only if x, y and p are collinear. From Theorem 2.2.2, this is equivalent to $\{x, y, p\}$ being a circuit.

Now we prove (ii). We first suppose that $\{p, x, y\}$ is not a circuit and one of $\{x, y, z\}$ and $\{p, x, y, z\}$ is a circuit. If $\{p, x, y, z\}$ is a circuit, then neither $\{p, y, z\}$ nor $\{p, x, z\}$ is a circuit. From (i), the three images of x, y and z are distinct. If $\{x, y, z\}$ is a circuit and $\{p, y, z\}$ were a circuit, then Condition **C4** would lead to the set $\{p, x, y\}$ being a circuit. As this is not so, $\{p, x, y\}$ is not a circuit. A similar argument shows that $\{p, x, z\}$ is not a circuit. Thus the images of each of x, y and z are distinct. In both cases the set of the three images is in the intersection of the plane $p \vee x \vee y$ and the plane of the perspective drawing. It is therefore a three-point circuit.

Conversely, from (i) we know that none of $\{p, x, y\}$, $\{p, y, z\}$ or $\{p, x, z\}$ is a circuit. The points x, y and z are coplanar with the line that contains the image of each and contains the point p. Therefore $\{p, x, y, z\}$ contains a circuit. This circuit can only be $\{x, y, z\}$ or $\{p, x, y, z\}$. □

Exercise 5.1.1. Figure 5.5 shows three perspective renditions of a tetrahedron. Let the vertex-set of the tetrahedron be $\{1, 2, 3, 4\}$. In each case list the circuits that contain the viewpoint and are subsets of the 5-point union of $\{1, 2, 3, 4\}$ and the viewpoint.

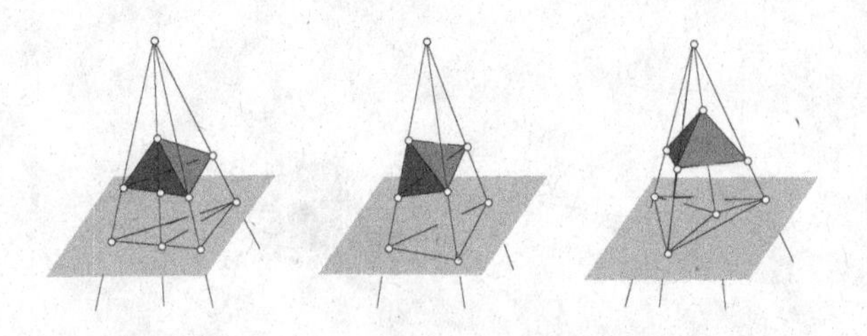

Fig. 5.5 Three perspective renditions of a tetrahedron

Proposition 5.1.1 gives the required conditions for a perspective drawing to accurately reflect the three-point circuit structure of its subject. It suggests the following definition.

Definition 5.1.2. *Let E be a figure. A point p that does not belong to any three-point or four-point circuit of $E \cup \{p\}$ is a* **general viewpoint** *of E.*

Example 5.1.2. The question of which of the six figures in Figure 1.13 is a perspective drawing of a model of a 4-gon is equivalent to the question asked in Exercise 1.2.1. The first is drawn from a general viewpoint of the 4-point figure of the vertices of the 4-gon but not of the model (remembering that the model consists of four segments). The viewpoint of each of the second, third, and fourth perspective drawings is neither a general viewpoint of the 4-point figure nor of the model. Neither the fifth nor sixth figure is a perspective drawing. Perspective renditions of the first and second figure are shown in Figure 5.6.

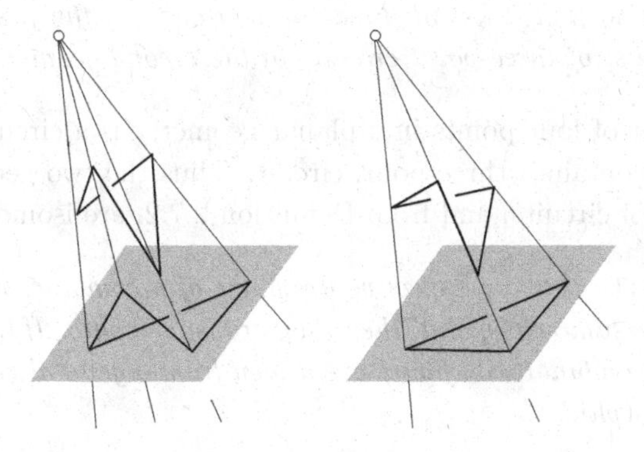

Fig. 5.6 Two perspective renditions of a model 4-gon

We now prove that perspective rendition from a general viewpoint preserves the three-point circuit structure of a figure. This leads us to two questions. First, what difference is there between perspective drawings of the same figure? Second, under what circumstances is a figure isomorphic to its perspective drawing?

Proposition 5.1.2. *The perspective rendition of a combinatorial figure from a general viewpoint is a pairing of the points of the figure with the points of its perspective drawing. If paired points are given the same label, then the list of three-point circuits of the figure is identical to the list of three-point circuits of the perspective drawing, and the list of lines of the figure is identical to the list of lines of the perspective drawing.*

Proof. From condition (i) of Proposition 5.1.1 we have that the association of each point with its image is the required pairing. It follows from condition (ii) of Proposition 5.1.1 that the two lists of 3-point circuits are identical. It then follows from Theorem 2.2.2 that the two lists of lines are identical. □

Corollary 5.1.1. *Any perspective drawing of a non-planar Desargues figure, drawn from a general viewpoint, is a planar Desargues figure.*

Proof. From Proposition 5.1.2 we know that the collection of circuits of the perspective drawing satisfies the requirements of Definition 2.3.3. □

Lemma 5.1.1. *Two planar geometries are isomorphic if the points of the first can be paired with the points of the second, and paired points given the same label so that a list of three-point circuits of the first geometry is identical to a list of three-point circuits of the second geometry.*

Proof. A set of four points in a planar geometry is a circuit if and only if it does not contain a three-point circuit. Thus the two geometries have identical lists of circuits, and from Definition 2.7.2, are isomorphic. □

Theorem 5.1.1. *If two perspective drawings of a combinatorial figure are drawn from the same viewpoint, then they are isomorphic. If two perspective drawings of a combinatorial figure are drawn from a general viewpoint, then they are isomorphic.*

Proof. A perspective drawing of a figure is also a perspective drawing of any other perspective drawing of the original figure that is drawn from the same viewpoint. Applying Proposition 5.1.2 and Lemma 5.1.1 to this second perspective rendition, and noting that any viewpoint of a planar figure that is not in the plane of the figure is a general viewpoint, we have the first isomorphism.

Applying Proposition 5.1.2 in turn to each of two perspective renditions of a figure from general viewpoints gives labelings of the perspective drawings for which the two perspective drawings have identical lists of 3-point circuits. Lemma 5.1.1 guarantees that the two perspective drawings are isomorphic. □

Next we prove that a figure and any perspective drawing of it that is drawn from a general viewpoint are as "alike" as possible. Unintended impressions of collinearity are avoided, no points are "lost" in the image, and a good impression of the pictured figure is given. However the result is not perfect in every case.

Proposition 5.1.3. *Let a perspective rendition of a combinatorial figure E from a general viewpoint be given. Then any restriction of the perspective rendition to a subgeometry of E is an isomorphism if and only if the subgeometry is planar.*

Proof. The viewpoint is also a general viewpoint for the restriction of the rendition to any subgeometry E_1 of E. If E_1 is planar, then we apply Proposition 5.1.2 and Lemma 5.1.1 to prove E_1 isomorphic to its image. Conversely, if E_1 is isomorphic to its image then every four-point subset of E_1 contains a circuit and by Theorem 2.2.4 the set E_1 is planar. □

We are now able to state exactly the requirement of a perspective drawing for it to accurately portray the geometric structure of the figure drawn.

Theorem 5.1.2. *Let a perspective rendition of a combinatorial figure be given. Then the perspective rendition is an isomorphism between the figure and its perspective drawing if and only if both the figure is planar and it is drawn from a general viewpoint.*

Proof. If the figure is drawn from a general viewpoint, then Proposition 5.1.3 implies that the figure is planar if and only if it is isomorphic to its perspective drawing. To complete the proof we need only note that Proposition 5.1.1 guarantees that only a rendition from a general viewpoint can be an isomorphism. □

Consequently, the most useful perspective drawings are those drawn from general viewpoints, shown by Theorem 5.1.2 to contain as much as possible of the geometric structure of their subject. Unfortunately it is not always feasible to restrict our discussion to perspective drawings from general viewpoints. The following lemma proves that a figure containing the union of two skew lines has no general viewpoint.

Lemma 5.1.2. *If A and B are two skew lines and p is a point on neither, then there is a unique three-point circuit that contains p, a point of A, and a point of B.*

Proof. The plane $p \vee A$ meets the line B in a point b. The line $p \vee b$ is the required line as it is coplanar with A and therefore meets A in a point a. The linear set $\{a, b, c\}$ is the required circuit. □

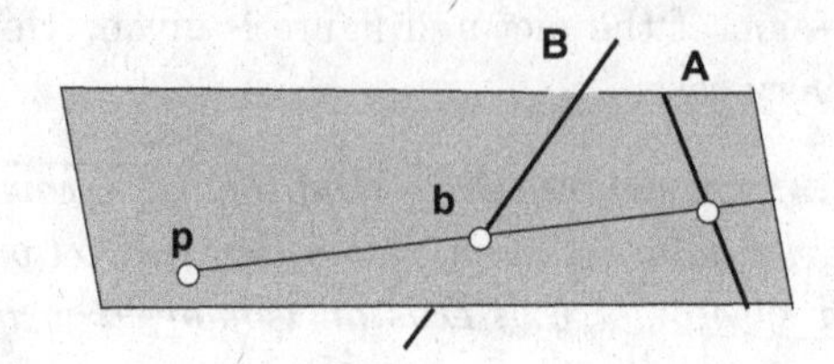

Fig. 5.7 A line intersecting each of two skew lines

5.2 Perspective drawings in a combinatorial geometry

In this section, we briefly indicate how perspective drawings in geometries that are not figures might be defined. Our use of a non-planar Desargues figure in the second part of the proof of Theorem 4.3.1 suggests that the notion of perspective drawing may be applicable in any projective space. Indeed, Definition 5.1.1 can be used in any projective space, and consequent results hold in that more general context.

In fact our definition of a geometry in terms of its circuits allows the notion of a "perspective drawing" of a geometry $\mathbf{G}$ to be defined even if $\mathbf{G}$ is not a subgeometry of a projective space. The following result shows how this is achieved. In it the planar geometry $(E', \mathbf{C}')$ plays the role of the perspective drawing of $\mathbf{G}$.

Theorem 5.2.1. *Let* $\mathbf{G}$ *be a geometry, where* $E(\mathbf{G}) = E' \cup \{p\}$ *and* p *is not in* E'. *If no three-point circuit of* $\mathbf{G}$ *contains* p, *then the collection of minimal members of the set* $\mathcal{C}' = \{C - \{p\} : C \in \mathcal{C}(G)\}$ *is the collection of circuits of a planar geometry on* E'.

Proof. The proof is left as a long but straightforward exercise in which the validity of the four conditions **C1** to **C4** of Definition 2.1.1 for the set $\mathcal{C}(G)$ enables one to prove their validity for the set $\mathcal{C}'$. □

However we will not pursue this idea further here, instead restricting our attention to perspective drawings in our world, that is, in extended Euclidean space. In the remainder of this chapter our discussions concern only figures, the subgeometries of extended Euclidean space.

5.3 Practical perspective drawing methods

In this section, we develop some practical methods of constructing perspective drawings of some simple figures. As the artist cannot, in practice, peer through his canvas with one eye at his proposed viewpoint, the other closed, and trace on it what he sees he must use other techniques based on Lemmas 5.3.1 and 5.3.2 for locating image points.

Fig. 5.8 Artist draws a Lute

Mechanical aids were used, as we see from Albrecht Dürer's "*Artist draws a Lute*", but the process shown is cumbersome. Nowadays, we can make a photographic slide, project it, and copy its main outlines on the canvas - and many artists do this. It is not at all obvious, nor even necessarily true, that the result of this process is a perspective drawing.

We can also now use computer help, but we first need give the computer correct rules for carrying out its task. These are just the rules that artists and geometers have spent many of the last five hundred years developing and practicing. They include the simple rules we derive below. Using them is the only technique available when beginning the outlines of a physically possible scene drawn from the artist's imagination.

The following basic result is the origin of the term "linear perspective" [51] used in art books to describe perspective rendition. From it we know that we need only find the images of two points of a line in order to construct the image of the line.

Lemma 5.3.1. *Let E' be a perspective drawing of the figure E. The image* *of any linear set of points of E is a linear subset of E'. In particular, the image of a line $a \vee b$ that does not contain the viewpoint of the drawing is the line containing the image of a and the image of b.*

Proof. Suppose that E is drawn in the plane P from the viewpoint p and that the linear set A is a subset of $a \vee b$. If $p \in a \vee b$, then the image of A is the single point $(a \vee b) \cap P$. If p is not in $a \vee b$, then suppose that x is any point of $(a \vee b) - \{a, b\}$, ensuring that $\{x, a, b\}$ is a circuit. Then, from Proposition 5.1.1, a and b have distinct images a', b' respectively and the images of x, a and b form a circuit. Therefore the image of x belongs to the line $a' \vee b'$. Any point $x' \in a' \vee b'$ is the image of the point of intersection of the coplanar lines $x' \vee p$ and $a \vee b$, and consequently $a' \vee b'$ is the image of $a \vee b$. □

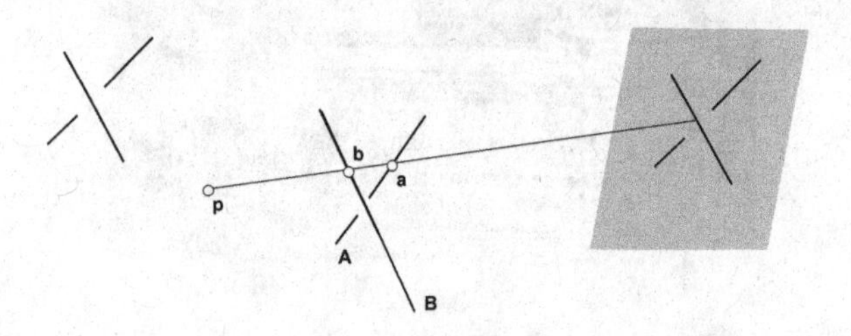

Fig. 5.9 A visibility convention for skew lines

We know from Lemma 5.3.1 above that the image of each of two skew lines in a perspective drawing is a line. When drawing in Chapter 1 we emphasized skew lines by drawing them as in the first figure of Figure 5.9. It is useful to refine this convention a little for use in a perspective drawing. Let the union of two skew lines A and B have a perspective drawing in which a point a of A and a point b of B have the same image. We draw them as in the second part of Figure 5.9 whenever we wish to emphasize the fact that b is between a and p. Such a visibility convention is useful for clarity, for example, in illustrating the answer to Exercise 1.2.1 or in illustrating the proof of Theorem 4.3.1, or even in Figure 5.9 itself.

It is clear that any planar figure is a perspective drawing of itself. Of more interest is the fundamental question of whether a given planar figure is a perspective drawing of some non-planar figure. We pursue this question with some success in Chapter 7 and later chapters.

Definition 5.3.1. *A* **scene** *is a perspective drawing of a non-planar figure.*

Exercise 5.3.1. Decide whether the visibility suggested in each of the figures of Figure 5.10 is compatible with each figure being a scene.

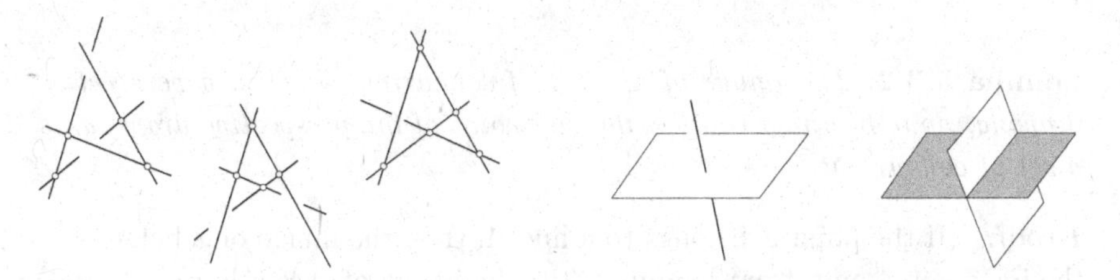

Fig. 5.10 Three possible scenes

Fig. 5.11 Visibility conventions for lines and planes

It is also sometimes useful, if part of a plane is farther from the viewpoint than part of a line or part of another plane (as in Figure 5.11).

Exercise 5.3.2. Decide whether the visibility suggested in each of the figures of Figure 5.12 is compatible with each being a perspective drawing of the union of three planes.

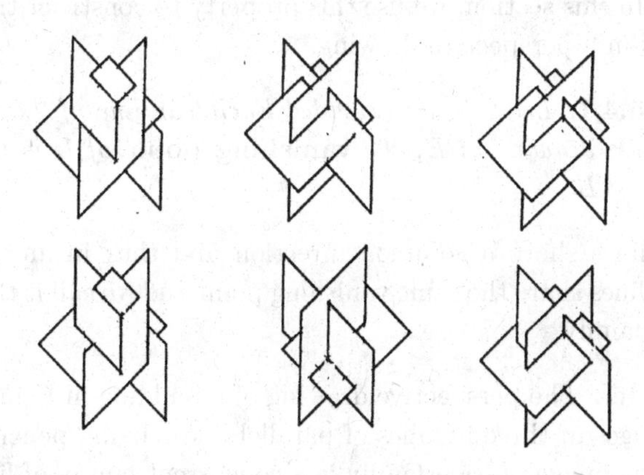

Fig. 5.12 Possible images of a set of three planes

We found in Exercise 5.3.1 that, even for a simple figure, it may not be easy to decide whether a given visibility is compatible with the figure being a scene.

Exercise 5.3.3. We know that both planar and non-planar Desargues figures exist. Try to devise a visibility for a planar Desargues figure that would be consistent with it being a scene. Either devise a visibility for a Pappus figure that would be consistent with it being a scene or prove that it is not a scene.

Lemma 5.3.2. *The image of any set of concurrent lines in a perspective drawing, none of which contain the viewpoint of the perspective drawing, is a set of concurrent lines.*

Proof. If the point x belongs to a line A, then the image of x belongs to the image of A and, from Lemma 5.3.1, the image of A is a line. □

Exercise 5.3.4. Do you have any comments on the illustration of a saw-horse from the Hobart, Tasmania "Mercury" newspaper reproduced in Figure 5.14?

5.4 Vanishing points

In the context of extended Euclidean space a family of parallel lines is concurrent. In this section, we use this property to construct the images of parallel lines in a perspective drawing.

Definition 5.4.1. *Let L be an extended Euclidean line of the figure E. In any perspective drawing of E, the* **vanishing point** *of L is the image of the direction of L.*

Parallel lines share a common direction and thus in any perspective drawing the lines share the same vanishing point and we call it the vanishing point of the family.

Exercise 5.4.1. The perspective drawing of a building in Figure 5.13 contains the images of three families of parallel lines. Using pencil and roller, check that the image of each family is a concurrent family of lines.

Exercise 5.4.2. One family of parallel lines drawn in "The Last Supper", shown in Figure 5.1, does not have a well-defined vanishing point. Find images of three of the lines of the family. Is the painting a perspective drawing?

As the world abounds with objects containing parts of parallel lines, an artist needs a quick and easy method of drawing images of a family

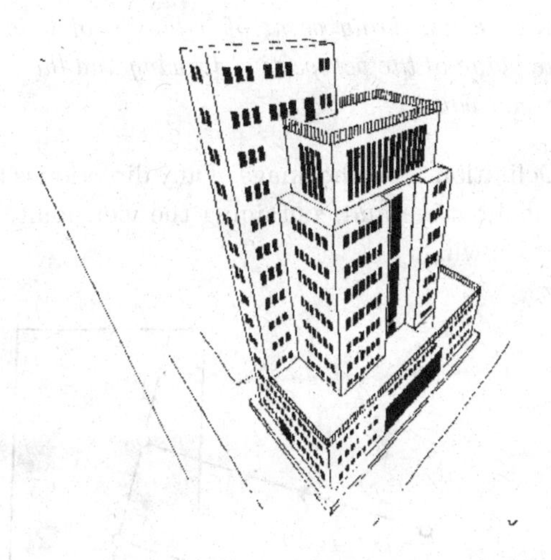

Fig. 5.13 A perspective drawing of a building

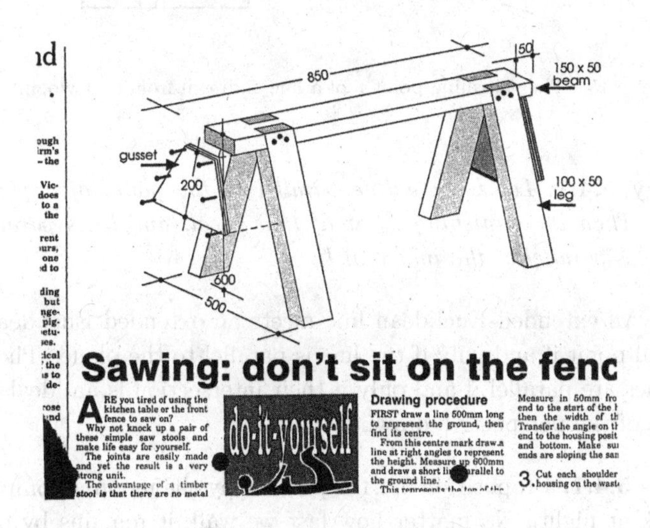

Fig. 5.14 A sawhorse

of parallel lines. The following theorem provides a practical method for locating the vanishing point of such a family. This point is on the image of each line of the family. Consequently, using Lemma 5.3.1, an artist needs only the image of one other point to be able to construct the image of a line of the family.

Theorem 5.4.1. *The vanishing point of a family of parallel lines is the intersection of the plane of the perspective drawing and the line of the family that contains the viewpoint.*

Proof. From Definition 5.1.1 the image of any direction is the intersection of the line, in that direction and containing the viewpoint, with the plane of the perspective drawing. □

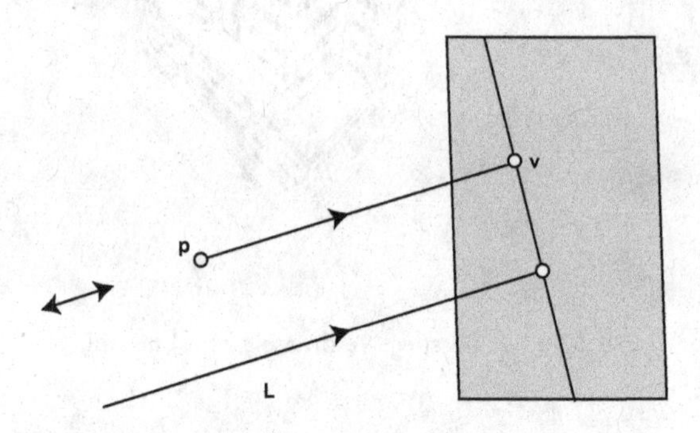

Fig. 5.15 The vanishing point v of a line L drawn from a viewpoint p

Corollary 5.4.1. *Let L be a line parallel to the plane of a perspective drawing. Then the vanishing point of L is ideal, and lines parallel to L have images parallel to the image of L.*

Proof. An extended Euclidean line meets an extended Euclidean plane in an ideal point if and only if the line is parallel to the plane. The images of two lines are parallel if and only if their intersection is an ideal point of the plane of the perspective drawing. □

Example 5.4.1. Proposition 5.4.1 explains why the moon accompanies us on a walk at night. No matter how fast we walk it remains by our side! All the light rays from it that reach us are approximately parallel, thus it would not matter where we chose to sit and draw, we would place the moon in the same part of our canvas.

Construction 5.4.1. Locate the vanishing point of the horizontal lines on the right hand side of the building drawn in Figure 5.13. Is any side of the building parallel to the plane of the perspective drawing?

Example 5.4.2. The plane of the following preliminary sketch for da Vinci's "Adoration of the Kings" is vertical and parallel to the front steps.

Fig. 5.16 Adoration of the Kings

Often the plane of a perspective drawing is vertical, as can be seen from the usual artist's easel. This is convenient for viewing, as pictures are usually hung vertically, and vertical lines then have vertical images.

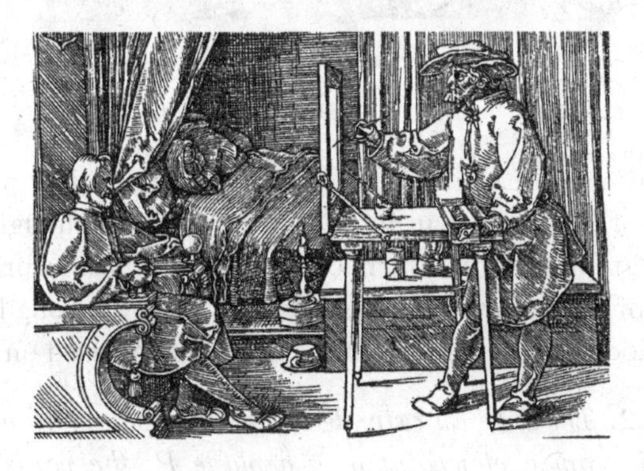

Fig. 5.17 An easel

Example 5.4.3. The plane of the perspective drawing in Figure 5.13 is not vertical, but the plane of Dürer's "Melencolia" in Figure 5.18 is vertical.

A photograph taken by a visitor to a city may not be in a vertical plane when exposed. This is one reason why it may look "distorted" - we usually "expect" vertical lines to have parallel images. How should we hold and look at a photograph of a tall building that is shot from ground level?

Fig. 5.18 Melencolia

There are many uses of vanishing points, and we are in a position to understand most of those given in instructional books, although we may find them couched in cumbersome terms. This should not surprise us as they are for the use of artists who do not necessarily have a strong background in geometry. We look at one more common application based on the following:

Theorem 5.4.2. *Let Q be an extended Euclidean plane. In any perspective drawing, drawn from a viewpoint p in a plane P, the vanishing points of lines parallel to Q are collinear, each belonging to the line of intersection of the plane P and the plane parallel to Q that contains the point p.*

Proof. The lines in question are exactly those whose directions belong to Q. A direction is in Q if and only if it is in a line parallel to Q and containing p. The image of each of these directions is therefore both in the plane parallel to Q that contains the point p and in the plane P. □

Corollary 5.4.2. *In any perspective drawing, the vanishing point of a horizontal line is in the intersection of the plane of the perspective drawing*

Fig. 5.19 A photograph of tall buildings

with the horizontal plane containing the viewpoint.

Artists call this line the *horizon line* of a perspective drawing. It is often the first construction line drawn to help them set out their picture.

Exercise 5.4.3. Find a drawing in which the horizon line plays a valuable role. If the picture shows the sea, then where is the horizon line? Why?

A sequence of five applications of Lemma 5.4.2 enables us to complete the perspective drawing of the roof in Figure 5.20. Referring to Figure 5.21, we give the required construction.

Construction 5.4.2. The point 1 is a vanishing point of a horizontal line, as is the point 2, each being the intersection of two images of parallel horizontal lines. Hence the line $1 \vee 2$ is the line of vanishing points of horizontal lines. Also the point 1 is a vanishing point of a line in the plane P, as is the point 3. Hence $1 \vee 3$ is the line of vanishing points of lines in the plane P. Similarly the line $2 \vee 3$ is the line of vanishing points of lines in the plane Q. The line B has vanishing point 2. The line A is in plane P and therefore has vanishing point 4. Each of A and B is in plane R. Consequently the line $2 \vee 4$ is the line of vanishing points of lines of R. Similarly the line $1 \vee 5$ is the line of vanishing points of lines contained in the plane S. The required line C is in both R and S. Therefore its vanishing point is $(2 \vee 4) \cap (1 \vee 5)$. Knowing two points of the line C enables us to draw the line.

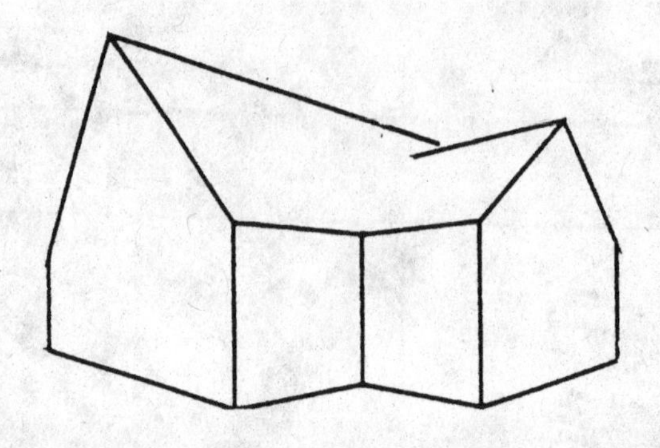

Fig. 5.20 A partially drawn house

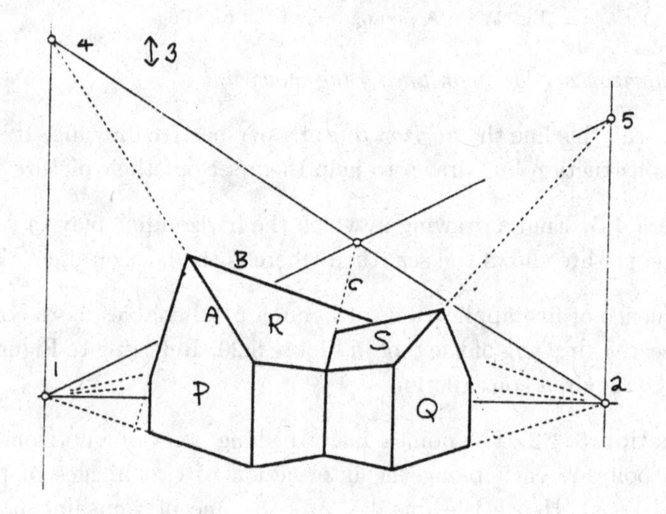

Fig. 5.21 Completion of the house roof

Construction 5.4.3. Let us test our mastery of perspective! Draw a book standing on an end, and having its pages fanned open.

5.5 Perspective drawings of a box

Many man-made constructions can be thought of as made up of several boxes. For example, the building shown in Figure 5.13 can be imagined as shown in Figure 5.22. Consequently perspective rendition of boxes is a basic artistic skill, and we investigate it in some detail in this section.

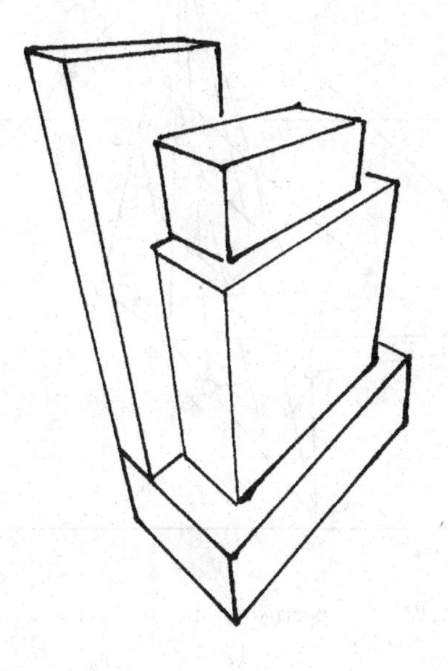

Fig. 5.22 A collection of boxes

Exercise 5.5.1. If you are able to, give a young child a piece of paper, pencil, and building block or some other box and ask for a drawing of the model. Repeat this procedure every six months or so. The resulting collection of drawings will arouse your curiosity.

To make a perspective drawing of a box, choose a viewpoint p and locate the vanishing point of each of the three families of parallel lines that contain edges of the box. Each vanishing point is on a line through the viewpoint that is parallel to an edge of the box. It is the intersection of such a line with the plane containing the page. Let $\{0, 1, 2, 3, 4, 5, 6, 7\}$ be the set of corners, or vertices, of the box. We label the vanishing point of the line $0 \vee i$ by v_i , for each $i = 1, 2, 3$.

Construction 5.5.1. We draw the box in the plane P containing the page, drawing the image $0' = (p \vee 0) \cap P$ of the point 0, drawing the image $1' = (p \vee 1) \cap P$ of of the point 1, drawing the image $2' = (p \vee 2) \cap P$ of the point 2, drawing the image $3' = (p \vee 3) \cap P$ of the point 3, and so on.

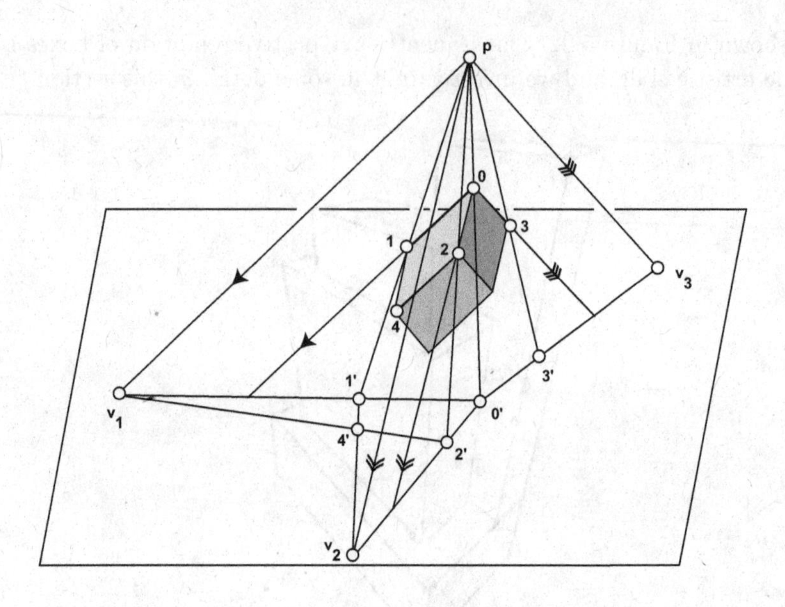

Fig. 5.23 Perspective rendition of a box

By using the three vanishing points and the points $0'$, $1'$, $2'$, and $3'$ we can save some time. As $2 \vee 4$ is parallel to $0 \vee 1$, then $2'$, $4'$ and v_1 are collinear. Similarly $1'$, $4'$ and v_2 are collinear, giving $4' = (v_1 \vee 2') \cap (v_2 \vee 1')$. Similar arguments give $5' = (v_3 \vee 1') \cap (v_1 \vee 3')$, $6' = (v_2 \vee 3') \cap (v_3 \vee 2')$ and $7' = (v_1 \vee 6') \cap (v_2 \vee 5')$.

Completing the image, we notice that the point $7'$ is on the three lines $v_1 \vee 6'$, $v_2 \vee 5'$ and $v_3 \vee 4'$. Consequently the resultant image is automatically a closed planar figure and we have the following result:

Lemma 5.5.1. *The eight-point figure consisting of the images of the vertices of a box in a perspective drawing, drawn from a general viewpoint of the eight-point set of vertices, is a closed planar figure.*

The above construction produces the images $0', 1', 2', 3', 4', 5', 6', 7'$, of the vertices 0, 1, 2, 3, 4, 5, 6, 7 of the box. We adopt the time and notation saving habit of labeling the image of a point by the point, so rather than using i' as a label for the image of i in the above perspective drawing we label the image of i by i, for all $i = 0, 1, 2, \ldots, 7$. As we mentioned following Construction 1.2.2, it is common practice to speak of "a box" when it is understood that we mean "a perspective drawing of a box".

When drawing a box we have several choices of presentation, as for example in Figure 5.24. Do we show just the visible surfaces, do we highlight the vertices, do we show all the edges, or do we highlight some edges according to the visibility conventions we discussed earlier? The answer depends on the context and the purpose of making the drawing. Is it part of an artistic scene, is it part of a construction blueprint? The answer determines the detail of the presentation.

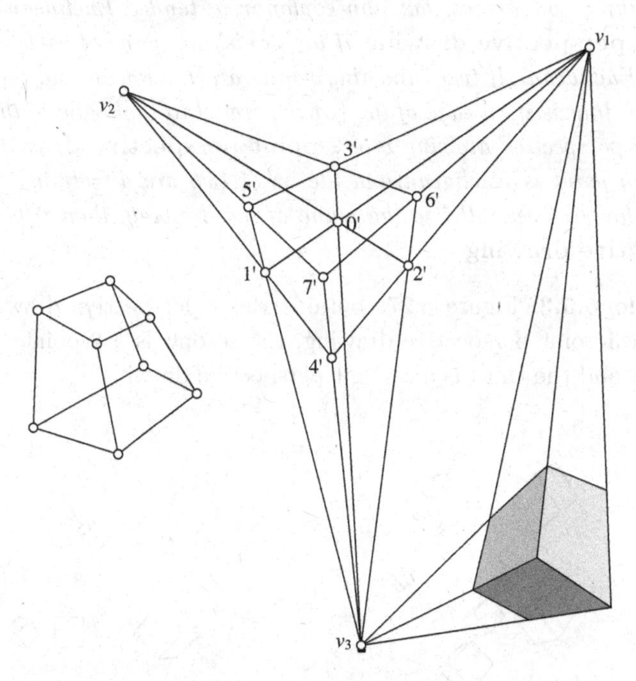

Fig. 5.24 Different drawing styles

Construction 5.5.2. Using a roller and a pencil, draw three boxes, that each have the same vanishing points, from the one viewpoint. Shade some faces, and then erase parts of the construction, leaving only the surfaces of boxes that are visible from the viewpoint.

Example 5.5.1. Figure 5.25 is a perspective drawing of several boxes. Each of the boxes has the same three vanishing points.

Example 5.5.2. Figure 5.26 is a perspective drawing of several boxes, no two boxes having the same vanishing points.

There is a striking difference between, for example, the perspective drawings in Figure 5.13 and in Figure 5.16, or between the various boxes in Figure 5.26, or between the three boxes in Figure 5.27. This is explained by taking note of the following distinction between perspective drawings of any figure that contains three important families of parallel lines:

Definition 5.5.1. *A perspective drawing of a box, or of any figure containing three concurrent, but non-coplanar, extended Euclidean lines, is a* **3-point perspective drawing** *if the vanishing point of each of the three lines is Euclidean. If two vanishing points are Euclidean and the third is a direction, that is if an edge of the box is parallel to the plane of the drawing, then the perspective drawing is a* **2-point perspective drawing**. *If one vanishing point is Euclidean and the other two are directions, that is if a face of the box is parallel to the plane of the drawing, then it is a* **1-point perspective drawing.**

Example 5.5.3. Figure 5.27 contains three perspective drawings. The first is a 3-point perspective drawing, the second is a 2-point perspective drawing, and the third is a 1-point perspective drawing.

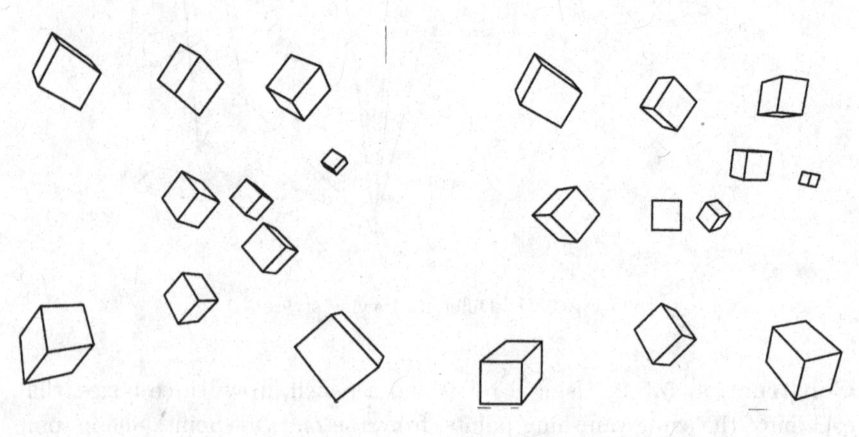

Fig. 5.25 A perspective drawing of several boxes

Fig. 5.26 A perspective drawing of more boxes

An advantage of using extended Euclidean space as the setting for our discussions is that 1-point, 2-point and 3-point perspective drawings can all be drawn with roller and pencil, starting with an appropriate triangle of vanishing points in the extended Euclidean image plane, and using Construction 5.5.1.

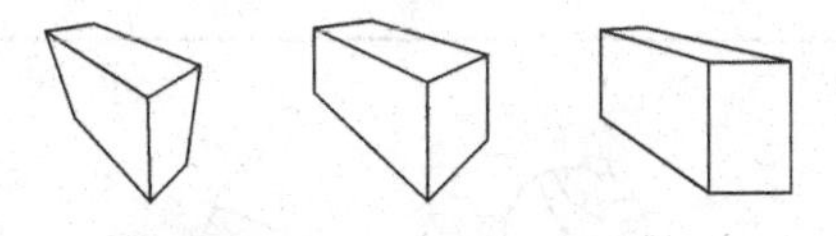

Fig. 5.27 Three perspective drawings of a box

Construction 5.5.3. Using a pencil and a roller, or a computer drawing package, place a box on a sheet of paper so that you may make a 3-point perspective drawing of it. Re-arrange the box in order to make a 2-point, and then a 1-point, perspective drawing of it.

Exercise 5.5.2. Decide whether the perspective drawing in Figure 5.28 is a 1-point, 2-point or 3-point perspective drawing.

Fig. 5.28 A scene containing images of parallel lines

Construction 5.5.4. Perhaps we should again test our mastery of perspective drawing. Let us make a 2-point perspective drawing of a small staircase of three steps.

Not every painting is drawn according to perspective rules. For example, "*Breakfast*" by the Cubist Juan Gris may not reward any attempt to find a preferred viewpoint. How do we explain the impression we have of the sides of a road converging into the distance away from us in both directions? Thinking about such questions led artists to wonder whether the rules of perspective are appropriate guides for their work. There is a practical difference between an artist's perspective drawing and the image in a pinhole camera. In applying Definition 5.5.1 an artist usually has the image between the artist's eye and the subject of the drawing, but the pinhole separates the image taken by a pinhole camera from the subject of

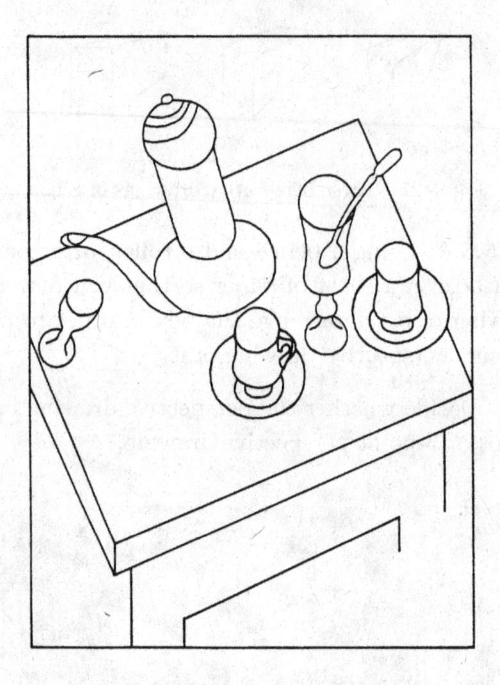

Fig. 5.29 Breakfast

the photograph. We have not had much experience examining images of points having the viewpoint between them and the picture plane! Let us experiment with such a perspective rendition.

Problem 5.5.1. *Imagine your head enclosed in a translucent box with one face parallel to a sheet of paper. Draw the box, including all its edges, on the paper.*

Exercise 5.5.3. Just in case we feel that we have mastered ideas that retained the attention of Renaissance giants, perhaps we could ask ourselves where we should sit to watch television or a movie?

Again we stress that artists try to affect our emotions and give us messages in whatever way they feel appropriate. Illusion and uncertainty may be part of the message. What, for example, do we make of Boring's "*Mother-in-Law*" (see Figure 5.31)?

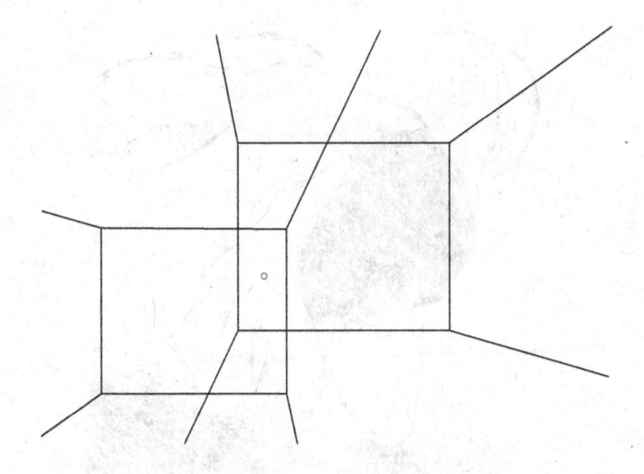

Fig. 5.30 A translucent box enclosing the head of the artist

5.6 Perspective rendition from an ideal viewpoint

In our discussions so far we have tacitly thought of viewpoints as Euclidean points, but our results hold for any point of extended Euclidean space as viewpoint. In fact it is common to use a direction as viewpoint in technical illustrations. In this section, we show some advantages of choosing a non-Euclidean viewpoint.

Definition 5.6.1. *An* **affine projection** *is a perspective drawing drawn from an ideal viewpoint.*

An affine projection image approximates to the photographs obtained using a telephoto lens, or the view through a telescope, of a distant scene. Clearly an affine projection image is best viewed from some distance! It is quickly drawn using a roller and pencil because of the following consequence of Theorem 5.4.1.

Theorem 5.6.1. *In an affine projection the vanishing point of each extended Euclidean line is a direction. Parallel lines have parallel images in an affine projection.*

Proof. A line containing the viewpoint and another direction is ideal and consequently the image of each direction is a direction. Thus, from Theorem 5.4.1, we have that each vanishing point is ideal. The images of the members of a family of parallel lines are lines containing the vanishing

Fig. 5.31 Mother-in-Law

point of the family. Thus in an affine projection parallel lines have parallel images. □

As a consequence, in any affine projection a rectangle has an image that is a parallelogram. This is the explanation for the convention we agreed on in Chapter 4 to denote planes. The illustrations in this book are often affine projections and typical rectangular pieces of pictured planes are drawn as parallelograms. For example, the four possible scenes of Exercise 5.3.2 are affine projections and are quickly drawn using a roller and a pencil.

The choice of a direction rather than a Euclidean point as viewpoint is much more convenient for small projective drawings. The choice of a Euclidean viewpoint that is far enough from the page to allow our eyes to comfortably focus on the page could give at least one Euclidean vanishing point some distance beyond the edge of the page. It is much easier to draw parallel lines with a roller and pencil, than to use the technique of Construction 3.5.4, to draw lines meeting, say, one meter off the page. But the difference between two such lines and two parallel lines is slight, and so a perspective drawing with an ideal viewpoint and a perspective drawing with a convenient Euclidean viewpoint may appear very similar.

Any box drawn in affine projection will have all three vanishing points ideal. In fact there is a very simple test to determine whether or not a particular perspective drawing is an affine projection.

Theorem 5.6.2. *A perspective drawing is an affine projection if and only if the vanishing points of three concurrent, but non-coplanar, extended Euclidean lines are ideal.*

Proof. Suppose that the viewpoint is Euclidean. Then at most two of the three lines are parallel to the plane of the perspective drawing, and so by Corollary 5.4.1, at most two of their vanishing points are ideal.

Conversely, if the viewpoint is ideal, by Proposition 5.6.1, all vanishing points are ideal. □

A quick examination of the saw-horse of Exercise 5.3.4, using a roller, suggests that it is an attempted affine projection.

Exercise 5.6.1. Figure 5.32 contains some perspective drawings of boxes. Decide which are 1-point perspective drawings, 2-point perspective drawings, 3-point perspective drawings, or affine projections.

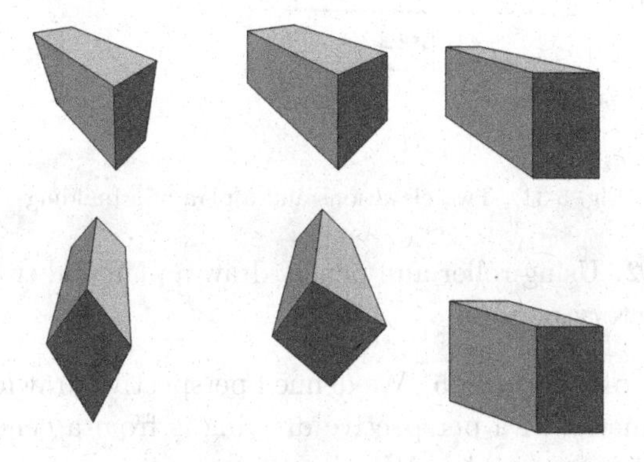

Fig. 5.32 Six possible perspective drawings of boxes

If an affine projection has, as its viewpoint, the direction orthogonal to its plane the projection is called an *orthographic* projection. Engineering and architectural drawings consist of, typically, three orthographic projections. The projection from the vertical direction is a *plan*, and the projections from horizontal directions are *elevations*.

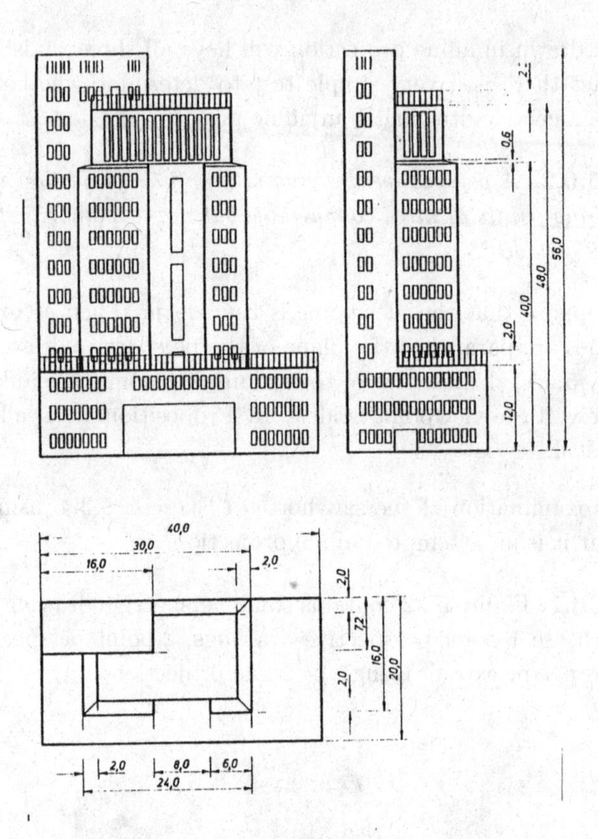

Fig. 5.33 Two elevations and a plan of a building

Exercise 5.6.2. Using roller and pencil, draw a plan and two elevations of an Olympic victory podium.

Summary of Chapter 5. We defined perspective drawings and examined their geometry. If a perspective drawing is from a general viewpoint it is as realistic as possible. We also gave conditions for a figure to be isomorphic to its perspective drawing.

We developed several practical techniques for drawing simple figures. Our use of extended Euclidean space enabled us to make different styles of perspective drawings in a unified way. We saw the simplicity of the rules for drawing from an ideal viewpoint. But we must remind ourselves that art remains more than the formulation of a collection of rules - no matter how mathematically appealing they may be.

Chapter 6

BINOCULAR VISION AND SINGLE IMAGE STEREOGRAMS

As useful as perspective drawing may be as a way of recording information about the world, it cannot exactly duplicate our vision, as we each see with two eyes. In this chapter we begin to unravel the mystery of how the human brain interprets simultaneous messages from the left eye and the right eye.

6.1 Binocular vision

We next analyze binocular vision and carry out some experiments. Each eye collects information, and this information is combined and interpreted by our brain. The process is complicated, but each of us continually and successfully carries it out. Let us begin to unravel a little of its mystery.

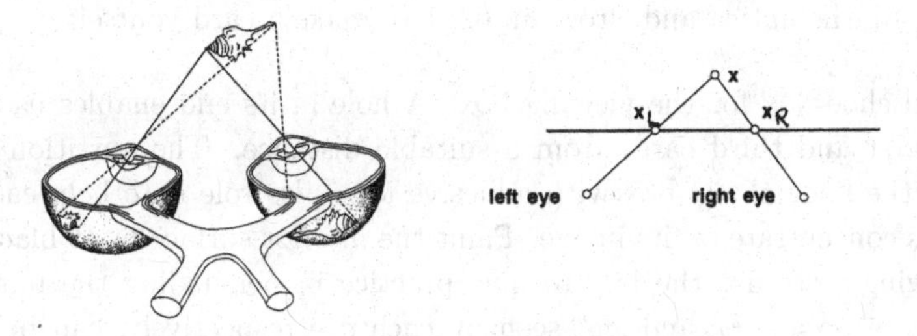

Fig. 6.1 Deceiving the Eyes

Looking at a point x, each eye turns towards x. If, instead of x, the left eye saw x_L and the right eye saw x_R as in Figure 6.1, could we be tricked into believing that we still saw the point x? There are some obstacles to such a deception. For example, each eye turns and changes the shape of its lens to focus on an object. We are aware of these physical states of our eyes and extract information from them about any viewed object. We will try a series of experiments to indicate whether or not we can trick our eyes and mind. A better way to look at the experiments is as a test of whether, and how, our minds create a non-planar world from only planar information.

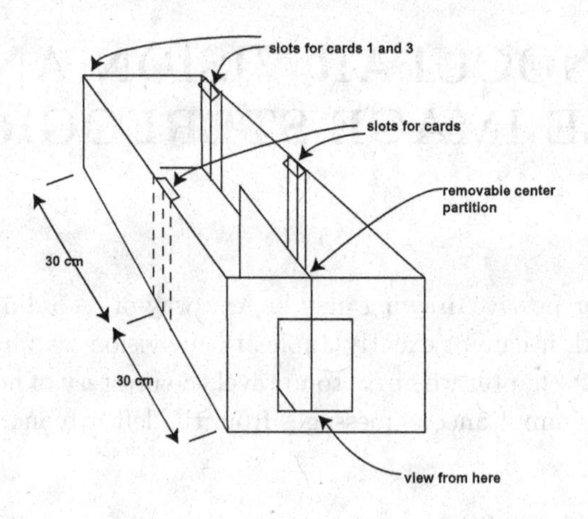

Fig. 6.2 A stereoscopic viewing box

Construction 6.1.1. Make a viewing box as indicated in Figure 6.2. Using a copier, enlarge each of the six cards shown in Figure 6.3 to the appropriate size or use Mathematica and Program 6.2.1 to make a card yourself.

Use old shoe-box for the viewing box. A hole in its end enables us to view the first and third cards from a suitable distance. The partition is hinged to the floor of the box with adhesive tape. Its role is to help each of our eyes concentrate on its image. Paint the interior surface matt black. The following tests use the box to give practice in persuading the mind that separate points x_L and x_R, seen by each eye respectively, can be a satisfactory substitute for a single point x seen by both eyes.

Exercise 6.1.1. Test whether you can distinguish between the first two cards, placing Card 1 60cm from the viewer, and then placing Card 2 30cm from the viewer as suggested in Figure 6.2.

Exercise 6.1.2. Repeat the experiment using Cards 3 and 4. Then look at each of the Cards 5 and 6, placing each 30cm from the viewer.

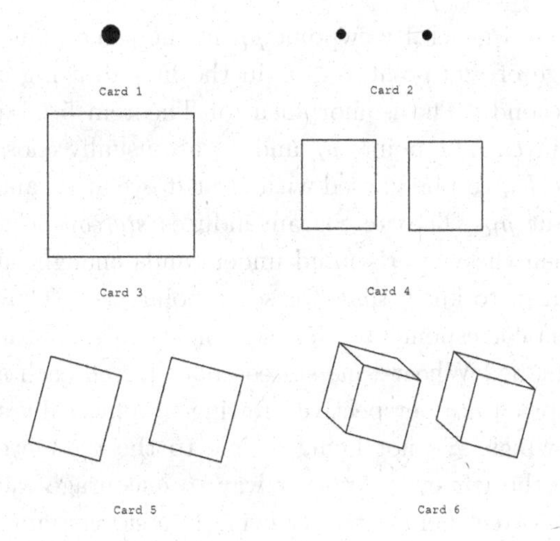

Fig. 6.3 Stereograms

Our viewing box is a rudimentary example of the once-popular stereoscope, invented by Charles Wheatstone [60] in the nineteenth Century. Cards 2, 4, 5, and 6 are simple examples of the photograph-pairs used in them, in which the two photographs are taken from slightly different positions. Their erstwhile popularity suggests that it must be possible to partially persuade our minds that when using them we are looking at a non-planar figure. Each of the photograph-pairs of the Wheatstone stereoscope and each card that we use in our viewing box is the result of combining two perspective drawings. We formalize this concept in the following definition.

Definition 6.1.1. *A* **stereoscope** *of a figure E is a pair of perspective renditions of E, each drawn from a general viewpoint, drawn in the same plane. The union of the two perspective drawings is a stereogram of the figure E.*

Just as a perspective drawing is planar, so is a stereogram. Both give information about non-planar figures in terms of planar figures.

Theorem 6.1.1. *The two perspective drawings in a stereogram are isomorphic.*

Proof. The result is an immediate consequence of Theorem 5.1.1. □

Let E_L be a perspective drawing of the figure E, drawn from a general viewpoint p_L in a plane P. Suppose also that E_R is a perspective drawing of E, drawn from a general viewpoint p_R in the same plane P. We write x_L for the image of any point $x \in E$ in the first drawing and x_R for its image in the second. The isomorphism of Theorem 5.1.1 pairs x_L with x_R, for each x in E. The points p_L and p_R are usually chosen so that the stereogram $E_L \cup E_R$ can be viewed with the left eye at p_L and the right eye simultaneously at p_R. The stereogram induces *stereopsis* (the perception of distance) when the viewer's mind understands enough, at least, of the above isomorphism to know that, for some points $x \in E$, an image of x_L on the left retina corresponds to an image of x_R on the right retina.

The old-fashioned Wheatstone stereoscopes forced each eye to concentrate on the appropriate perspective drawing by physically separating the perspective drawings, E_L not being visible to the right eye and E_R not being visible to the left eye. Another way to encourage each eye to concentrate on the correct perspective drawing in a stereogram $E_L \cup E_R$ is to color the figure E_L red and the figure E_R blue. Viewing through spectacles having a blue left lens and a red right lens lets each eye see only the appropriate drawing, avoiding the need for a viewing box. The blue lens absorbs the red light from E_L, causing its outline to appear black to the viewer. All light allowed through this lens is blue and so E_R is barely noticeable. Similarly the right lens encourages the right eye to concentrate on E_R. It is difficult to exactly match the colors of the lenses and figures in practice.

Exercise 6.1.3. Copy Card 6 of Figure 6.3 in red and blue, coloring the figure on the left-hand side red and the figure on the right-hand side blue. View it through spectacles with cellophane lenses, one red and one blue. Is the result more or less convincing than that experienced in Exercise 6.1.2?

With practice, we can view a card without using a viewing box or coloring images. Each eye of the viewer then sees both images on the card. If the eyes turn in slightly, then the viewer's mind identifies the inner image from each eye and interprets the result as a non-planar subject which in the case of Card 6, for example, is a box. The viewer then "sees" a box (flanked by a pair of flat drawings which are the images E_L and E_R seen by the right and left eye, respectively, and not not used in creating the central impression of the box). We might wonder how our minds cope with the problem, for example, of an Euclidean point x_L corresponding to a direction x_R. But, as we now see, in practice this does not arise.

Lemma 6.1.1. *If the line containing the viewpoints of a stereoscope is par-*

allel to the plane of the stereogram, then the isomorphism of Theorem 5.1.1 pairs Euclidean points with Euclidean points and directions with directions.

Proof. If x is a point collinear with p_L and p_R, then $x_L = x_R$. Otherwise the line $p_R \vee x$ is parallel to the plane of the stereogram if and only if the plane $x \vee p_L \vee p_R$ is parallel to the plane of the stereogram. An analogous argument shows that this is the case if and only if the line $p_L \vee x$ is parallel to P. Thus in each case x_L is Euclidean if and only if x_R is Euclidean. □

6.2 Binocular vision and random-dot stereograms

It has been argued that the process of believing that we are seeing non-planar objects when viewing a stereogram relies only on our prior knowledge that each image is an outline of a familiar non-planar object, in the above case a box, and does not rely on binocular vision at all. In this section, we investigate this claim. In a famous experiment, Adalbert Ames, Jr. covered the interior surfaces of a room by leaves. In monocular vision, the shape of the room was not strongly evident, but an observer was easily aware of its shape when viewing with both eyes. In 1960 the American psychologist and engineer Bela Julesz [39, 40] took this idea a step further and replaced any recognizable outline in a stereogram by a collection of pairs, x_L and x_R, of dots, each pair to represent a single point x, as in Figure 6.1. In this way no recognizable outline was available in E_L or in E_R to prompt the mind of the viewer of a stereogram $E_L \cup E_R$.

Problem 6.2.1. *Can all the information our brains receive during binocular vision be obtained by monocular vision alone?*

Solution. Place a copy of Card 7, shown in Figure 6.4, in the viewing box and examine it.

If you see something that you recognize as a simple non-planar figure, then you have disproved the argument mentioned above, and shown that you make unavoidable use of your binocular vision to observe the world around you in a way which cannot be duplicated by monocular vision. When you succeed you will be in no doubt. The clarity is quite stunning. Consequently, perspective drawing alone cannot be a fully satisfactory method of information transfer, in planar form, about our world.

If you have difficulty with this experiment there are a variety of suggestions, all designed to encourage you to turn your eyes in a little, so that the

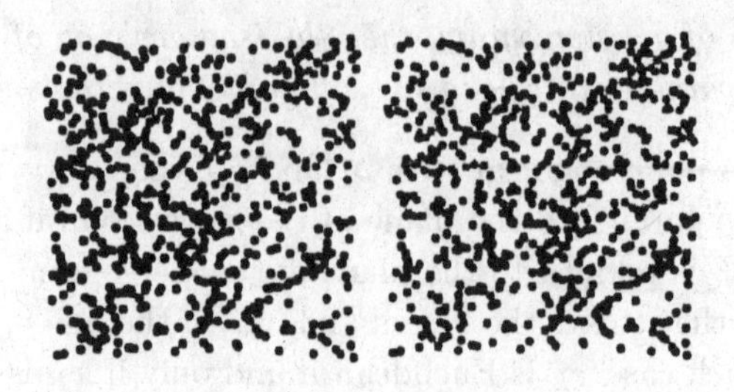

Fig. 6.4 Card 7

left looks directly at x_L and the right at x_R, for some point x beyond the card. It may help if you imagine that you are looking through the box to a point x beyond and let your eyes "glaze" a little, that is let the focus of their lenses wander. If still in difficulty, viewing Cards 8 and 9, above, in the viewing box may help you interpret stereograms containing randomly chosen points. The box outlines of Card 8 may help fuse the corresponding dots of the left and right images. Alternate it with Card 9 until you feel at home with the latter. Then try Card 7 in the viewing box again.

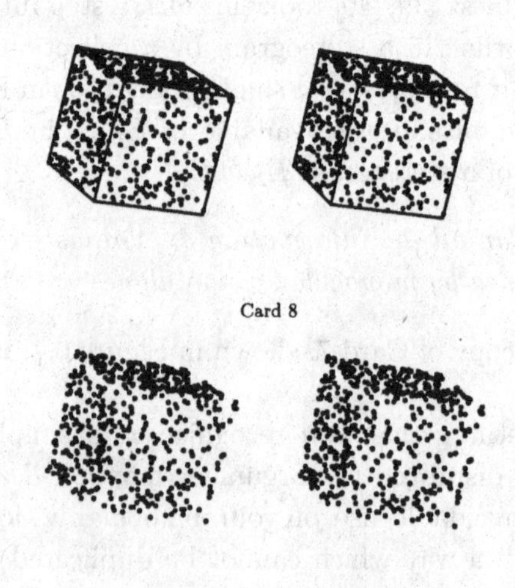

Fig. 6.5 Cards 8 and 9

Definition 6.2.1. *Let a figure E be chosen, the points of E forming a reasonably random collection on the surface of an object. Then any stereogram* *of the figure E is a* **random-dot stereogram** *of the object.*

The stereogram on Card 7 is an example of a random-dot stereogram. It was created by the following Mathematica program. Vary the function f in order to give different cards for the viewing box.

Program 6.2.1. Random Dot Stereogram of a Surface $f(x, z) = y$

```
leftViewPt={1,-3,-30,0}; rightViewPt={1,3,-30,0};
f[pair_]:=40+pair[[1]];
subjDomain=
Table[{Random[Real,{-3,3}],Random[Real,{-3,3}]},{700}];
stereoPt[m_]:={1,m[[1]],f[m],m[[2]]};
stereoSubject=Map[stereoPt, subjDomain];
leftShadow[subj_]:=Module[{u},u=m*subj+n*leftViewPt;
First[u/. Solve [{u[[1]]==1, u[[3]]==0}, {m,n}]]];
rightShadow[subj_]:=Module[{v},v=r*subj + s*rightViewPt;
First[v/. Solve[{v[[1]]==1, v[[3]]==0}, {r,s}]]];
dot[image_]:={image[[2]], image[[4]]};
leftImage=Chop[N[Map[leftShadow, stereoSubject], 3]];
leftStereo=Map[dot, leftImage];
rightImage=Chop[N[Map[rightShadow, stereoSubject], 3]];
rightStereo=Map[dot, rightImage];
RDStereogram=Graphics[{Map[Point, leftStereo],
Map[Point,rightStereo],
Line[{{-5,-5}, {5,-5}, {5,5}, {-5,5}, {-5,-5}}]}];
Show[RDStereogram, AspectRatio-> 1, PlotRange -> All]
```

The code "leftViewPt" and "rightViewPt" specify the two viewpoints from which the perspective drawings that constitute the stereogram are drawn. The surface being examined has equation $f(x, z) = y$. "SubjectDomain" is a square containing a random collection of pairs (x, z). The code "stereo-Subject" is the corresponding collection of points $(x, f(x, z), z)$. This collection is E, that part of the surface defined by f, the subject of the drawing. The function "leftShadow" specifies the perspective drawing "leftStereo" of "stereoSubject" that is drawn from the viewpoint of the left eye in coordinate form. Similarly, "rightShadow" specifies the perspective drawing that is drawn from the viewpoint of the right eye in coordinate form. The function "RDStereogram" collects the two drawings to form the required stereogram, and "Show" prints the stereogram. The "AspectRatio" code ensures that no distortion takes place in the setting out of the output.

6.3 Single-image stereograms

In this section, we show how to construct "random-dot stereograms". In Figure 6.6, when making stereograms of a figure E that is too big or too close, the two images E_L and E_R overlap in $E_L \cup E_R$ and their points are jumbled with each other. In other words, some points of the stereogram have two labels, one as a member of E_L and another as a member of E_R. Exercise 6.3.1 suggests that the only way to view such a stereogram is with colored spectacles.

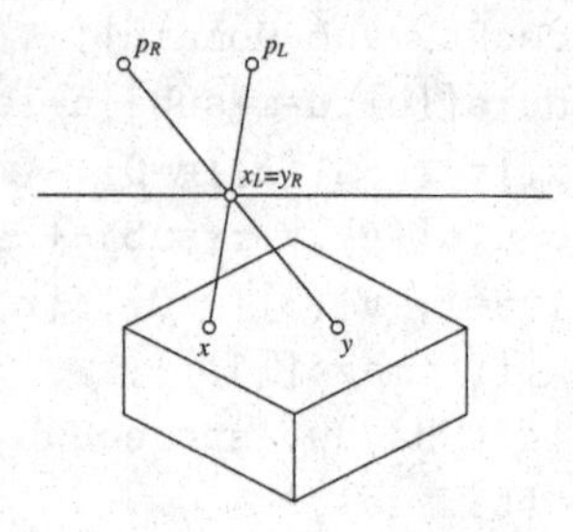

Fig. 6.6 A point of a stereogram that is seen by both eyes

Exercise 6.3.1. Copy the card shown in Figure 6.7, coloring the left-displaced outline red and the right-displaced outline blue. View it through spectacles having a blue left lens and a red right lens.

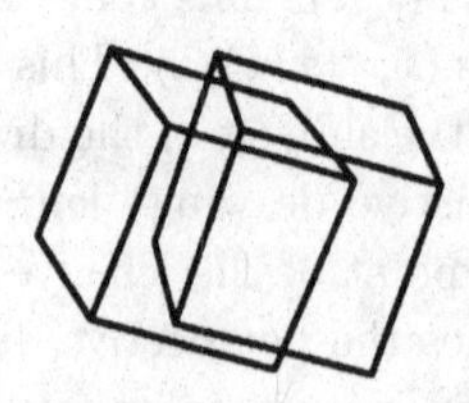

Fig. 6.7 A colored stereogram

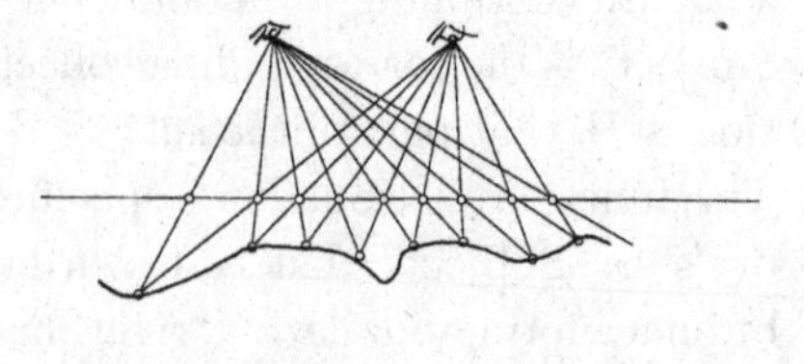

Fig. 6.8 Many image points that are seen by both eyes

But perhaps by a suitable choice of points on the surface of an object we may construct a stereogram of a figure E that could be understood without the need to resort to different colors. If we could select the points of E, as for example in Figure 6.8, and train our eyes to move in partnership, then many points of the stereogram would be used twice, once by each eye. The plethora of coffee-table books containing so-called random-dot stereograms is testament to the fact that we can successfully do this.

The points of these stereograms are obtained in the following way. Suppose that we wish to visualize a figure E_0 by means of a stereogram. We choose a plane P and two points p_L and p_R, neither in P. We define two functions f_L and f_R as follows: Suppose that e is in E_0. We write $e_L = (p_L \vee e) \cap P$ and $e_R = (p_R \vee e) \cap P$, and define $f_R(e_L) = e_R$ and $f_L(e_R) = e_L$. Thus $f_R \circ f_L$ is the identity function defined on some subset of the plane P. We construct a set E_L and a set E_R as follows. We select a point x of P. Each point $f_R^n(x)$, for which the composition $f_R^{n+1}(x) = f_R \circ f_R^n(x)$ is defined, is included in E_L. Each point $f_L^n(x)$, for which the composition $f_L^{n+1}(x) = f_L \circ f_L^n(x)$ is defined, is included in E_R. We note that these points are collinear, each being in the line of intersection of P and the plane $p_L \vee p_R \vee x$. We repeat this process, choosing points until P is reasonably densely covered. It follows that $E_R = \{f_R(x) : x \in E_L\}$ and $E_L = \{f_L(x) : x \in E_R\}$. Then $E_L \cup E_R$ is a stereogram of the figure $E = \{e \in E_0 : e = (p_L \vee x) \cap E_0, x \in E_L\}$. The set of points of the figure E is a reasonably representative subset of the set E_0.

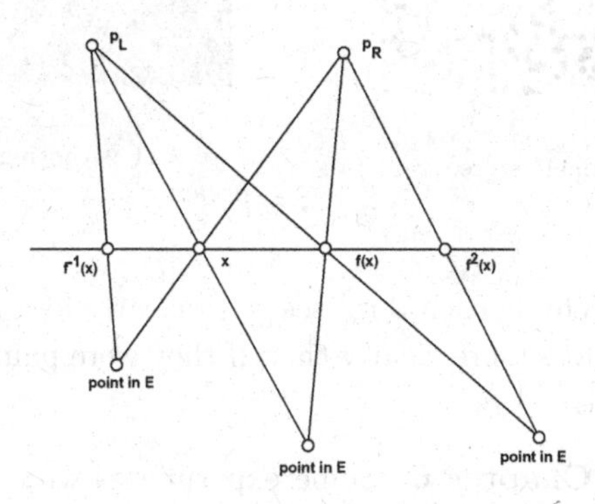

Fig. 6.9 Constructing a single image stereogram

It is not really appropriate to call the resultant stereogram a random dot stereogram. Tyler and Clark [17], the originators of the idea, coined the phrase *auto random dot stereogram.* Perhaps a more appropriate name is *single-image stereogram.* This encapsulates the distinguishing feature, the almost total overlap of left and right perspective drawings. This removes the annoying restrictions, on size and location, of the objects viewed in traditional stereoscopes. A single-image stereogram can be a large object, and is restricted only by inability of the mind to cope with unexpected changes in the distance between successive dots in the stereogram. There are a number of programs available for constructing single-image stereograms. A good example is the **SIS** Mathematica package [45].

Exercise 6.3.2. Examine the single-image stereograms in Figures 6.10 and 6.11 in the light of the above discussion. If in difficulty, use any of the techniques which helped when using the viewing box.

Fig. 6.10 A single image stereogram

Fig. 6.11 Another single image stereogram

Exercise 6.3.3. Why is each dot "seen" so clearly in a stereogram? Is it easier to "see" all dots more clearly than if they were painted on the actual surface represented?

Summary of Chapter 6. Some experiments with stereopsis proved that binocular vision is essentially different to monocular vision. In particular, we constructed and investigated four methods of creating the effect of non-planar figures in our minds from planar information, starting with old-fashioned stereoscopes and ending with the currently popular single-image stereograms.

Chapter 7

SCENE ANALYSIS

In Chapter 5 we developed some techniques that enable us to make perspective drawings of a figure. We now tackle the converse problem of determining whether a planar figure is a scene, that is, a perspective drawing of a non-planar figure. *Scene analysis* is the identification of any given planar figure as a perspective drawing of a non-planar figure. We use our results both to investigate planar figures that are claimed to be scenes and to make perspective drawings of subjects that exist "in our minds".

Our methods would enable "flat folk" with no experience of a non-planar world to theoretically deal with non-planar figures by means of their images, deciding whether a plane figure does represent something beyond their experience but theoretically possible. This is analogous to what we would do when discussing geometry of four dimensions. We would look at the images or "shadows" that its objects cast in our world.

When drawing figures that contain parts of planes we often draw "outlines", that is, parts of lines that are the images of the intersections of these planes. The nature of our drawing implements makes this a natural procedure. Shading or coloring does help - but outlines are perhaps the main clue to subject recognition.

In Chapter 5 we drew boxes. That was quite straight forward, and we commented in Lemma 5.5.1 that the resultant perspective drawings are closed figures. We know from our experiments in Chapter 1 that drawing closed figures may be fraught with difficulty. So we begin our scene analysis with the task of drawing "imaginary" boxes in mind. We broaden the scope of, and at the same time simplify, our analysis by thinking of boxes, and many other figures, as parts of sets of planes.

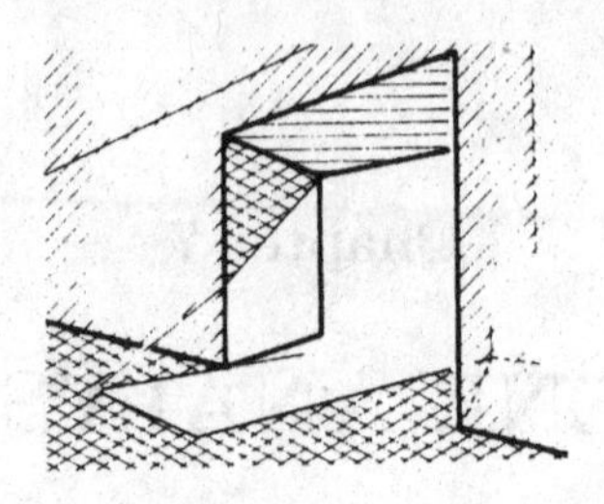

Fig. 7.1 Outlines and shading

7.1 Perspective drawings of intersections of planes

We begin by examining a planar set of lines and prove that, under certain conditions, it is a scene. We do this by constructing a set of planes. The union of the given lines is a perspective drawing of the union of the pairwise intersections of these planes. We first prove the following simple lemma:

Lemma 7.1.1. *Let p be a point and P_1 and P_2 be two planes, neither containing p. Suppose that M and N are two lines in P_1 concurrent with the line of intersection of P_1 and P_2. Then the lines M and $(p \vee N) \cap P_2$ are coplanar, say belonging to the plane P. The images of $P \cap P_1$ and $P \cap P_2$, in any perspective drawing in the plane P_1 drawn from the viewpoint p, are M and N, respectively.*

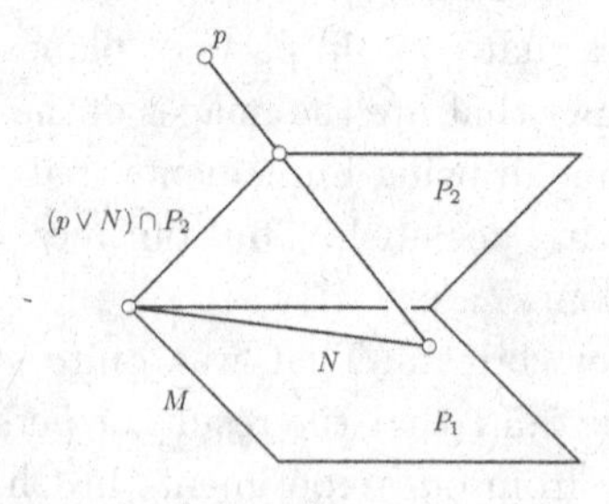

Fig. 7.2 A plane containing two given lines

Proof. The common point of the three planes P_1, P_2 and $p \vee N$ is in each of the lines $(p \vee N) \cap P_1 = N$, $P_1 \cap P_2$ and $P_2 \cap (p \vee N)$. Therefore it is the point of concurrence of $P_1 \cap P_2$, N and M. Therefore the lines M and

$P_2 \cap (p \vee N)$ are coplanar. □

We now derive a condition sufficient to ensure that a given planar set **L** of lines is a perspective drawing of the pairwise intersections of a set of planes. The condition enables us to label each line of **L** by ij , so that we are able to construct a suitable set $\mathbf{P} = \{P_i : i = 1, 2, \ldots, n\}$ of planes. The planar union of the lines of **L** is a perspective drawing of the non-planar union of the lines of the set $\{P_i \cap P_j : i, j = 1, 2, \ldots, n\}$.

Theorem 7.1.1. *Let* **L** *be a planar set of finitely many lines with the property that if* **L** *contains more than three lines, then they are not all concurrent. Then the following two conditions are equivalent.*

1. *The set* **L** *is a perspective drawing, drawn from any given point p as viewpoint, of the pairwise intersections of a set* **P** *of planes. No plane of* **P** *contains the viewpoint.*
2. *There is a natural number n so that, for each $\{i, j\} \subseteq \{1, 2, \ldots, n\}$, the symbol ij labels a member of* **L***; the two symbols ij and ji label the same member of* **L***; three lines labeled ij, jk and ki are concurrent; and if ij and jk label the same line, then ik is also a label of this line.*

Proof. First suppose the Condition 1 to be true. Let $\mathbf{P} = \{P_1, P_2, \ldots, P_n\}$. We label the image of each intersection $P_i \cap P_j$ in the perspective drawing by ij and ji. Lemma 4.1.3 guarantees that $P_i \cap P_j$, $P_j \cap P_k$ and $P_k \cap P_i$ are concurrent. Lemma 5.3.2 ensures that their images are also concurrent. If a line is labeled by both ij and jk, then p is coplanar with each of the lines $P_i \cap P_j$ and $P_j \cap P_k$. As p is not in P_j the two lines are equal and therefore equal to $P_i \cap P_k$. Thus Condition 2 holds.

Conversely, suppose that a labeling of the lines of **L** satisfies Condition 2. We choose an arbitrary viewpoint p not in the plane containing the lines of **L**. We choose this plane as P_1. If **L** is not empty, then $n \geq 2$. As P_2 we choose a plane containing the line labeled 12 but not the point p. For each $i = 3, 4, \ldots, n$ we apply Lemma 7.1.1, with the line labeled by 12 as $P_1 \cap P_2$, the line labeled by $1i$ as M, and the line labeled by $2i$ as N, in order to define a plane P_i that contains both the lines M and $(p \vee N) \cap P_2$. It may be helpful to think of each P_i being "hinged" at the line labeled by $1i$ and swung about this hinge out of the plane P_1 until its intersection with P_2 has an image, viewed from p, of the line labeled by $2i$. Thus we have a set $\mathbf{P} = \{P_1, P_2, \ldots, P_n\}$ of planes.

An ambiguity arises in defining P_i if M and $(N \vee p) \cap P_2$ are the same line. This occurs only if $1i$ and $2i$ label the same line, and consequently

12 also labels this line. Suppose that the line labeled by 12 and 21 has another label. On the one hand, there may be another member of **L** that has only two labels. In this case a permutation of $\{1, 2, \ldots, n\}$ gives a re-labeling of the members of **L** for which the line re-labeled by 12 has 12 and 21 as its only labels. This re-labeling satisfies Condition 2. Suppose, on the other hand, that each line of **L** has at least three labels. In this case let r be any member of $\{1, 2, \ldots, n\}$. If a line is labeled by ir and jr, then it is also labeled by ij. Therefore each line of **L** has a label not containing r. We delete all labels containing r, leaving each line labeled at least twice. Continuing to delete labels in this way until some line has only two labels, we are left with a labeling for which ij labels a member of **L**, for each $\{i, j\} \subseteq \{r_1, r_2, \ldots, r_m\}$. We replace each r_i by i, and permuting $\{1, 2, \ldots, m\}$ if necessary, we obtain a labeling that satisfies Condition 2 and in which the line labeled by 12 has only 12 and 21 as labels.

Thus, without loss of generality, there is a labeling for which the above procedure gives a well-defined set of planes.

If the point p were in one of the planes, say P_i, then $1i$ and $2i$ would label the same line. Then 12 would label this line, contradicting the fact that this line has only two labels. Therefore p is not in any of the planes of **P**.

We can be sure that each intersection $P_i \cap P_j$ is a line. If not, then the planes P_i and P_j would not be distinct and $1i$ and $1j$, and therefore ij, would label the same member L of **L**. As well, $2i$ and $2j$, and therefore ij, would label the same member M of **L**. Thus $L = M$ and $1i$ and $2i$ would both label this line. Consequently 12 would also label the line. But this could not be so as 12 and 21 label a line that has only two labels.

From our choice of planes we have that the image of $P_1 \cap P_i$ is the line labeled by $1i$, and that the image of $P_2 \cap P_i$ is the line labeled by $2i$ in the perspective drawing of the non-planar union of the set $\{P_i \cap P_j \mid i, j = 1, 2, \ldots, n\}$ of lines from the viewpoint p.

We need to check that the intersection of any two planes P_i and P_j, where neither i nor j is in $\{1, 2\}$, has the required image. The line $P_i \cap P_j$ contains the point $P_1 \cap P_i \cap P_j = (P_1 \cap P_i) \cap (P_1 \cap P_j)$. This point is the intersection of the line labeled by $1i$ and the line labeled by $1j$, and so belongs to the line labeled by ij. Also $P_i \cap P_j$ contains the point $P_2 \cap P_i \cap P_j = (P_2 \cap P_i) \cap (P_2 \cap P_j)$. This point has an image belonging to the line labeled by $2i$ and to the line labeled by $2j$. This point also belongs to the line labeled by ij.

Thus $P_i \cap P_j$ has the line labeled by ij as its image, unless possibly the image of $P_1 \cap P_i \cap P_j$ and the image of $P_2 \cap P_i \cap P_j$ is the same point. As p is not in P_1, Lemma 2.2.2 ensures that this could only occur if P_1, P_2, P_i, and P_j were concurrent. But then certainly $1i$, $1j$, and 12 would label concurrent lines. We know that not all lines of $\mathbf{L}$ are concurrent. But the concurrence of lines labeled by $1r$, for all r, would force the concurrence of all lines of $\mathbf{L}$ as, for each r, three lines labeled by $1r$, $1s$ and rs are concurrent. So, without loss of generality, there is a line labeled by 13 that is not concurrent with the lines labeled by $1i$ and ij. We replace P_2 in the above argument by P_3 and thereby prove that $P_i \cap P_j$ has the line labeled by ij as its image as required. □

We now refer to each line of $\mathbf{L}$ by any of its labels, saying "the line ij" rather than the more tedious "the line labeled by ij". We take care not to confuse a label ij for a line with the more common usage $i \vee j$ for a line containing points i and j. The label ij reflects the fact that the line is an image of the intersection of two planes in some perspective drawing.

Corollary 7.1.1. *Let the lines of* $\mathbf{L}$ *be labeled so that* $1i \neq 1j$. *Then the planes* P_i *and* P_j *are distinct.*

Proof. If the planes P_i and P_j were not distinct, then $P_1 \cap P_i$ and $P_1 \cap P_j$ would be the same line and have the same image in any perspective drawing. □

The following example proves that Theorem 7.1.1 may fail should we allow all the lines of $\mathbf{L}$ to be concurrent.

Example 7.1.1. The union of the seven lines in Figure 7.3 is not a perspective drawing of the pairwise intersections of seven planes P_1, P_2, $\ldots$, P_7 in which the image of $P_i \cap P_j$ is the line ij, for all $i, j = 1, 2, \ldots, 7$.

If Figure 7.3 were such a scene, by intersecting the seven planes with an eighth plane P_8 that is not concurrent with them, we would obtain the planar figure shown in Figure 7.4. This supposed figure is a Fano geometry which, we know from Corollary 3.7.1, does not exist as a figure. Therefore the set of seven planes does not exist.

We have restricted ourselves so far to discussing possible scenes drawn from a viewpoint that is not in any of the planes of $\mathbf{P}$. Precise necessary and sufficient conditions can be given for the union of the lines of $\mathbf{L}$ to be a scene drawn from a viewpoint belonging to some of the planes of $\mathbf{P}$. In this case certain intersection lines may have points as images.

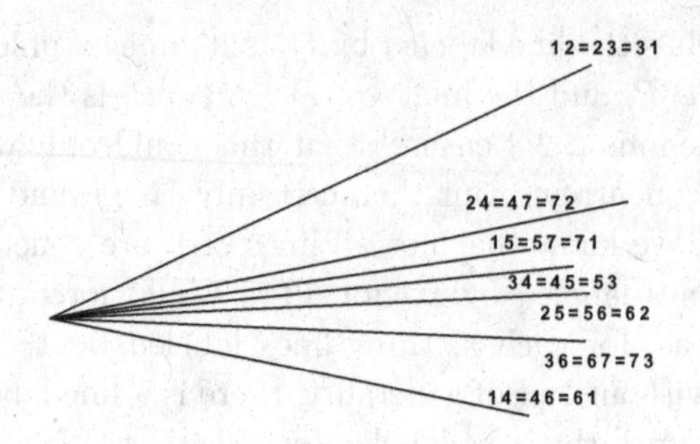

Fig. 7.3 Incorrectly labeled lines

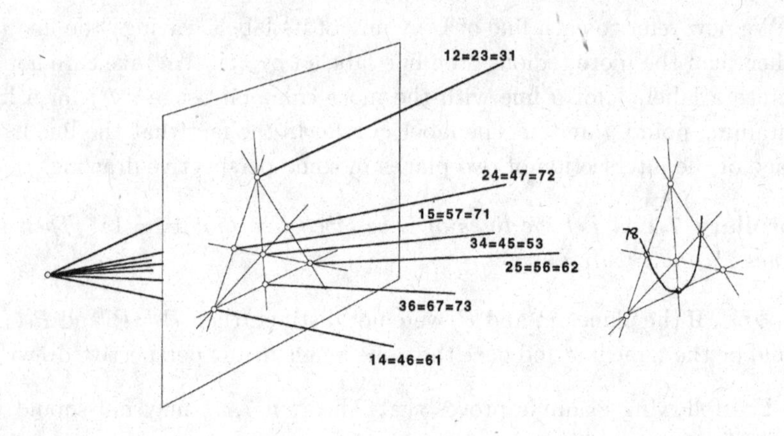

Fig. 7.4 The intersections of P_8 with seven planes

As a representative result we give the conditions for the union of a planar set **L** of lines to be a scene drawn from a viewpoint in one of the planes.

Theorem 7.1.2. *Let* **L** *be a planar set of finitely many lines with the property that if* **L** *contains more than three lines, they are not all concurrent. Then the following two conditions are equivalent:*

1. *The set* **L** *is a perspective drawing, drawn from any given viewpoint, of the pairwise intersections of a set* **P** *of planes. Exactly one member of* **P** *contains the viewpoint.*

2. *There is a natural number n so that, for each* $\{i, j\} \subseteq \{1, 2, \ldots, n\}$, *the symbol ij labels a member of* **L**; *the two symbols ij and ji label the same member of* **L**; *three lines labeled ij, jk and ki are concurrent; the two*

symbols $3i$ and $3k$ label the same line; and if $k \neq 3$ and ij and jk label the same line then ki is also a label of this line.

Proof. The proof follows that of Theorem 7.1.1, except that P_3 is chosen to be the plane containing both p and the line labeled 13. □

Exercise 7.1.1. Figure 7.5 contains a perspective drawing of a box, drawn in 3-point perspective from a viewpoint coplanar with one face of the box. Draw extra lines and assign labels to the lines of the resulting figure that support the conclusion that it is indeed a perspective drawing of a box.

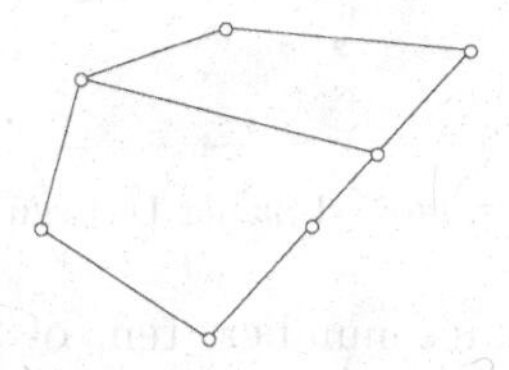

Fig. 7.5 A box drawn from a viewpoint coplanar with a face

7.2 Scenes and other planar figures

In this section, we give some varied applications of Theorem 7.1.1. A first application verifies the following intuitively satisfying result. It enables us to be certain, for example, that our illustration for Theorem 4.3.1 in Figure 4.4 is a perspective drawing.

Theorem 7.2.1. *Any planar Desargues figure is a perspective drawing of a non-planar Desargues figure, drawn from a general viewpoint. Conversely, a perspective drawing of any non-planar Desargues figure, drawn from a general viewpoint, is a planar Desargues figure.*

Proof. Applying Theorem 7.1.1 to any planar Desargues figure labeled as in Figure 7.6 gives a set of five planes. We note, as we did in Exercise 4.4.2, that the pairwise intersections of these five planes are the lines of a non-planar Desargues figure that has the planar Desargues figure as its perspective drawing.

As both figures contain the same number, ten, of points the viewpoint is not in a 3-point circuit of the non-planar Desargues figure. As well,

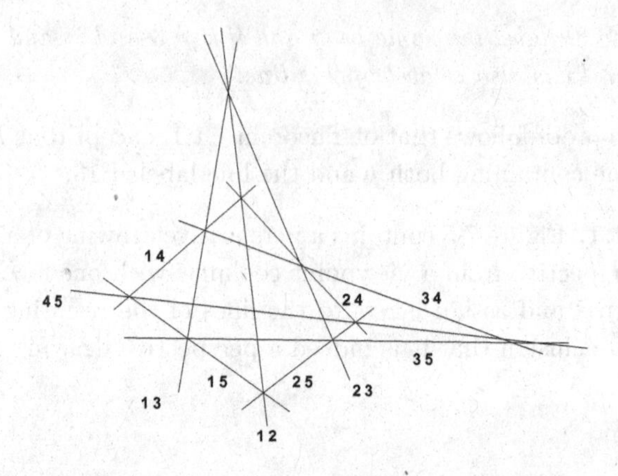

Fig. 7.6 A labeled planar Desargues figure

both figures contain the same number, ten, of 3-point circuits, ensuring that the viewpoint is not in any 4-point circuit of the non-planar Desargues figure. From Proposition 5.1.1 we now have that the viewpoint is a general viewpoint. The converse statement is Lemma 5.1.1. □

We apply Theorem 7.1.1 to the Penrose triangle.

Definition 7.2.1. *The* **Penrose triangle**, *described by L.S. Penrose and the British mathematical physicist Roger Penrose* [50] *in* 1958, *and used to great effect by Maurits Escher in several lithographs* [21, 54], *is the planar figure shown in Figure 7.7*

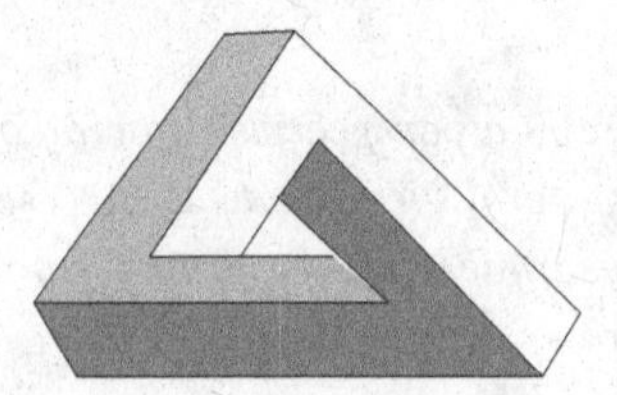

Fig. 7.7 The Penrose "triangle"

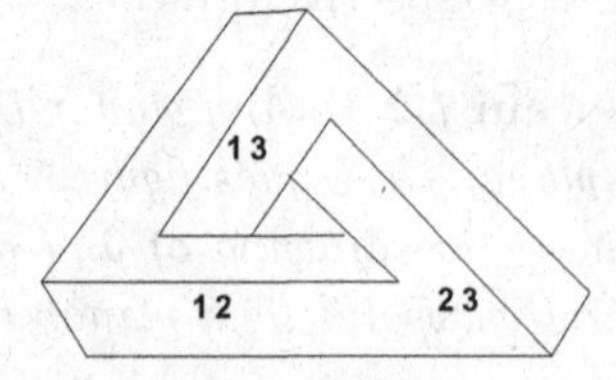

Fig. 7.8 A labeled Penrose "triangle"

Problem 7.2.1. *Is the Penrose "triangle" a scene in which each shaded region is the image of a planar set?*

Solution. If Figure 7.4 were to be a scene then the three labeled lines in Figure 7.8 would be the images of pairwise intersections of three planes.

But they are not concurrent, and consequently from Theorem 7.1.1 we have that the Penrose triangle is not a scene.

The following easily stated but non-trivial question of plane projective geometry would seem to have little to do with non-planar matters.

Problem 7.2.2. *Let P be a plane. If A, B, and C are three concurrent lines in P, and a, b, and c are three non-collinear points in P, then is there a triangle with vertices on each of A, B, and C, respectively, - and sides through each of a, b, and c, respectively?*

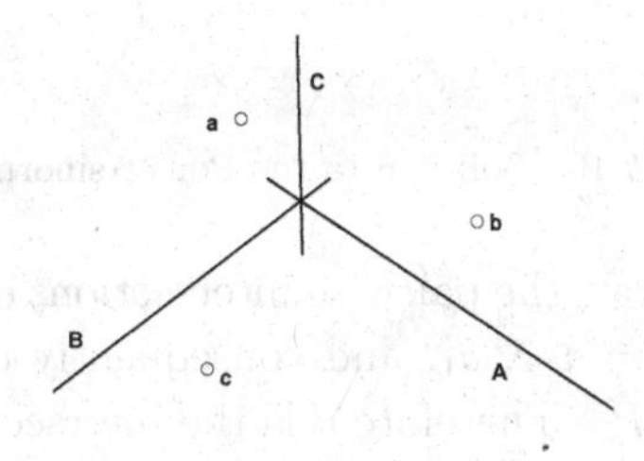

Fig. 7.9 A figure containing three points and three lines

Solution. A few attempts to draw such a triangle will rapidly lead to the conclusion that the question is not a trivial one. Even through the problem concerns a planar figure, Theorem 7.1.1 allows us to gain insight by thinking of the figure as a scene.

We select a point p. We know that $\{A, B, C\}$ can be labeled $A = 23$, $B = 13$, and $C = 23$, thus ensuring that their union is a perspective drawing of the pairwise intersections of three planes P_1, P_2, P_3 from the viewpoint p. We think of $a_1 = P_1 \cap (p \vee a)$, $b_1 = P_2 \cap (p \vee b)$, and $c_1 = P_3 \cap (p \vee c)$ as points on the planes P_1, P_2, P_3, respectively.

If we define $P_4 = a_1 \vee b_1 \vee c_1$, then, in any perspective drawing of the family $\{P_1, P_2, P_3, P_4\}$ drawn from the viewpoint p, the images of $P_1 \cap P_4$, $P_2 \cap P_4$, and $P_3 \cap P_4$ are the sides of the required triangle.

Conversely, any solution to Problem 7.2.2 would be the perspective drawing of the pairwise intersections of the members of $\{P_1, P_2, P_3, P_4\}$ drawn from the viewpoint p. Thus we deduce that there is exactly one triangle solving the problem.

But how can we construct this unique triangle? Suppose that d is a point of C. Then it is the image of a point d_1 on $P_1 \cap P_2$. Let P_5 be the

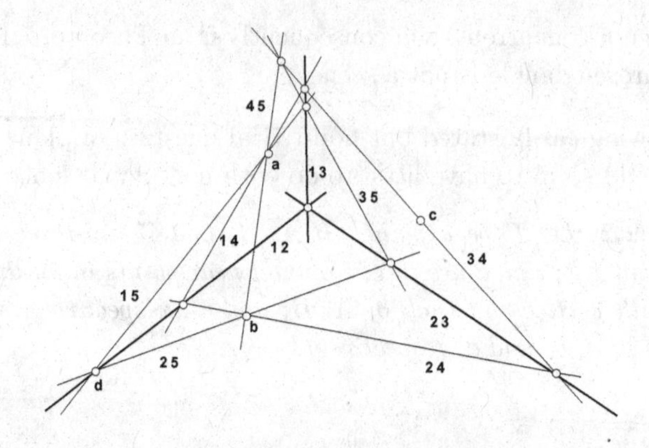

Fig. 7.10 Solution to the Pons Asinorum

plane $a_1 \vee b_1 \vee d_1$. We draw the pairwise intersections of $\{P_1, P_2, P_3, P_4, P_5\}$. We have that $P_4 \cap P_5 = a_1 \vee b_1$, and consequently $a \vee b = 45$. Also d is the image of $P_1 \cap P_2 \cap P_5$. Therefore d is the intersection of 12, 25 and 51. Thus $a \vee d$ is labeled 15, and similarly $b \vee d$ is labeled 25. In this scene, 35 must be concurrent with 13 and 15, so $13 \cap 15 \in 35$. Similarly, 35, 13, and 15 are concurrent, giving $23 \cap 25 \in 35$. These two points belong to the line 35. Also $35 \cap 45 \in 34$, but as well $c \in 34$, ensuring that $c \vee (35 \cap 45)$ is the line 34. A similar argument ensures that 24 is the line $b \vee (34 \cap 23)$. The line 14 contains the two points $24 \cap 12$ and $34 \cap 13$. We know that the three points a, $24 \cap 12$, and $34 \cap 13$ are each on the image of $P_1 \cap P_4$ and therefore the line 14 contains the point a. We are therefore able to construct the (unique) solution as follows:

Construction 7.2.1. Choose any point d in the line C. Draw the line $a \vee d = 15$, draw the line $b \vee d = 25$. Draw the line $(B \cap (a \vee d)) \vee (A \cap (b \vee d)) = 35$. Draw $a \vee b = 45$. Draw the lines $(35 \cap 45) \vee c = 34$ and $(34 \cap 23) \vee b = 24$. Finally, draw the line $(34 \cap 13) \vee a = 14$. Then the required triangle has sides 14, 24, and 34.

7.3 The existence of boxes

Think back to your work in Construction 5.5.2. Did you draw a box from your imagination, or did you draw an existing physical model? If you drew from your imagination, how can we be sure that a box having that perspective drawing could really exist? We are asked to believe that the existence of a planar figure guarantees the existence of a special non-planar figure. In this section, we prove this claim by examining a likely construction for the perspective drawing, labeling the resultant figure, and applying Theorem 7.1.1 to it. We first examine the requirements of the set $\{v_1, v_2, v_3\}$ of proposed vanishing points of the scene in the light of Definition 5.5.1 and Theorem 5.6.2.

Lemma 7.3.1. *Let L_1, L_2 and L_3 be three concurrent, but non-coplanar extended Euclidean lines. Suppose that the lines are drawn from the viewpoint p and that, for each i in $\{1,2,3\}$, v_i is the vanishing point of L_i. If p is Euclidean, then $\{v_1, v_2, v_3\}$ is a set of three non-collinear points. If p is ideal, then $\{v_1, v_2, v_3\}$ is a set of two or three collinear ideal points.*

Proof. Suppose that the viewpoint p is Euclidean. Then the set of directions of each of L_1, L_2, and L_3 is non-linear and the set of these directions together with p is non-planar. From the first conclusion of Proposition 5.1.1 we therefore have that no two of the vanishing points are equal, and from the second conclusion, $\{v_1, v_2, v_3\}$ is a set of three non-collinear points. Alternatively, if p is ideal then, from Theorem 5.6.1, the set of vanishing points is a set of ideal points belonging to an extended Euclidean plane, and consequently linear. The point p can be collinear with at most two of the set of three non-collinear directions, ensuring that at least two of the vanishing points are distinct. □

Therefore if we hope to draw a box from a Euclidean viewpoint, then we must begin by choosing three non-collinear points v_1, v_2 and v_3 to be vanishing points and follow the steps of Construction 5.5.1.

Construction 7.3.1. Let v_1, v_2 and v_3 be three non-collinear points of an extended Euclidean plane. Choose a point 0 in the plane. Draw the line $v_i \vee 0$ and choose a point i on $0 \vee v_i$, for each $i = 1, 2, 3$. Draw the lines $v_2 \vee 1$ and $v_1 \vee 2$, and construct 4, their point of intersection. Similarly, draw the points $5 = (v_3 \vee 1) \cap (v_1 \vee 3)$, $6 = (v_3 \vee 2) \cap (v_2 \vee 3)$, and $7 = (v_1 \vee 6) \cap (v_2 \vee 5)$.

We noted in Lemma 5.5.1 that when drawing an existing box our rules of perspective drawing, based on Lemmas 5.3.1 and 5.3.2, ensure that the three points 4, 7, and v_3 are collinear. This collinearity guarantees that Construction 7.3.1 results in a closed figure. It is only when drawing a box from imagination, as artists constantly do, that the existence of the associated closed figure comes into question. The following Proposition deals satisfactorily with this question and enables us to derive a labeling that satisfies the requirements of Theorem 7.1.1.

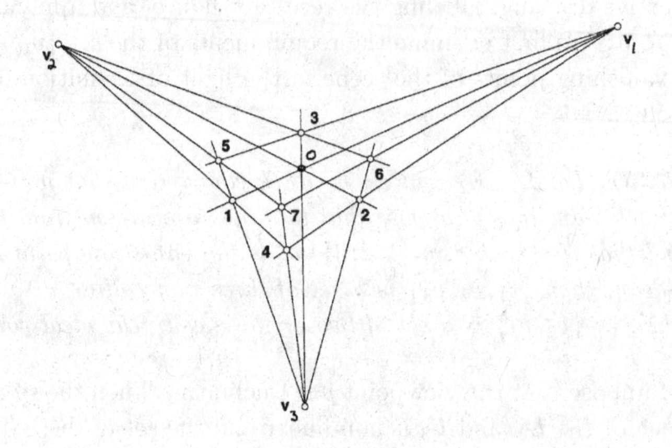

Fig. 7.11 A possible scene that is a closed figure

Proposition 7.3.1. *Construction 7.3.1 results in a closed figure.*

Proof. Apply Desargues' Theorem to the two triangles 306 and 517 of Construction 7.3.1 to ensure that $(0 \vee 6) \cap (v_2 \vee v_3) = (1 \vee 7) \cap (v_2 \vee v_3)$. The converse of Desargues' Theorem applied to the two triangles 062 and 174 ensures that $v_3 \in 4 \vee 7$. This ensures that the figure on $\{v_1, v_2, v_3, 0, 1, 2, 3, 4, 5, 6, 7\}$ satisfies the requirements of Definition 3.3.2 and is closed. □

We now concentrate particularly on the set **L** consisting of the non-trivial lines of the construction, together with the three lines $v_1 \vee v_2$, $v_2 \vee v_3$, and $v_1 \vee v_3$. We label each of these lines as in Figure 7.12.

The labeling satisfies the requirements of Condition 2 of Theorem 7.1.1. From this theorem and Corollary 7.1.1 we have the existence of seven distinct planes P_1, P_2, P_3, P_4, P_5, P_6, and P_7 with the property that the union of the fifteen distinct lines of pairwise intersection of these planes has the union of the lines of **L** as a perspective drawing. There are eight points that are each the intersection of exactly three of the planes. The figure of these eight points has the figure on $\{0, 1, 2, 3, 4, 5, 6, 7\}$ as its perspective drawing. Already we have succeeded in proving that Construction 7.3.1 gives a scene. The question now is: How special do we need this scene to be before we are happy to call it the image of a box?

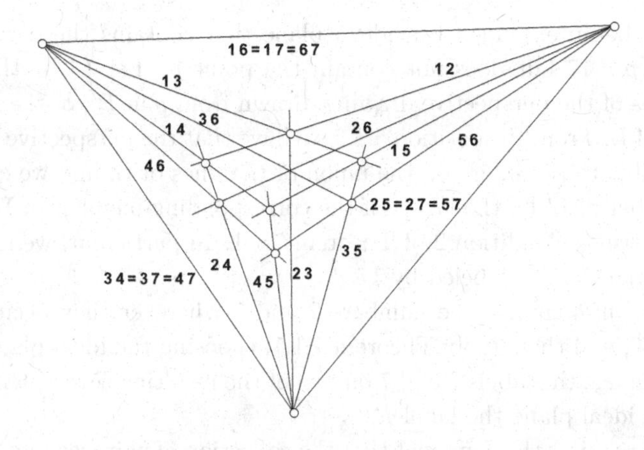

Fig. 7.12 The possible scene of labeled lines

Lemma 7.3.2. *Let P be any extended Euclidean plane and p any Euclidean point not in P. Then the figure on $\{0,1,2,3,4,5,6,7\}$ that results from an application of Construction 7.3.1 in the plane P is the perspective drawing of a combinatorial cube, drawn from p. Each set having an image that is one of $\{1,4,5,7\}$, $\{2,4,6,7\}$, $\{0,2,3,6\}$, $\{0,1,3,5\}$, $\{0,1,2,4\}$, and $\{1,2,4,7\}$ is planar.*

Proof. The points with images 1, 4, 5, and 7 are on the plane P_4 and no subset of three is collinear. Therefore $\{1,4,5,7\}$ is a circuit. From similar arguments we have that each of the other five 4-point sets is planar as required, and the requirements of Definition 2.3.5 are satisfied. □

It is clear that if the plane P_7 were chosen to be ideal, then the planes P_1 and P_6 would be parallel, the planes P_3 and P_4 would be parallel, and the planes P_2 and P_5 would be parallel. We will call the resulting cube of Euclidean points a *box*, or *parallelepiped*, thinking of the six planar 4-point sets above as the *faces* of the box. Each two opposite faces would then belong to parallel planes. We call the two-point intersection of each two non-opposite faces an *edge* of the box.

We now prove that we are able to choose the plane P_7 to be the ideal plane in the application of Theorem 7.1.1 to Figure 7.12. Let p be any Euclidean point, and let P be any extended Euclidean plane not containing p. Suppose that $\mathbf{L}$ is any set of lines obtained by Construction 7.3.1 in P. Label $\mathbf{L}$ as in Figure 7.12.

Let P' be an extended Euclidean plane that contains the ideal line of the plane $p \vee 17$ but does not contain the point p. Let $\mathbf{L}'$ be the set of image-lines of the perspective drawing, drawn from p in P', of the union of the lines of $\mathbf{L}$. From Proposition 5.1.2 we have that the perspective drawing is isomorphic to the figure on the union of the lines of $\mathbf{L}$, and we can label each member of $\mathbf{L}'$ by the labels of the corresponding member of $\mathbf{L}$ so that $\mathbf{L}'$ then satisfies Condition 2 of Theorem 7.1.1. In particular, we note that the ideal line of P' is labeled by 17.

We now interchange the numbers 2 and 7 wherever they occur in the labels of $\mathbf{L}'$, and then apply Theorem 7.1.1, choosing the ideal plane as P_2. We interchange the labels 2 and 7 on two of the resulting seven planes, thus giving the ideal plane the label P_7.

It only requires the comment that the collection of pairwise intersections of these planes has both the union of the lines of $\mathbf{L}'$ and the union of the lines of $\mathbf{L}$ as perspective drawings, to complete the proof of the following theorem:

Theorem 7.3.1. *Let p be any Euclidean point and P be any extended Euclidean plane not containing p. Then any figure obtained by Construction 7.3.1 in P is a perspective drawing of a box drawn from the general viewpoint p.*

This is quite a remarkable result. We have argued that the existence of a planar figure implies the existence of a corresponding non-planar figure. We have successfully shown that the existence of a "shadow" proves the existence of something non-planar having that "shadow". Think how useful this method of arguing would be were we to contemplate the existence of a four-dimensional world!

As we mentioned in Chapter 5, an advantage of using extended Euclidean space as the setting for our discussions is that 1-point, 2-point and 3-point perspective drawings can each be constructed using only roller and pencil, by applying Construction 7.3.1 to an appropriate choice of non-collinear vanishing points v_1, v_2, and v_3.

In Chapter 8 we complete our analysis of perspective drawings of boxes from a Euclidean viewpoint by specifying in Theorems 8.3.1, 8.4.1, and 8.5.1 the exact conditions under which any result of Construction 7.3.1 is the perspective drawing of a rectangular box. In Theorem 8.6.1 we also detail conditions for Construction 7.3.1 to be the image of an affine projection of a rectangular box.

Construction 7.3.2. Using a computer drawing program such as The Geometer's Sketchpad, follow the instructions of Construction 7.3.1 to make a 3-point perspective drawing of a box. Dragging the point 0, or dragging a point i along a line $0 \vee i$ for $i = 1, 2$ or 3, gives a perspective drawing of a parallel box. Dragging a vanishing point gives a perspective drawing of a differently oriented box. The comments below apply particularly to The Geometer's Sketchpad and, taken in conjunction with the hints following Construction 1.2.6, may be helpful.

Open a new **Sketch** and use the **Point** tool to construct four points v_1, v_2, v_3, and 0. Select **Ray** from the **Line** menu in the **Toolbox**. Click on 0, and drag to the point v_1 to produce a ray from 0. The reason for using the **Ray** tool rather than the **Line** tool is merely to avoid unnecessary parts of lines cluttering up the screen. Repeat the process to construct lines $0 \vee v_2$ and $0 \vee v_3$. Use the **Point** tool to construct i on $0 \vee v_i$, for $i = 1, 2, 3$. Use the **Ray** tool to construct the lines $j \vee v_k$, for the various values of j and k required by Construction 7.3.1, being sure to click on j first in each case. Then construct the points 4, 5, 6 and 7 as required.

You may like to use a computer drawing program with shading abilities, shading appropriate faces and then hiding the construction lines used. The following comments also apply particularly to The Geometer's Sketchpad and may be helpful.

Decide on the faces to be shaded and carry out the procedure described following Construction 1.3.1 for each of these faces. To hide the lines of the construction, choose **Ray**, from the **Edit** menu choose **Select All Rays**, then from the **Display** menu choose **Hide All Rays**. If you would also like to remove the points, choose **Point**, **Select All Points**, **Hide All Points**. Do not forget that the **Undo** option is available in the **Edit** menu if things get out of hand.

A computer-generated result of Construction 7.3.2, drawn from a viewpoint of the left eye of an observer, can be displayed on a small translucent screen that is fixed to a headset in front of the eye. Sensors in the headset inform the computer of the eye's position and the display is re-calculated many times each second. At any moment, it gives a perspective drawing of a fixed imaginary, or "virtual", box drawn from the position of the left eye. Thus the imaginary box remains fixed relative to the real world as the wearer of the headset moves. It appears to the observer as if the imaginary box is really present. More complex *virtual reality* may be used in order to investigate the appearance and effect of a proposed building in a cityscape, thus giving city planners a better chance of discovering unexpected surprises before building is commenced. We discuss stereoscopic virtual reality, produced by simultaneously using this technique for each eye, in Chapter 6

7.4 Completing a partial scene

Often we see only part of a figure, although we suspect that the hidden part exists. In this section, we add lines to a planar figure and apply Theorem 7.1.1 to the augmented figure, in order to prove that the initial figure is itself a scene.

As an example we look again at the technique we used to complete, in Construction 5.4.2, the image of the house roof shown in Figure 5.21. The required line can be obtained by labeling the lines of the scene as shown in Figure 7.13.

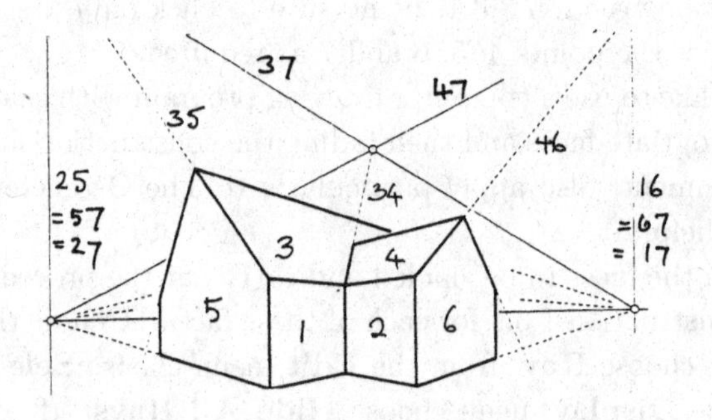

Fig. 7.13 Labeled lines of a house image

We use the concurrence of lines 13, 17, and 37, and the concurrence of 35, 57, and 37 to draw two points of 37. A similar argument constructs the line 47. Finally the concurrence of the lines 13, 14, and 34 and the concurrence of lines 34, 37, and 47 gives two points of the required line 34.

Our approach in general is as follows. We suspect that a set of lines forms a desired scene. To verify this we interpret the scene as the perspective drawing of some, but not all, of the pairwise intersections of a set **P** of planes. We add other lines to the scene to give the images of all pairwise intersections of the members of **P**, enabling us to apply Theorem 7.1.1.

For instance we can think of Figure 7.14 as an attempt at a perspective drawing of a triangular plateau with three sloping planar side cliffs. Applying Theorem 7.1.1 to Figure 7.14 we have that pairs of the cliffs must intersect in three concurrent lines for the figure to be the required scene. That is not the case, and so the figure is not a perspective drawing of a triangular plateau.

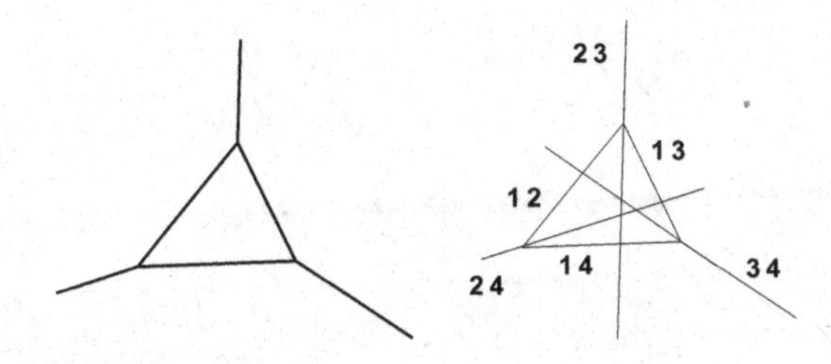

Fig. 7.14 A possible triangular plateau

Let us examine the figures in Figure 7.15 in this spirit. Each may loosely be thought of as a possible perspective drawing of a plateau with sloping planar cliff sides. The lines of the first figure can be labeled as in Figure 7.16.

We need to add lines 34, 15, and 25 in order to satisfy the requirements of Theorem 7.1.1. These lines, and the existing lines, must satisfy concurrency conditions. Looking at these conditions in detail, we see that the following triples need to be concurrent: $\{12, 23, 13\}^*$, $\{12, 24, 14\}^*$, $\{12, 25, 15\}$, $\{13, 34, 14\}$, $\{13, 35, 15\}$, $\{14, 45, 15\}$, $\{23, 34, 24\}$, $\{23, 35, 25\}$, $\{24, 45, 25\}$, and $\{34, 45, 35\}$. Those concurrences marked with an asterisk follow directly from our labeling.

In order to satisfy all requirements of Condition 2 of Theorem 7.1.1, we choose 25 through the two points $23 \cap 35$ and $24 \cap 45$, we choose 34 through the two points $23 \cap 24$ and $13 \cap 14$, and we choose 15 through the two points $12 \cap 25$ and $13 \cap 35 = 14 \cap 45$. This can always be done, as in Figure 7.16, for example. So the first figure is a scene, a perspective drawing of a flat-topped mountain.

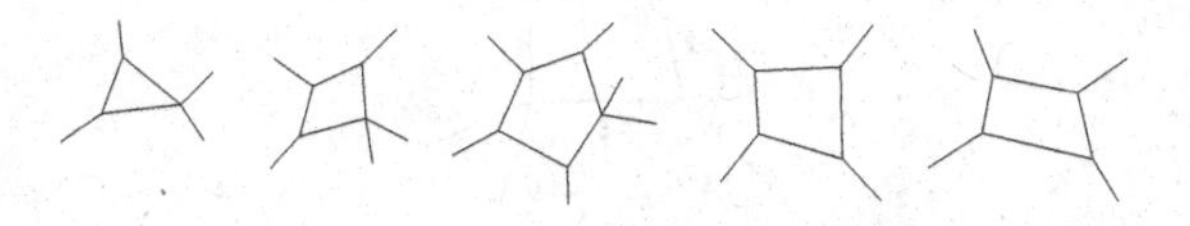

Fig. 7.15 Some plateau

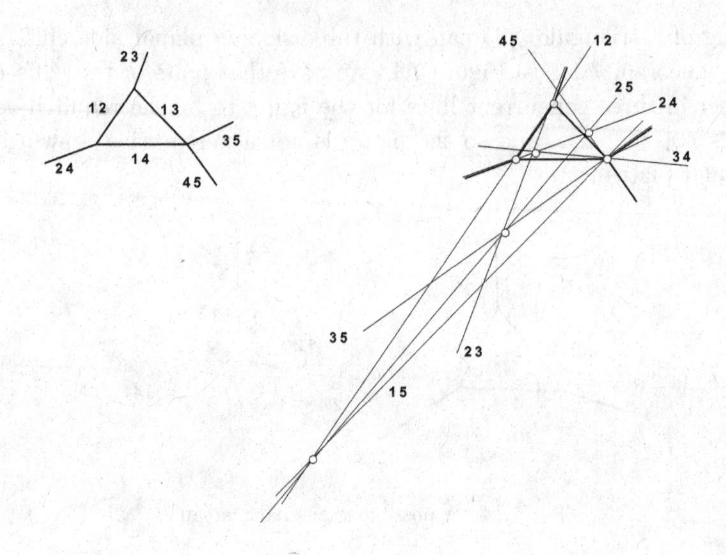

Fig. 7.16 Images of pairwise intersections of the first plateau and its cliffs

Similarly, we investigate the second figure of Figure 7.15. We label it as in Figure 7.17. Lines 16, 24, 26, 35, 36, and 45 are needed to augment the figure to a possible perspective drawing of the pairwise intersections of six planes. We draw $24 = (12 \cap 14) \vee (23 \cap 24)$, $35 = (13 \cap 15) \vee (23 \cap 25)$, and $26 = (24 \cap 46) \vee (25 \cap 56)$. To be able to draw the required line 45 we need the points $45 \cap 56$, $34 \cap 35$, and $24 \cap 25$ to be collinear. An application of Desargues' Theorem to the figure so far completed proves this to be true, enabling us to draw 45. Similarly we add lines 16 and 36 that satisfy all concurrency requirements of Condition 2 of Theorem 7.1.1, and consequently the original figure is a perspective drawing of a four-sided plateau.

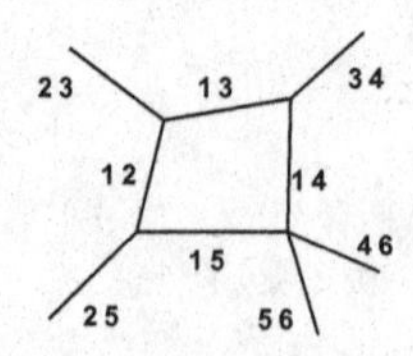

Fig. 7.17 A four-sided plateau

Exercise 7.4.1. Is the third figure of Figure 7.15 part of a scene?

We label each of the fourth and fifth figures of Figure 7.15 as in Figure 7.18. In order to prove that each is a scene we need only check the concurrences $\{12, 23, 13\}^*$, $\{12, 24, 14\}$, $\{12, 25, 15\}^*$, $\{13, 34, 14\}^*$, $\{13, 35, 15\}$, $\{14, 45, 15\}^*$, $\{23, 34, 24\}$, $\{23, 35, 25\}$, $\{24, 45, 25\}$, and $\{34, 45, 35\}$. Those marked with an asterisk are certainly correct. In each case we need the existence of lines 24 and 35, satisfying the above concurrences, to complete a scene. Clearly $(13 \cap 15) \vee (23 \cap 25)$ is the unique possibility for 35. We could find a suitable line 24 exactly if $12 \cap 14$, $23 \cap 34$ and $25 \cap 45$ were concurrent. If this were so, then Theorem 7.1.1 would guarantee that the figure was a scene. Thus the fourth figure is, and the fifth figure is not, a scene.

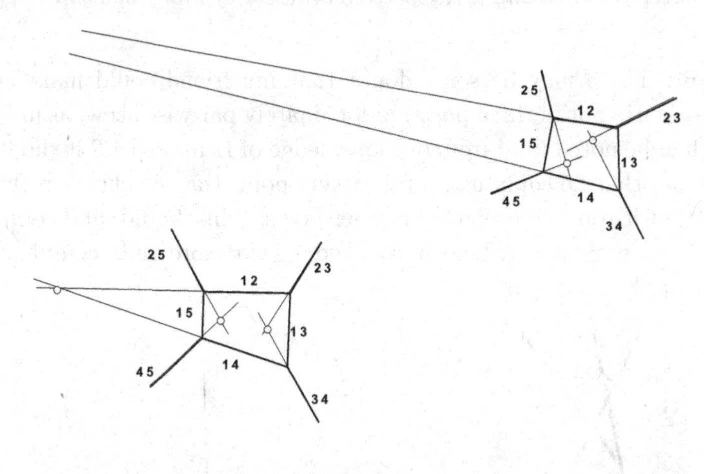

Fig. 7.18 A scene and a figure that is not a scene

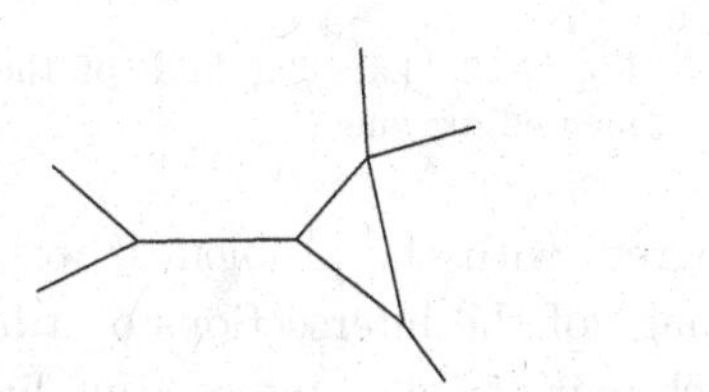

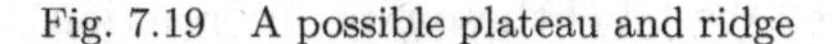

Fig. 7.19 A possible plateau and ridge

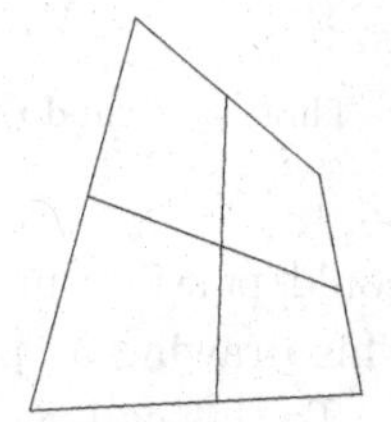

Fig. 7.20 A warped framework

Exercise 7.4.2. Is Figure 7.19 part of a scene?

7.5 Two problems

To complete this chapter we illustrate the wide-ranging use to which Theorem 7.1.1 may be put by applying it to two quite disparate problems, Problems 7.5.1 and 7.5.2. The latter, in particular, seems initially to have little in common with our work.

The first problem is a very practical one. Derelict garden fences are a common sight, many with only the frames of some panels remaining.

Problem 7.5.1. *A friend made a framework for a panel of a garden fence but it was warped - even through each of the six pieces of timber remained straight. He drew it for me. Figure 7.20 contains the result of his drawing. Did he succeed in making a perspective drawing of the framework?*

Solution. There may be some doubt that my friend could make such a framework. He put in three posts, unfortunately pairwise skew, as in Figure 7.21. Then he noted (and from our knowledge of Lemma 5.1.2 it should not surprise us) that he could nail a rail to any point that he chose on the first post so that it touched each of the other posts. This he did, and completed nailing it to each post where it touched. Two more rails completed his attempt at the fence panel.

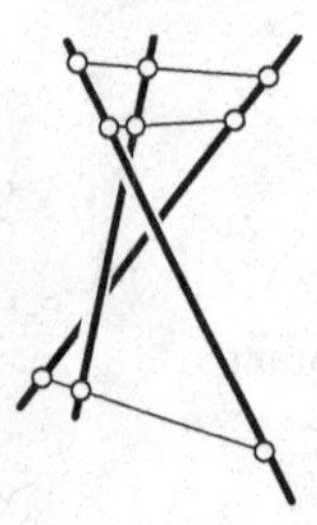

Fig. 7.21 Three posts and three rails

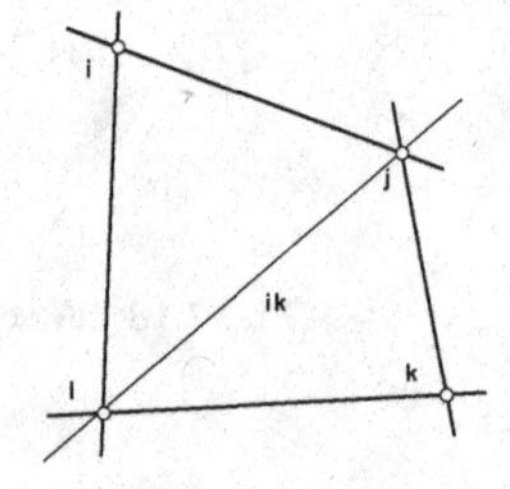

Fig. 7.22 Labeling lines of the augmented drawing

We could possibly make some progress with the problem if we could think of his drawing as part of the image of the intersections of a family of planes. To this end we note that each pair of nailed intersecting lines in the framework belongs to a unique plane. We label these nine intersection points as in Figure 7.23, and call the plane determined by the ith intersection P_i. After a little thought we see that the lines of my friend's drawing must be labeled as in Figure 7.23 if we are to apply Theorem 7.1.1.

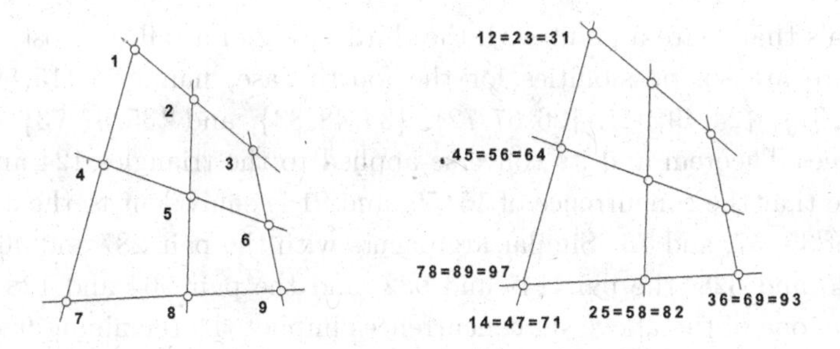

Fig. 7.23 Labeling intersections, posts, and rails

In order to apply Theorem 7.1.1, we need the image of the intersection of *each* pair of the nine planes P_1, P_2, ..., P_9. Examining P_1 and P_5 for example, the plane P_1 contains the line $1 \vee 2$, and P_5 contains the line $2 \vee 5$, ensuring that the point 2 is in $P_1 \cap P_5$. Similarly the point 4 is in $P_1 \cap P_5$, giving that $P_1 \cap P_5 = 2 \vee 4$. Therefore we should add a line through 2 and 4 to the drawing and label it by 15. We are imagining a skew 4-gon 1254 in $P_1 \cup P_5$ that has four pieces of wood as its sides. We repeat this argument for each of the seventeen other non-planar 4-gons with wooden sides that has a pair $\{i, j\}$ of nails as opposite vertices, labeling the line through the images of the other two vertices of the 4-gon by ij.

In order to use Theorem 7.1.1 we must establish the concurrence of the three lines ij, jk, and ki, for $\{i, j, k\} \subseteq \{1, 2, 3, 4, 5, 6, 7, 8, 9\}$. Four different cases arise, depending on the configuration of the points i, j, and k and shown in Figure 7.24.

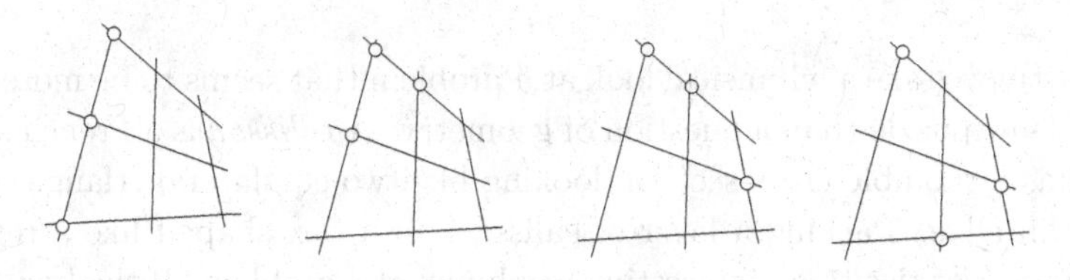

Fig. 7.24 Testing concurrent lines

In the first case, concurrence follows directly from the labeling of the images of the posts and rails. In the second case, the points are images of three of the vertices of a 4-gon contained in the framework; ij being the image of a post, jk the image of a rail, and ki the image of the diagonal of th 4-gon through j as required. In the third possibility, two of the lines are diagonals that share a point with the third image of a rail (or post).

There are six possibilities for the fourth case, namely: $\{15, 59, 91\}$, $\{16, 68, 81\}$, $\{24, 49, 92\}$, $\{26, 67, 72\}$, $\{34, 48, 83\}$, and $\{35, 57, 73\}$. From Desargues Theorem and its converse applied to the triangles 124 and 968 we have that the concurrence of 15, 59, and 91 is equivalent to the concurrence of 35, 57, and 73. Similar arguments with the pair 287 and 463, the pair 746 and 328, the pair 146 and 982, and the pair 964 and 128 prove that any one of the above six concurrences implies the remaining five.

Consequently the drawing is a perspective drawing of a skew framework if and only if we can choose points $\{i, j, k\} \subseteq \{1, 2, 3, 4, 5, 6, 7, 8, 9\}$, with no two on the same line of the original drawing, so that the added lines ij, jk, and ki are concurrent.

In Figure 7.25 we apply the test to my friend's illustration.

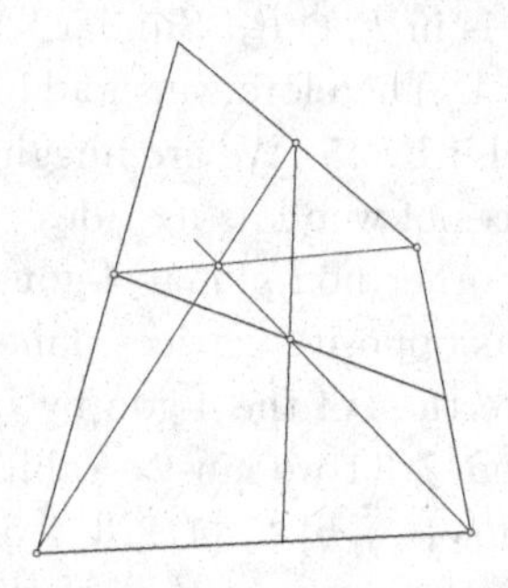

Fig. 7.25 Testing my friend's drawing

Lastly we take a whimsical look at a problem that seems to be more akin to a jigsaw puzzle than a question of geometry. A *calisson* is a French sweet that has a rhombic cross-section, looking like two equilateral triangles that share an edge. Packing a layer of calissons in a box shaped like a regular hexagon gives rise to an interesting combinatorial problem. If the box, with side-length n, is filled with sweets of side length 1, then each sweet will have one of three possible orientations.

Problem 7.5.2. *What can we say about the number of calissons in each of the three possible orientations in a filled hexagonal box?*

Solution. Let us examine the example of such a packing in Figure 7.26.

Coloring all sweets that share the same orientation the same color gives

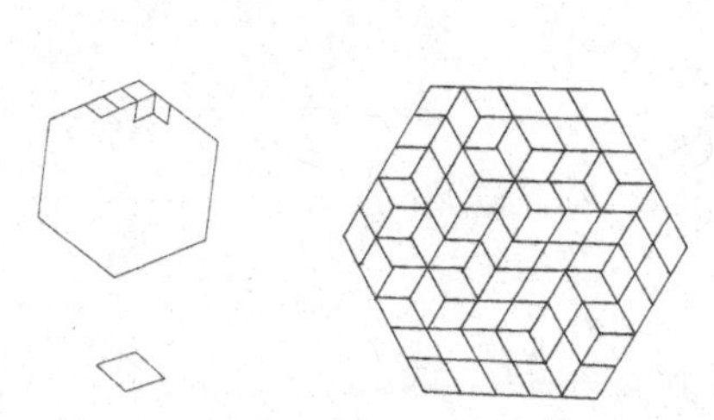

Fig. 7.26 A packing of calissons

Fig. 7.27 A colored packing of calissons

the very suggestive diagram in Figure 7.27. If this were a persplective drawing of boxes, then there would be twenty-five facing left, twenty-five facing right and twenty-five vertically up and the answer to our question would be clear. But are these the directions faced by the boxes? Often, looking at Figure 7.27 for a few moments will cause our perception of the cubes to reverse. Theorem 5.6.2 suggests that if we are indeed looking at a perspective drawing then it is an affine projection. Our later discussion of distortion in affine projections, following Figure 8.18 in Chapter 8, will lead us to expect such a lack of certainty in perception. In order to prove our argument we show that the illustration above is indeed a perspective drawing of the pairwise intersections of three parallel families of planes. Our intuition leads us to guess that the planes should be labeled as shown in Figure 7.28 for $i = 1, 4, 7, \ldots, 16$, $j = 2, 5, 8, \ldots, 17$, and $k = 3, 6, 9, \ldots, 18$.

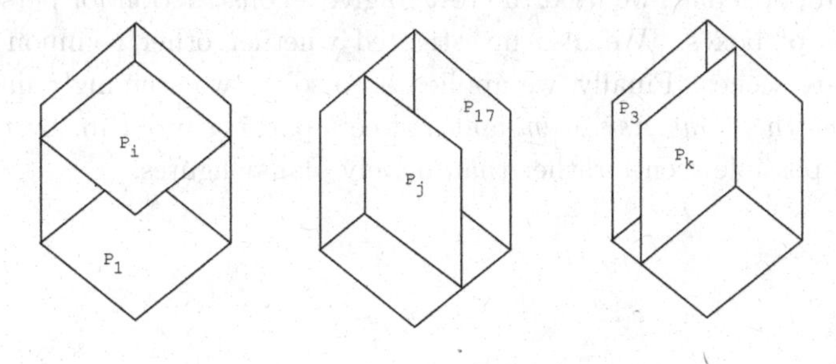

Fig. 7.28 Three families of parallel planes

Some consequent labeling of images of the $\binom{18}{2} = 153$ pairwise intersections of the planes is shown in Figure 7.29. For each i and j differing by a multiple of 3, the label ij is given to the ideal line of the page.

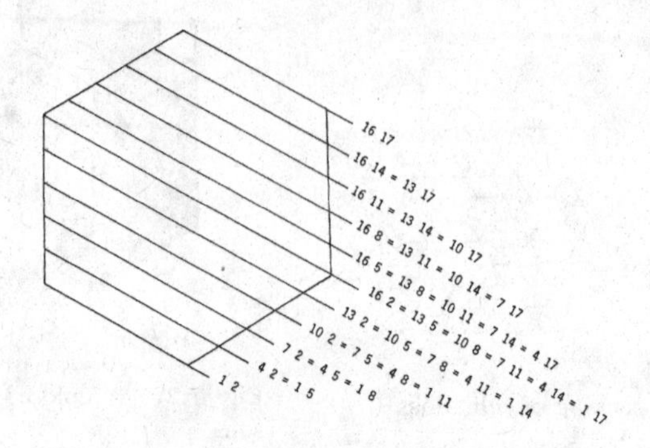

Fig. 7.29 Some labeled image lines

There are 18 choices for i, j and k, giving $\binom{18}{3} = 816$ concurrences to be checked. If each of us successfully verifies a representative sample of these 816 required concurrences we can be reasonably confident of any conclusion based on the interpretation of Figure 7.27 as a scene. We could carry out an analogous argument for a box of side-length n, for any natural number n, and we conclude that consequently, in any packing of calissons, the number with any given orientation is one third of the total number in the box.

Summary of Chapter 7. When given a set of labeled lines purporting to be the image of pairwise intersections of a set of planes, we used Theorem 7.1.1 as a simple internal test of the lines which determined the truth of the claim. In particular, we used the test to give a construction for perspective drawings of boxes. We also investigated whether other common planar figures are scenes. Finally, we applied it to solve two seemingly unrelated problems, the *Pons Asinorum* and a sweets packing problem, by treating them as possible scenes rather than merely planar figures.

Chapter 8

DISTORTION AND ANAMORPHIC ART

We have applied rules, derived from the definition of perspective drawing, in order to draw various simple subjects. But what is the justification for this particular form of planar representation of figures? Why do it? Are perspective drawings particularly easy representations to interpret? Are they intuitively convincing, or does it require an effort of intellect to use and understand them? Together with words, they constitute our major attempt to exchange information on paper about the world around us. But they sometimes produce an uneasy feeling in the viewer that they may be misleading, that they "just do not look right". We should find out why.

8.1 Viewing perspective drawings

We next decide how best to view a perspective drawing by drawing some more boxes. We have perspective drawings of boxes with the same and different vanishing points in Figures 5.25 and Figure 5.26, respectively.

Example 8.1.1. Figure 8.1 contains a collection of perspective drawings of the same box, each drawn from a different viewpoint.

The pictures we drew in Figures 5.24, 5.25, 5.26, and 8.1, and some others drawn in Chapter 5, look terrible! Surely five hundred years of European art can lead to something better. What is wrong? We can do no better than quote the following translation [52] of Leonardo da Vinci;

"... *If you want to represent an object near you which is to have the effect of nature, it is impossible that your perspective should not look wrong, with every false relation and disagreement of proportion that can be imagined in a wretched work, unless the spectator, when he looks at it, has his eye at the*

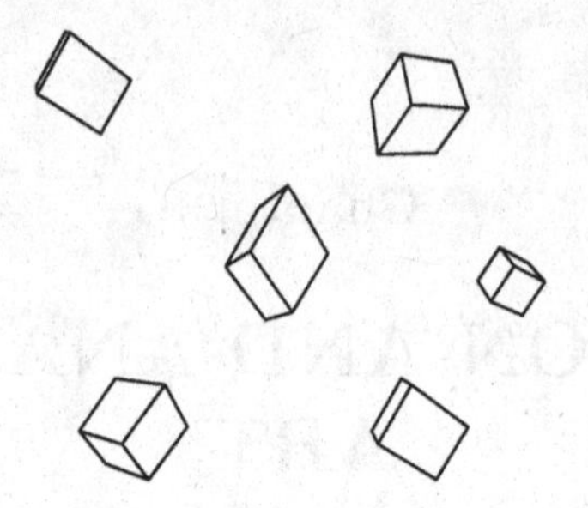

Fig. 8.1 Boxes drawn from different viewpoints

very distance and height and direction where the eye or the point of sight was placed in doing this perspective."

If we view a picture, with one eye, from the viewpoint of its creator we can be confident that it will not appear distorted to us. Definition 5.1.1 ensures that, seen from the viewpoint, the outlines on the canvas coincide exactly with those of the subject. Viewed from elsewhere, or with two eyes, we can have no such expectation about the appearance of a perspective drawing. Binocular vision - using both eyes - raises complex physiological and psychological questions which neither the scientific nor the artistic community are able to satisfactorily answer. We discussed some aspects of binocular vision in Chapter 6, and proved that it differs fundamentally from monocular vision. Here we concentrate on viewing with one eye, trying to determine where to place the eye. There seems to be no one spot from which all the boxes in Figure 8.1.1 "look right". This would agree with our hypothesis that each needs to be observed from its particular viewpoint.

Exercise 8.1.1. By viewing from different positions with one eye open, find a point from which the scene in Figure 5.13 "looks right". A little experimentation may suggest that it is very close to the page, perhaps less than 5 cm above it.

Looking at Figures 5.25 and 5.26, we may also find distortion is least when each is viewed from close to the page. A disadvantage of such a viewpoint is the difficulty of looking at the drawing as it is meant to be viewed, from its viewpoint, as the viewer's eye cannot easily focus on the page from a point so close. A disadvantage for the artist in using a viewpoint far from his canvas is the necessity of drawing lines that meet at vanishing points off the canvas. He then has the time-consuming need to use techniques like that of Construction 3.5.4, whereas we noticed in Construction 5.5.1 how easy it is to draw a box if the three vanishing points used are on the page. The conflict of these two requirements of a viewpoint is exacerbated whenever we look at small reproductions of paintings in art books.

Exercise 8.1.2. When next you visit an art gallery try to locate the viewpoint of the paintings by looking at each from several locations. (Caution: start from afar and move in. Curators are suspicious of close observers.)

8.2 Anamorphic art

In this section, we discuss the use that artists make of the fact that a perspective drawing has a preferred viewing position. Hidden meaning in figures, as for example we saw in Boring's "Mother-in-Law" in Figure 5.31, has been a constant theme of artists. In its most general usage the term *anamorphic art* includes figures drawn on non-planar surfaces, figures which need to be viewed via non-planar mirrors, and figures whose message is disguised in any of a variety of ways. In Chapter 6 we investigated a modern manifestation which needs binocular vision for its interpretation, the so-called "random-dot stereogram".

The simplest, and possibly earliest, technique of anamorphic art is that of making a perspective drawing from an unlikely viewpoint, so that it is rarely initially seen from this point. As the understanding of perspective developed only during the Renaissance it is not surprising that the earliest examples are found in Leonardo da Vinci's drawings.

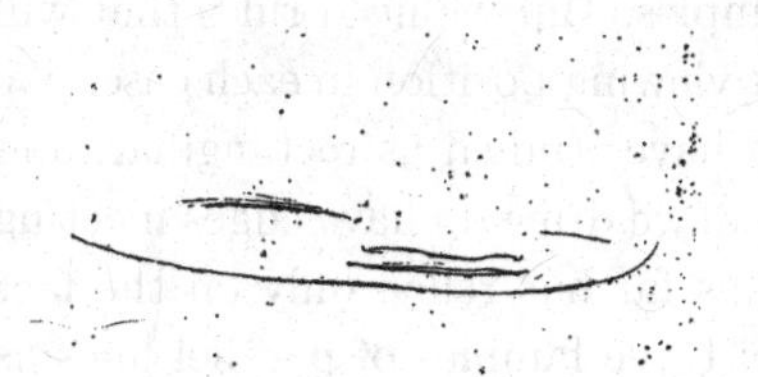

Fig. 8.2 Sketch of a baby's face

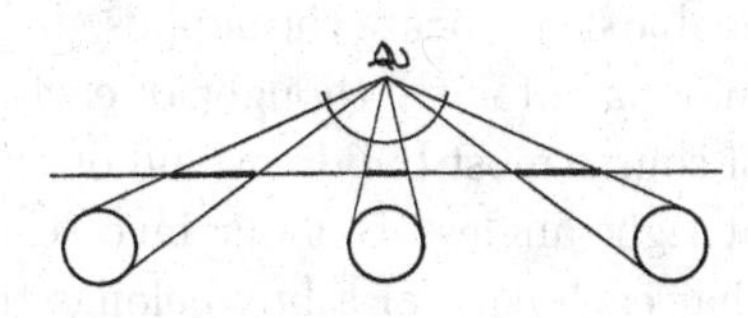

Fig. 8.3 Cross-sections of columns

Figure 8.2 contains a sketch, probably drawn before 1478, from folio 35, versa *a*, of the Codex Atlanticus [19]. Experiment suggests that the baby's face was drawn from a viewpoint about 2 cm above the right-hand edge of the page. In Figure 8.3 da Vinci demonstrates that images of cylindrical columns far from the viewpoint are bigger than images of closer columns, quite a surprising result.

Anamorphic techniques were a significant part of an artist's armory for some centuries. In 1639 Johann Heinrich Glaser engraved "*Christ with the Crown of Thorns, Flanked on the Right by the Fall from Grace and on the Left by the Expulsion from Paradise*". Christ is drawn anamorphically, and His "unseen" presence during normal viewing of the Fall and Expulsion is a significant part of the artist's message. We have earlier seen such examples of figures, parts of which are drawn from different viewpoints such as the three parts of Figure 5.27 or sections of Gris' "*Breakfast*" in Figure 5.29.

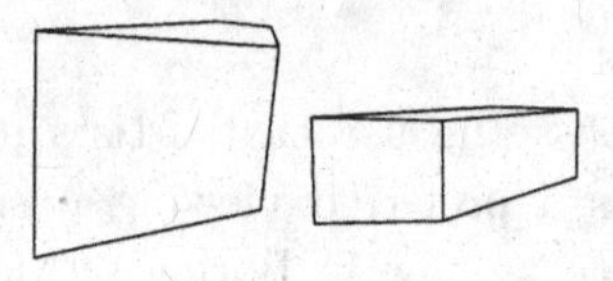

Fig. 8.4 Two perspective drawings of the same cubic box

Figure 8.4 contains two perspective drawings of the same cubic box. The viewpoint of the first is very close to a vertex of the box, the second is drawn from a viewpoint close to the right-hand edge of the page. We repeat the advice of da Vinci:

Instruction 8.2.1. If we view a perspective drawing, with one eye, from the viewpoint of its creator we can be confident that it will not appear distorted to us.

To test this advice we view some examples. But we need rules that will enable us to locate the supposedly correct viewing position in each case. We may have tacitly thought of each box we have studied as rectangular, and of course most buildings and other subjects we draw do have edges meeting at right-angles. But our investigation thus far has relied only on the fact that each edge of a box belongs to one of three families of parallel lines as defined following Lemma 7.3.2. When discussing whether the image of a particular box "looks right" we need to know more about the box itself.

Definition 8.2.1. *Each line that contains an edge of a box belongs to one of three families of parallel lines. The box is* **rectangular** *if the directions of these families are pairwise orthogonal. It is* **cubic** *if it is rectangular and the distance between the each pair of vertices of an edge is the same.*

In fact each box that we have drawn so far is rectangular. Figures 5.25, 5.26, and 8.1 contain perspective drawings of identical cubic boxes. These drawings and the ones that follow in this chapter will repay careful study in the light of Theorems 8.3.1, 8.4.1, 8.5.1, and 8.6.1. The theorems may also explain any dissatisfaction you felt with the results of Construction 5.5.3.

8.3 Distortion in 3-point perspective drawings

We now locate the viewpoint of a 3-point perspective drawing of a rectangular box, using only clues that are contained within the drawing. We then test the usefulness of da Vinci's Viewing Advice 8.2.1 on various examples.

Theorem 8.3.1. *A box is drawn in 3-point perspective (Figure 5.23).*
(1) If p is a Euclidean point not in the plane of the perspective drawing, then the scene is also a 3-point perspective drawing of a box drawn from the viewpoint p.
(2) If the triangle of vanishing points of the scene is not acute-angled, then the scene is not a perspective drawing of a rectangular box from any viewpoint.
(3) If the triangle of vanishing points is acute-angled, then the scene is the perspective drawing of a rectangular box drawn from the fourth vertex of the unique right-angled tetrahedron whose base is the triangle of vanishing points.

Proof. The perspective drawing is obtained by Construction 5.5.1. But this is exactly the process of Construction 7.3.1, and Theorem 7.3.1 guarantees that the resultant figure is the perspective drawing of a box that is drawn from any Euclidean viewpoint using the original vanishing points v_1, v_2 and v_3. Theorem 5.4.1 ensures that each edge of the box is parallel to one of $p \vee v_1$, $p \vee v_2$ and $p \vee v_3$. Consequently the box is rectangular if and only if the tetrahedron $pv_1v_2v_3$ is right-angled at p. From Theorem 4.8.1 we know that this happens if and only if the triangle of vanishing points is acute-angled. □

The vanishing points of the first scene in Figure 8.5 are the vertices of a triangle that is not acute-angled. The vanishing points of the second scene in Figure 8.5 are the vertices of an acute-angled triangle.

Example 8.3.1. The first scene in Figure 8.5 is a 3-point perspective drawing of a box, drawn from a viewpoint 8cm above the + sign. It is not a perspective drawing of a rectangular box from any viewpoint. The development of the box is also shown. The second scene is a 3-point perspective drawing of the same box, drawn from a viewpoint 8cm above the $*$ sign. This scene is also a perspective drawing of a rectangular box from some viewpoint.

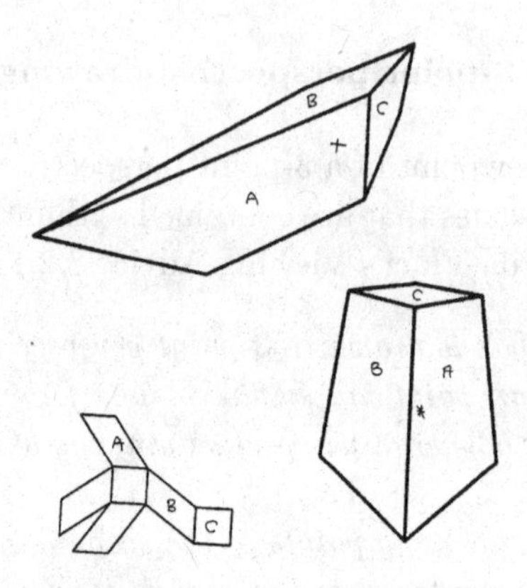

Fig. 8.5 Two views, and a development, of a parallelepiped

Of course in practice most "boxes" drawn are rectangular. In order to check the claim that it is best to look at a scene from its viewpoint it will be useful if we have a practical method of locating the viewpoint of such a scene. Consider any 3-point perspective drawing of a rectangular box, drawn from a viewpoint p, and having vanishing points v_1, v_2 and v_3. We saw in Theorem 4.8.1 that the viewpoint is on the line, perpendicular to the plane of the perspective drawing, that passes through the orthocenter of the triangle of vanishing points. As the line $p \vee v_1$ is perpendicular to each extended Euclidean line of $p \vee v_2 \vee v_3$ it is perpendicular to the line $p \vee a$. After drawing a semicircle on the segment $[av_1]$, and a perpendicular to the line $a \vee v_1$, both in the plane of the drawing, as in Figure 8.7, we can measure the distance between the intersection of the semi-circle and the perpendicular and the point c. This is the height of the point p above the perspective drawing plane. Thus to locate the unique viewpoint, above the page, of a 3-point perspective drawing of a rectangular box, we carry out the following process.

Construction 8.3.1. Locate the Euclidean vanishing points v_1, v_2, and v_3. Draw the orthocenter c of the triangle $v_1v_2v_3$. Draw a semicircle on an

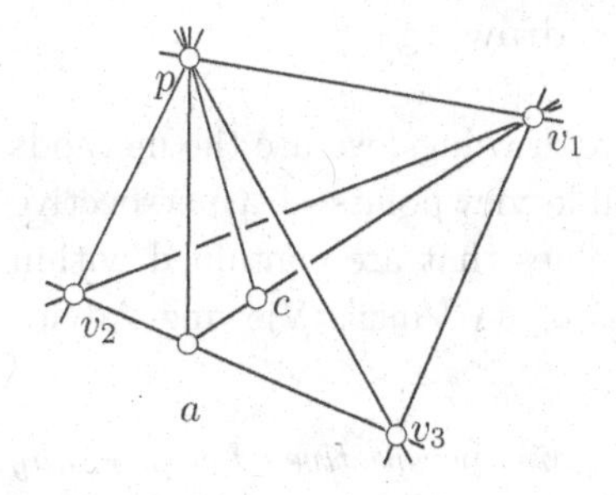

Fig. 8.6 Locating the viewpoint of a 3-point perspective drawing

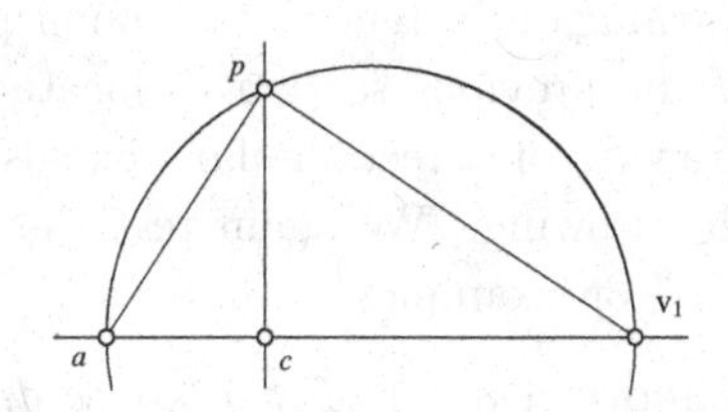

Fig. 8.7 Construction in the perspective drawing plane

altitude of the triangle $v_1v_2v_3$ in the plane of the drawing. Draw a perpendicular through c to this altitude. Measure the distance of the intersection of this perpendicular with the semicircle from c. The correct viewpoint is this distance above c, on the perpendicular to the page through c.

Exercise 8.3.1. Use the above method to find the viewpoint of the scene in Figure 5.13. Compare it with the results of Exercise 8.1.1.

Exercise 8.3.2. Find the viewpoint of the pinhole camera photograph in Figure 8.8.

Fig. 8.8 A pinhole camera photograph

8.4 Distortion in 2-point perspective drawings

Turning our attention to 2-point perspective drawings, we use the methods of the previous section to locate the possible viewpoints of a perspective drawing of a rectangular box, using only clues that are contained within the drawing. We again test the usefulness of da Vinci's Viewing Advice 8.2.1 on examples.

Theorem 8.4.1. *Let a box be drawn in 2-point perspective. Let p be any Euclidean point not in the plane of the perspective drawing. Then the scene is also a 2-point perspective drawing of a box drawn from the viewpoint p. Suppose that the direction v_1 is the ideal vanishing point, and v_2 and v_3 are the two Euclidean vanishing points, of the scene.*

(1) If the direction v_1 and the direction of $v_2 \vee v_3$ are not orthogonal, then the scene is not a perspective drawing of a rectangular box from any viewpoint.

(2) Suppose that v_1 and the direction of $v_2 \vee v_3$ are orthogonal, then suppose that V is the plane that contains $v_2 \vee v_3$ and is perpendicular to each line that has direction v_1. If p is any point of the circle that is in the plane V, is not in the plane of the scene, and has the segment $[v_2v_3]$ as diameter, then the scene is a perspective drawing of a rectangular box drawn from the viewpoint p.

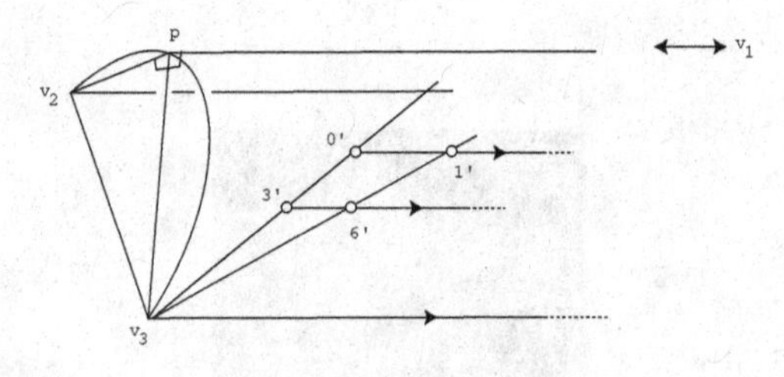

Fig. 8.9 A 2-point perspective rendition of a box

Proof. The first part of the proof is exactly as in Theorem 8.3.1, using Construction 7.3.1 and Theorem 7.3.1. We thus establish that the figure is a 2-point perspective drawing of a box, drawn from any Euclidean viewpoint.

If the scene is a perspective drawing of a rectangular box from a viewpoint p then each pair of the three lines $p \vee v_1$, $p \vee v_2$ and $p \vee v_3$ is perpendicular. Consequently, the line $p \vee v_1$ is perpendicular to the plane

$p \vee v_2 \vee v_3$, ensuring that the direction v_1 is orthogonal to the direction of the line $v_2 \vee v_3$.

Conversely, suppose that v_1 is orthogonal to the direction of the line $v_2 \vee v_3$. From Theorem 4.6.1 we have that there is a line L through v_3 and perpendicular to the plane of the perspective drawing. Any line having direction v_1 is perpendicular to the plane $V = v_2 \vee L$. Let p be a point of the circle that is in the plane V and has $[v_2v_3]$ as diameter. From Theorem 4.7.5 we have that the lines $p \vee v_2$ and $p \vee v_3$ are perpendicular, and each is perpendicular to the line $p \vee v_1$. Theorem 5.4.1 then ensures that the box drawn from the viewpoint p is a rectangular box. □

Example 8.4.1. The first scene in Figure 8.10 is a 2-point perspective drawing of a box, drawn from a viewpoint 10cm above the + sign. It is not a perspective drawing of a rectangular box from any viewpoint. The development of the box is also shown. The second scene is a 2-point perspective drawing of the same box, drawn from a viewpoint 8cm above the * sign. This scene is also a perspective drawing of a rectangular box from some viewpoint.

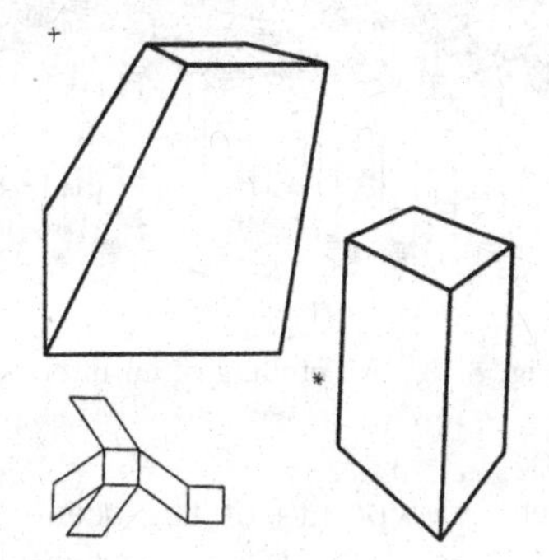

Fig. 8.10 Two views, and a development, of a parallelepiped

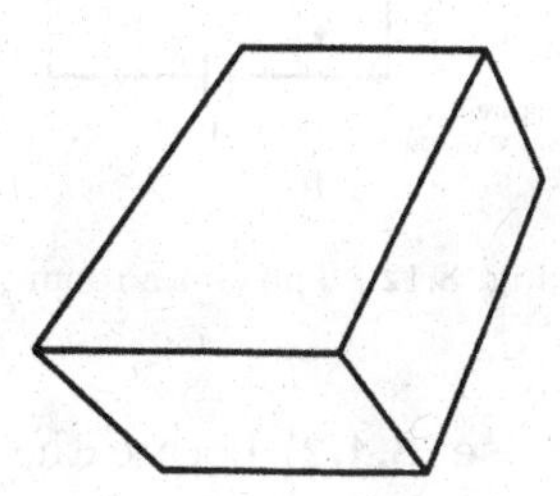

Fig. 8.11 A 2-point perspective drawing of a rectangular box

There is an inherent uncertainty about the location of the viewpoint of any 2-point perspective drawing of a rectangular box.

Construction 8.4.1. Locate the Euclidean vanishing points, v_2 and v_3. Any point p, of the semicircle that is on the segment $[v_2v_3]$ and in the plane perpendicular to the plane of the scene, is a correct viewpoint of the scene as a 2-point perspective drawing of a rectangular box.

Thus, for example, we cannot say whether Figure 8.11 is the perspective drawing of a shallow box, or a deeper box, without additional information.

Exercise 8.4.1. Choose, on the appropriate semi-circle above the page, the viewpoint from which you think a cubic box was drawn to give the 2-point perspective drawing in Figure 8.11

We often unconsciously use extra information to help us interpret drawings. For example, we may look at a human figure in a scene knowing approximately the relation of height to width of the person. This information cannot be derived from the scene alone. This is dramatically illustrated in the famous 1950's experiment of the American psychologist Adalbert Ames, Jr. in which a trapezoidal room, shown in plan and in a photograph in Figure 8.12, was built deliberately to mislead the viewer.

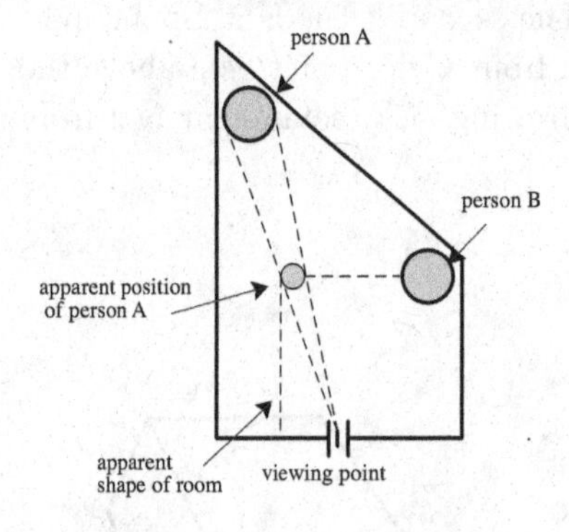

Fig. 8.12 The Ames room

Fig. 8.13 An etching of an interior

Exercise 8.4.2. Locate the possible correct viewpoints of the scene in Figure 8.13. Using any clues given by the impression that images of familiar objects make, choose the viewpoint which you think is the most appropriate.

8.5 Distortion in 1-point perspective drawings

In this section, we concentrate on 1-point perspective drawings, using the methods of the previous sections to locate the possible viewpoints of a perspective drawing of a rectangular box, using only clues within the drawing. We again test the usefulness of da Vinci's Viewing Advice 8.2.1 on examples.

Theorem 8.5.1. *Let a box be drawn in 1-point perspective and p be any Euclidean point not in the plane of the perspective drawing. Then the scene is also the perspective drawing of a box, drawn in 1-point perspective from the viewpoint p. Suppose that the directions v_1 and v_2 are the ideal vanishing points and v_3 is the Euclidean vanishing point.*

(1) If the directions v_1 and v_2 are not orthogonal, then the scene is not a perspective drawing of a rectangular box drawn from any viewpoint.

(2) Suppose the directions v_1 and v_2 are orthogonal and L is the line through the point v_3 that is perpendicular to the plane of the scene. If p is any Euclidean point of the line L other than v_3, then the scene is a perspective drawing of a rectangular box drawn from the point p.

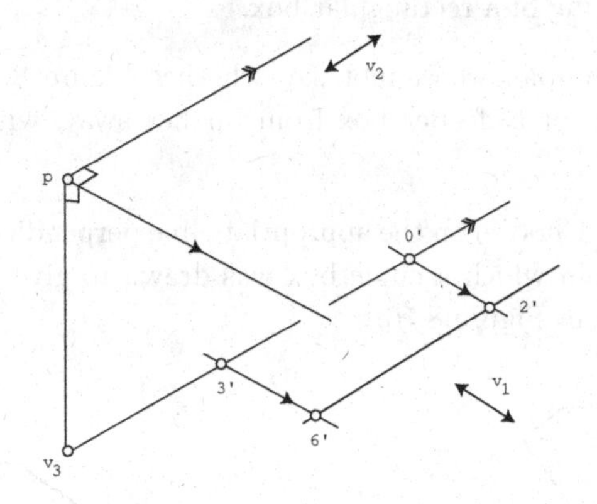

Fig. 8.14 A 1-point perspective rendition of a box

Proof. We use Construction 7.3.1 and Theorem 7.3.1 as in Theorem 8.3.1 in order to show that the figure is a perspective drawing of a box, drawn from any Euclidean viewpoint and using the original viewpoints. If the scene is the perspective drawing of a rectangular box from some viewpoint p, then the lines $p \vee v_1$, $p \vee v_2$, and $p \vee v_3$ are pairwise perpendicular. Hence v_1 and v_2 are orthogonal directions.

Conversely, suppose that v_1 and v_2 are orthogonal directions. Then v_1 and v_2 are directions in the plane of the scene. If p is a point on the line L, then the lines $p \vee v_1$, $p \vee v_2$ and $p \vee v_3$ are pairwise perpendicular. From Theorem 5.4.1 we have that the box drawn from p is rectangular. □

Example 8.5.1. The first scene in Figure 8.15 is a 1-point perspective drawing of a box, drawn from a point 6cm above the + sign. It is not a perspective drawing of any rectangular box. The development of the box is also shown. The same box is drawn again, from a viewpoint 6cm above the $*$ sign. This scene is a perspective drawing of a rectangular box from some viewpoint.

As in the 2-point perspective case, there is uncertainty about the location of the viewpoint of any 1-point perspective drawing of a rectangular box.

Construction 8.5.1. Locate the Euclidean vanishing point v_3. Any point p on the perpendicular to the plane of the perspective drawing through this vanishing point is a correct viewpoint of the scene as a perspective 1-point perspective drawing of a rectangular box.

Thus, for example, we cannot say whether Figure 8.16 is the image of a shallow box, or a deeper box from further away, without additional information.

Exercise 8.5.1. Choose, on the appropriate line perpendicular to the page, the viewpoint from which a cubic box was drawn to give the 1-point perspective drawing in Figure 8.16.

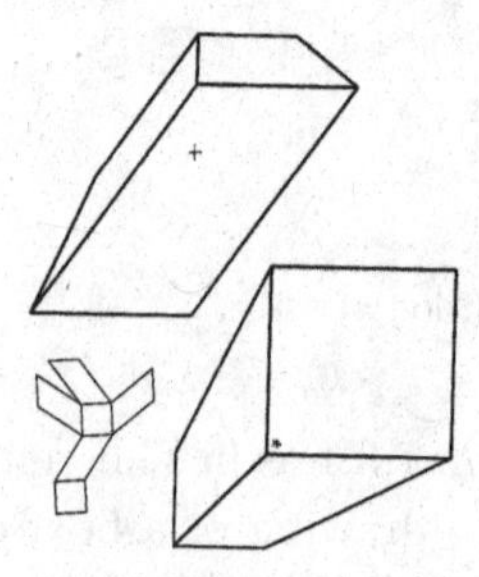

Fig. 8.15 Two views, and a development, of a parallelepiped

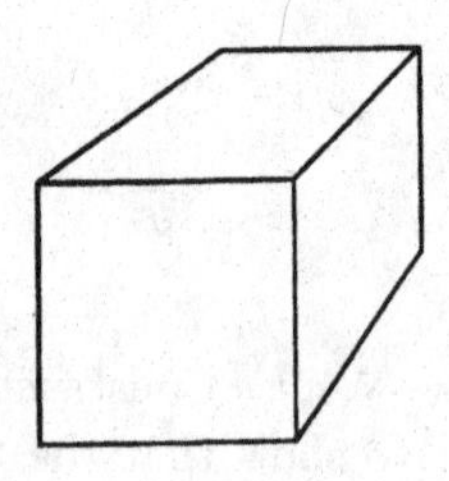

Fig. 8.16 A 1-point perspective drawing of a rectangular box

Exercise 8.5.2. Locate the possible correct viewpoints of "Adoration of the Kings" in Figure 5.16. Choose the viewpoint which you think is the most appropriate.

Exercise 8.5.3. Calculate the position of the viewpoint for some "coffee-table art book" old-master reproduction. Does the representation seem more convincing when viewed, by one eye, from this point?

Exercise 8.5.4. Locate the possible correct viewpoints of the scene in Figure 8.17. Using any clues given by the impression that images of familiar objects make, choose the viewpoint which you think is the most appropriate.

Fig. 8.17 A perspective drawing of a folly

8.6 Distortion in affine projections

We have investigated scenes drawn from Euclidean viewpoints and seen how important it is to look at them from their viewpoint in order to avoid false impressions.

It is obviously not possible to view an image of an affine projection from its viewpoint. However the choice of a direction as viewpoint is much more convenient for drawing small pictures than any Euclidean viewpoint which is conveniently far enough from the page to allow our eye to focus on the page. We commented earlier that it might be best to view an affine projection from afar, and again we quote Leonardo da da Vinci in translation:

"... unless indeed you make your view at least twenty times as far off as the greatest width or height of the objects represented, and this will satisfy any spectator placed anywhere opposite to the picture".

Many perspective drawings contain images of parallel lines. From Theorem 5.6.1 we have that these images are themselves parallel in any affine projection, whereas when drawn from an Euclidean viewpoint the images are parallel only when the lines themselves are parallel to the plane of the drawing. We might feel that the distortion observed by the viewer of an affine projection image would be slight. It is certainly much easier to draw parallel lines, using a roller and pencil, than it is to draw lines meeting, say, one meter off the page using the technique of Construction 3.5.4. The difference between the two images is slight, and so a small affine projection and a small perspective drawing drawn from a viewpoint 1 or 2 meters from the page appear very similar.

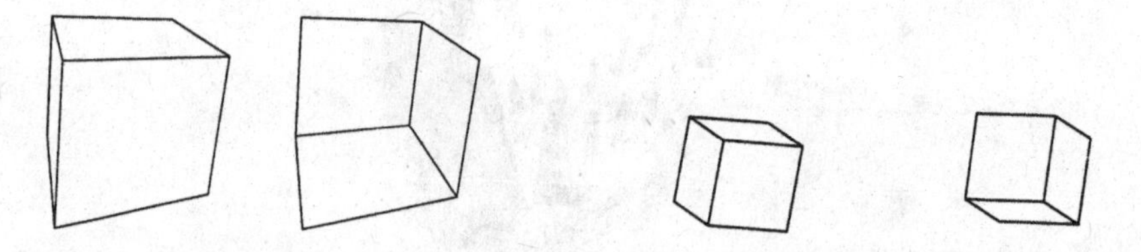

Fig. 8.18 Two 3-point perspective drawings

Fig. 8.19 Uncertainty inherent in affine projections

However there is one aesthetic disadvantage of an affine projection not shared by a perspective drawing drawn from an Euclidean viewpoint - an uncertainty it induces in the viewer. Even though distortion may be low when viewed from a reasonable distance, the eye lacks clues as to whether one is looking "up into" or "down onto" the subject. The reason for this is our inability to associate a direction with any one particular "end" of a Euclidean line in our construction of extended Euclidean space.

The first scene in Figure 8.18 is the image of a cubic box drawn in 3-point perspective, or more precisely, it is the image of those three panels of the box visible from the viewpoint. The second scene is the image of the same box with those three panels removed.

Figure 8.19 contains an affine projection of each subject of the scenes in Figure 8.18. It is possible to tell which of the two scenes in Figure 8.18 is which. This is not the case with the two scenes in Figure 8.19. Each of the two shells shown in Figure 8.20 has the same affine projection from the direction p. Without external clues, such as a shadow, it is not possible to distinguish between the two from the projection. The brain of a viewer is uncertain which interpretation to choose. This results in unexpected difficulties when scenes, such as that in Figure 8.21 for example, are viewed without interruption for a minute or so. Of course, uncertainties are sometimes sought by the artist. Boring, in his "*Mother-in-law*" shown in Figure 5.31, wants us to ponder on the flickering transition between

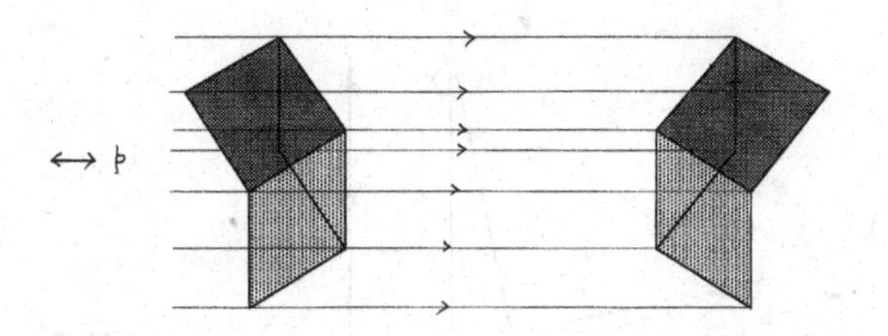

Fig. 8.20 Two shells sharing the same affine projection

young girl and old woman. The question of from which direction an affine projection is drawn is, as we might expect by now, difficult.

Fig. 8.21 Looking down on, or looking up to, some boxes?

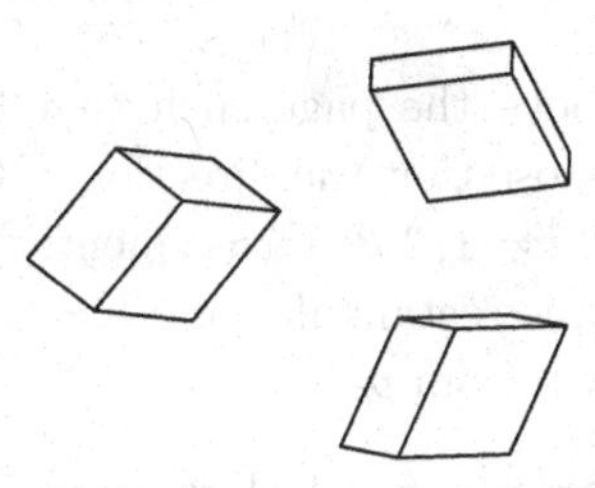

Fig. 8.22 Affine projections of three boxes

Theorem 8.6.1. *Let a box be drawn in affine projection, and suppose that p is any direction not in the plane of the drawing. Then the scene is also the affine projection of a box drawn from the viewpoint p.*

There exists a direction so that the scene is the affine projection of a rectangular box drawn from that direction.

Proof. We vary Construction 7.3.1 by choosing the vanishing points v_1, v_2, and v_3 to be (necessarily collinear) directions, leading to a variant of Theorem 7.3.1 in which the point p is ideal. Using this result, the first part of the proof follows the first part of the proof of Theorem 8.3.1.

In order to complete the proof, in the affine projection we choose an internal vertex image and label it, without loss of generality, by $0'$. We draw an acute-angled triangle $1_1 2_1 3_1$, whose vertex i_1 is on the edge $0' \vee i'$, for each $i = 1, 2, 3$. From Theorem 4.8.1 we have that there is a point

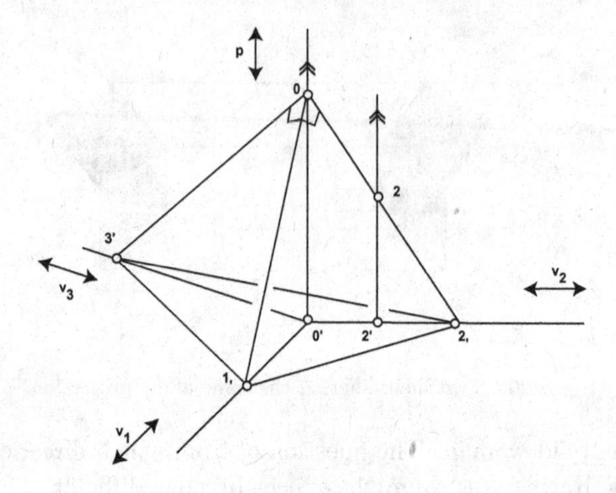

Fig. 8.23 An affine rendition of a box

0 above the page such that $01_12_13_1$ is a tetrahedron, right-angled at 0. Suppose that the direction of $0 \vee 0'$ is p. We define $i = (0 \vee i_1) \cap (p \vee i')$, for each $i = 1, 2, 3$. Consequently any box, with corners 1,2 and 3 each adjacent to 0, is rectangular and has the original scene as its affine projection when drawn from p. □

Example 8.6.1. Each figure in Figure 8.22 is an affine projection of a box. Each figure is also an affine projection of a rectangular box from some direction.

Summary of Chapter 8. We specified the exact circumstances that allow a perspective drawing of a box to also be a perspective drawing of a rectangular box. We proved that each 3-point perspective drawing of a box is also a 3-point perspective drawing of a box drawn from any Euclidean viewpoint. From information contained within the perspective drawing we determined whether the image is a perspective drawing of a rectangular box. If so, the information enabled us to locate the viewpoint of this particular rendition of a rectangular box. We were then able to view the drawing from the point recommended by Leonardo da Vinci, and test his advice. We performed similar analyzes of 2-point and 1-point perspective drawings. In these cases additional information is needed in order to locate the desired point from which to view the drawings. We discussed the advantages and disadvantages of affine projection, in particular the existence of an inherent uncertainty in our interpretation of any affine projection.

Chapter 9

PLANAR BAR-AND-JOINT MECHANISMS

Throughout our geometric investigations we have drawn and modeled figures in order to help us visualize them and understand their properties. In Chapter 5 we began an analysis of perspective drawings of figures. We examined the similarities between a figure and its drawings and developed rules for perspective rendition. In this chapter and in Chapter 10 we pay more attention to models of figures. Our models are simple examples of the grander works that engineers and builders engage in. This guarantees us a plentiful supply of examples to engage our attention.

We focus on some properties of the models that make them useful to engineers. In particular, we look at the question of the rigidity [14] of models. Designers of structures such as the Tacoma Narrows Bridge shown in Figure 9.1, and the Forth Railway Bridge re-built in 1890 and shown in Figure 9.2, are interested in the rigidity of bar-and-joint models.

Fig. 9.1 The Tacoma Narrows bridge in a strong wind

Fig. 9.2 The Forth Railway Bridge

The models we make, and man-made constructions in general, fall naturally into two types - those in which parts may move relative to other parts, and those in which the distance between any two parts remains unchanged. A *mechanism* is a model in which the rigid components (bars or panels, say) are able to move relative to one another. A model in which no part can move relative to another is *rigid*.

The model of the cube in Construction 2.5.1 was designed to have a planar unassembled format for convenience of storage. As it is folded into its assembled position, the panels move relative to one another and their corners no longer model a planar set. In its assembled form the corners model the points of the designated cube. Generally, one mechanism may model a family of distinct figures as the parts move relative to one another.

We examine models from an engineer's point of view, using results proved about mechanisms to make practical use of examples. A structural engineer requires a building to be rigid, but a mechanical engineer wants his engine to be a mechanism.

9.1 Bar-and-joint models

There are two common engineers' constructions, those held together by their "skin" and those in which a "skeleton" plays the main structural role. A cardboard box and a typical skyscraper of the 1940's are examples of these differing approaches to structural integrity. In some ways it is more enlightening to draw the first type of structure, and to model the second. In this section, we examine models of the second type.

Strong leverage forces sometimes make it impractical for the joins of rigid bars of such "skeletons" to be designed to resist rotational movement of one bar relative to another. We examine models that are designed to accept this limitation.

Definition 9.1.1. *A* **bar-and-joint model** *consists of rigid bars, their only interaction being that members of a subset of bars may each have an end held together. Such a connection is often called a universal joint.*

The adjective "universal" emphasizes the fact that bars are attached as freely as possible, able to move relative to one another subject only to the condition that the ends remain together. Even though any two bars may share only a universal joint, the overall rigidity of a bar-and-joint model may be difficult to determine.

Definition 9.1.2. *A* **bar-and-joint mechanism** *is a model in which the* *rigid bars are able to move relative to one another. A model in which no bar can move relative to another is rigid.*

Simple bar-and-joint models are widely used in our every-day lives. The simplest interesting bar-and-joint model is shown in Figure 9.3. It is usually called a flail and is a mechanism. It can be made of two pieces of wood hinged by a leather strip and used to thresh grain. Unfortunately it also has a manifestation as the *nunchuk* of martial arts notoriety. It is also the only successful model of the golf swing, duplicating very closely the swings of great players. One bar represents the arms, and the other models the club. The joint models the wrists. The model's success is due to the fact that at the important part of a golf swing, at contact with the ball, the wrists hinge freely. In Figure 9.3, taken from "*The Dynamics of the Golf Swing*" by D. Williams [64], the motion of the flail derived from these assumptions closely resembles stroboscopic photographs of actual swings.

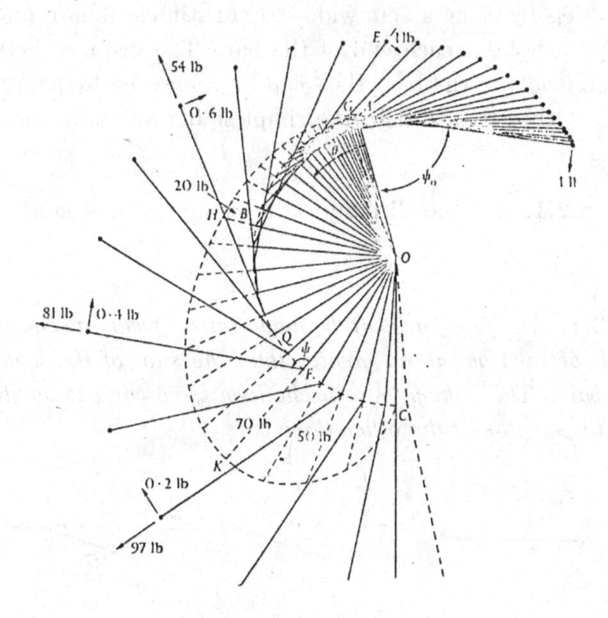

Fig. 9.3 A flail as a golf swing model

Another simple bar-and-joint model is a triangular framework. From Lemma 4.7.3 we have that a triangular framework exists exactly if the length of each bar is no greater than the sum of the other two bar lengths. It is not a mechanism and is the basic shape in most rigid engineering bar-and-joint frameworks. It is repeatedly present in the Forth Bridge, in scaffolding, and in the octahedral-tetrahedral trusses that make up space-frames, for example.

The difficulty of constructing freely movable universal joints in practice makes non-planar models more of interest to structural engineers than to mechanical engineers. It is easier to make a non-planar model that stays rigid than a satisfactory long-lasting machine with moving universal joints.

9.2 Four-bar linkages

In this section, we investigate bar-and-joint models in which the models themselves always remain planar. Any penknife, or pair of scissors, or pair of pliers serves as a reminder that joints which allow movement freely in a plane are commonplace and reliable. We can easily make planar bar-and-joint models by using a 2cm wide strip of stiff cardboard for each bar, and punching a hole at each end of the bar. The distance between hole centers is the *length* of the bar. Bars can be connected by paper fasteners in the holes. Other than a flail, the simplest bar-and-joint mechanism is the following:

Definition 9.2.1. *A* **4-bar linkage** *is a planar bar-and-joint model of a 4-gon.*

Lemma 9.2.1. *Any four bars can be made into a 4-bar linkage if and only if the length of each bar is no greater than the sum of the lengths of the other three bars. The linkage is a mechanism if no bar has length equal to the sum of the lengths of the other three bars.*

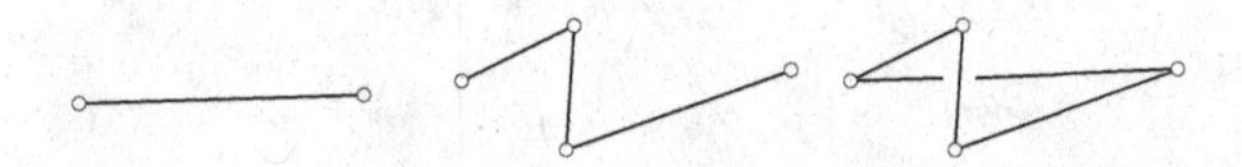

Fig. 9.4 Assembling a 4-bar linkage

Proof. From Theorem 4.7.3 we have that the separation of the ends of a flail is at most the sum of the lengths of its two bars. In turn this ensures that the separation of the ends of a chain of three bars is at most the sum of the lengths of the three. So such a chain could not be completed to a

4-bar linkage by a fourth bar of length greater than the sum of the lengths of the members of the chain.

Conversely if we choose a longest bar, assemble the others into a chain where ends are separated by the length of this largest bar, then the linkage can be completed. From our discussion following the Triangle Inequality in Chapter 4, we have that an assembled linkage remains linear if the length of the longest bar is equal to the sum of the lengths of the remaining bars. ☐

Construction 9.2.1. Make a 4-bar linkage, composed of bars of length 5cm, 6cm, 7cm, and 7.5cm respectively. Make a second linkage modeling a parallelogram with two 8cm bars and two 14cmbars.

Experimenting when assembling the linkages confirms that each of the two models is a mechanism. Examples of 4-bar linkages are all about us, sometimes well disguised. With a little practice we will be able to find them in surprising places. Let us examine one such example.

Example 9.2.1. Water from the hose spins the driving crank of the sprinkler [32] in Figure 9.5 via a little water-wheel connected by gearing to the crank. As the crank turns it moves the spraying tube and distributes the water in a rectangular pattern. Each setting of the clamping screw gives a different pattern of coverage as the spraying tube moves back and forth. The length of one bar of the 4-bar linkage underlying this mechanism is determined by the screw setting, the other lengths not being adjustable.

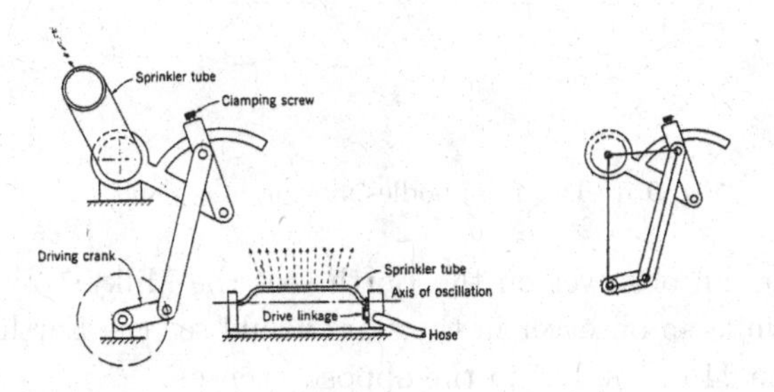

Fig. 9.5 A garden sprinkler

Exercise 9.2.1. Locate a 4-bar linkage in each of the crane and auto-hood hinge in Figure 9.6.

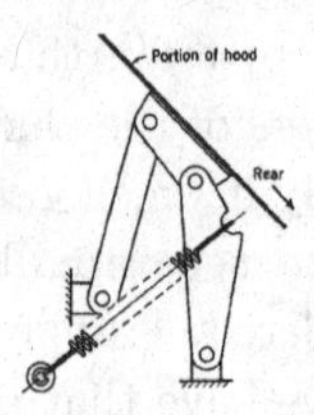

Fig. 9.6 A crane and an auto-hood hinge

In order to analyze the behaviour of 4-bar linkages we observe the motion of one bar relative to another. Observed change in an observed object depends also on any change in the observer. Thus two observers, one standing and one spinning like a top, would differ in their descriptions of the sun's motion. Let us agree, without loss of generality, that we take clockwise as a direction of positive rotation in the page. We may sum up the situation in the following lemma:

Lemma 9.2.2. *Let two rigid bodies be confined to a plane. If the first turns through an angle θ, as seen from the second, then the second turns through an angle $-\theta$, as seen from the first. In particular, one makes a full turn relative to the second exactly if the second makes a full turn relative to the first.*

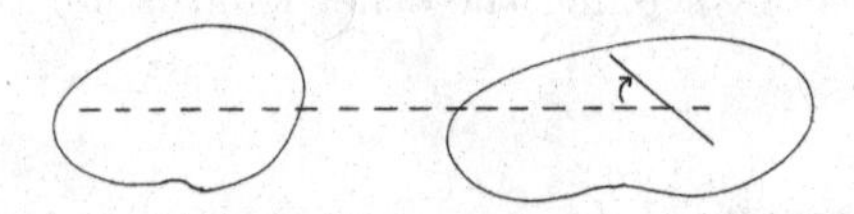

Fig. 9.7 Two rigid bodies moving in a plane

For example, an observer on the Earth sees the Milky Way turn fully in 24 hours - while an observer in the stars would see the Earth turn once during the same 24 hours, but in the opposite sense.

Exercise 9.2.2. Confirm that we can move the bars of the parallelogram model of Construction 9.2.1 so that the linkage remains a model of a parallelogram, the short sides staying parallel to each other as they turn fully relative to each of the longer bars. Rephrasing this in the light of Lemma 9.2.2, we say that each long bar is able to rotate completely relative to each short bar.

9.3 Rocking and rotation in 4-bar linkages

Linkages are typically used to transform one type of input motion (rocking or rotation) into an output motion (again, rocking or rotation). In this section, we classify 4-bar linkages by the possible transformations they can carry out. The classification reflects the uses to which they are put. Jacob Leupold, an early 18th century engineer, was perhaps the first to attempt such a systematic classification and the following theorem, proved in 1883 by F. Grashof [29], is the key to it.

Theorem 9.3.1. (Grashof's Theorem) *A bar of a 4-bar linkage can rotate completely, relative to each of the other bars, exactly when it is a shortest bar and the sum of its length and the length of a longest bar is no greater than the sum of the lengths of the other two bars of the linkage.*

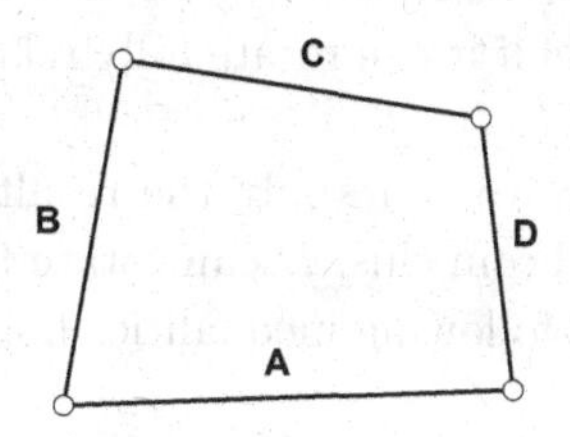

Fig. 9.8 A 4-bar linkage

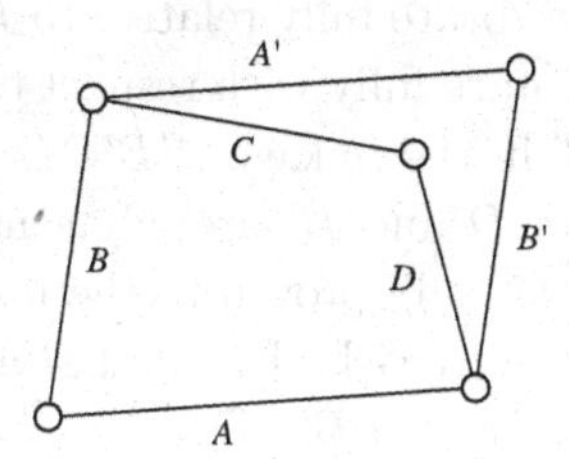

Fig. 9.9 Adding bars to the original linkage

Proof. We denote the bars of the linkage by A, B, C and D, and their lengths by a, b, c and d respectively.

Consider two adjacent bars A and D such that, without loss of generality, $d \leq a$. Then the bar D can rotate fully relative to A if and only if the figures in Figure 9.10 exist.

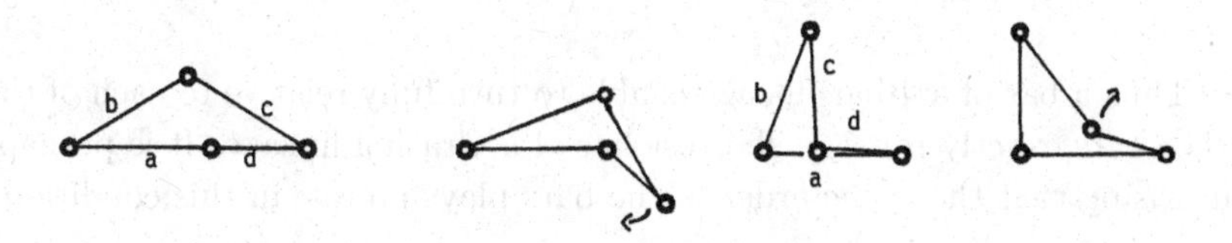

Fig. 9.10 Stages in a full turn of D relative to A

From Lemma 4.7.3 this requirement is equivalent to the set $a+d \leq b+c$, $b \leq (a+d)+c$, $c \leq (a+d)+b$, and $c \leq b+(a-d)$, $b \leq c+(a-d)$, $a-d \leq b+c$ of six conditions.

This set is equivalent to the set $a+d \leq b+c$, $c+d \leq a+b$, and $b+d \leq a+c$, of three conditions. This set can be restated as:

$d+$ length of any bar $\leq$ sum of the lengths of the remaining two bars.

Using a similar argument with c in place of a, we prove that this condition determines also whether D is able to rotate fully relative to C.

In order to answer the question of whether D can rotate fully relative to B we need to consider the model $\mathbf{M}$ in Figure 9.9 where A' has length a and B' has length b. As we saw in our experiments with the parallelogram model in Construction 9.2.1, the parallelogram linkage $A'B'AB$ allows B' to rotate fully relative to A, with B remaining parallel to B' during the motion.

Thus D can rotate fully relative to B in the linkage $ABCD$ exactly if it can rotate fully relative to B in the 6-bar model $\mathbf{M}$ of Figure 9.9. But D can rotate fully with respect to B in this model if it can rotate fully relative to B' in the linkage $A'B'CD$.

As D and B' are adjacent in this linkage we can apply the result we have already proved to the linkage $A'B'CD$. From this, D can rotate fully relative to each of B' and B if and only if the following inequalities hold in the linkage $A'B'CD$:

$d+$ length of any bar $\leq$ sum of the lengths of the remaining two bars.

As the bar lengths are a, b, c and d this is the condition we have met twice before, once enabling D to rotate relative to A and once enabling D to rotate relative to C in the original linkage. To finish the proof we need only note that for this condition to hold, D must be a shortest bar. □

Definition 9.3.1. *The* **Grashof condition** *for a 4-bar linkage is the requirement that the sum of the lengths of a shortest and a longest bar does not exceed the sum of the lengths of the other two bars. A linkage satisfying this condition is a Grashof linkage.*

Thus a bar of a 4-bar linkage is able to turn fully relative to each of the other bars exactly if it is a shortest bar of a Grashof linkage. It is perhaps surprising that the cyclic order of the bars plays no role in this condition.

Exercise 9.3.1. Decide by measurement if the linkage in Example 9.2.1 is a Grashof linkage.

Decide by measurement if each linkage of Construction 9.2.1 is a Grashof linkage. Does this confirm the results of your experimentation with the parallelogram linkage in Exercise 9.2.2? Test the other linkage by holding each bar in your hand in turn, and attempt to rotate the others of the model relative to it.

Definition 9.3.2. *The* **base,** *or* **frame-link,** *of a 4-bar linkage is a selected bar of the linkage. One bar adjacent to the base is called the input, and the other is called the output of the linkage. The remaining bar is the coupler, or connecting-rod, of the linkage.*

This nomenclature reflects practical applications in which the base of a 4-bar linkage is fixed, and the input bar moved, thereby giving motion in a convenient form to the output bar via the coupler. Unless otherwise stated, motion of parts of a linkage is usually taken to mean relative to the base of the linkage. A standard pictorial notation is shown in Figure 9.11.

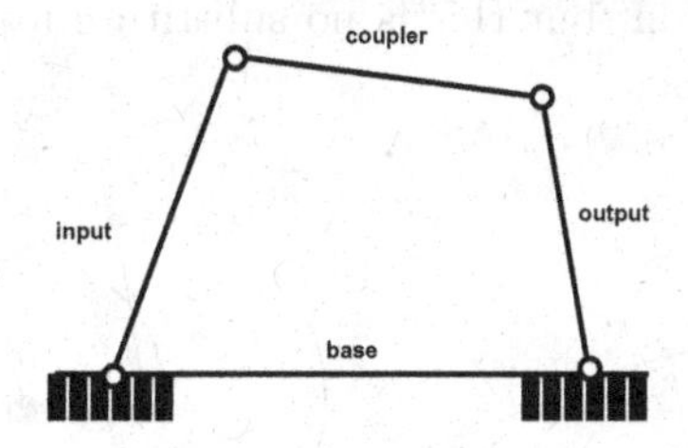

Fig. 9.11 A notation that distinguishes bars of a 4-bar linkage

Definition 9.3.3. *A* **double-crank,** *or* **drag-link,** *mechanism is a Grashof linkage in which the base is a shortest bar of the linkage. Both input and output rods are called cranks of the mechanism.*

A crank/rocker mechanism is a Grashof linkage in which the input is a shortest bar. The input is the crank, and the output is the rocker, of the mechanism.

A rocker/crank mechanism is a Grashof linkage in which the output is a shortest bar. The input is the rocker, and the output is the crank, of this mechanism.

A double-rocker mechanism is either a non-Grashof linkage, or a Grashof linkage in which the coupler is the shortest bar. Both input and output bars are called rockers.

We sum up the behavior of these mechanisms in the following lemma:

Lemma 9.3.1. *A crank of any 4-bar linkage is free to fully rotate relative to the base of the linkage, but a rocker is restricted to oscillatory motion relative to the base.*

Proof. An input or an output bar of a Grashof linkage is a crank if either it or the base is a shortest bar of the linkage. From Theorem 9.3.1 we have that this is exactly the condition enabling it to rotate fully relative to the base. □

Exercise 9.3.2. In which of the 4-bar linkage types of Definition 9.3.3 is the coupler able to rotate fully relative to the base?

We draw arcs of circles, as in Figure 9.12, to show the possible positions of the coupler joints relative to the base of a linkage and help us visualize linkage motion, but this is no substitute for making models and experimenting with them.

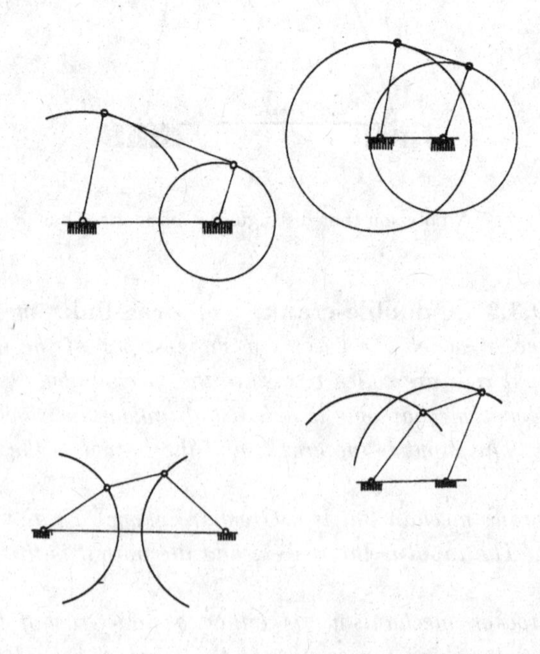

Fig. 9.12 Possible ranges of coupler joint movements

Exercise 9.3.3. Choose the 7cm bar as the base in the first model of Construction 9.2.1. Remove the fasteners of the two joints of the base (and the bar itself). Place a sheet of paper between the model and a reasonably soft backing board. Replace the fasteners by thumb-tacks, pushed into the backing board. You can now conveniently investigate the motion of the model. Trace, with a pencil through a joint of the coupler, the possible positions of the joint. Repeat the experiment with the other coupler joint. Do the results agree with those expected from Theorem 9.3.1?

Repeat these experiments with the 5cm bar chosen as base. Does this experiment lead you to doubt the possibility of wide-spread practical use of the full range of motions of a double-crank mechanism?

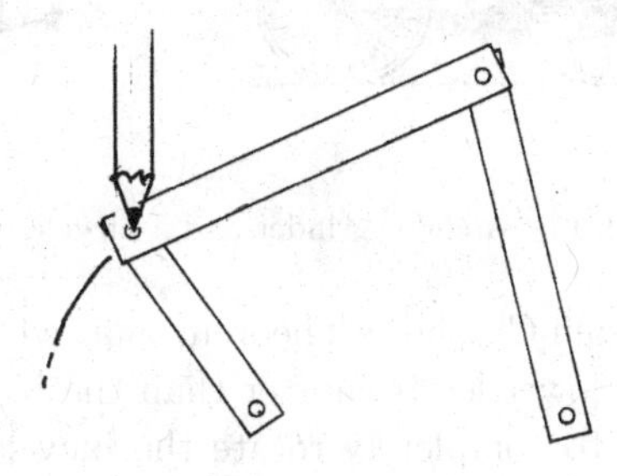

Fig. 9.13 Plotting the positions of coupler joints on backing paper

Example 9.3.1. The lawn sprinkler of Figure 9.5 is a crank/rocker mechanism - the rotary motion of the water-driven crank being transformed into an oscillating motion of the spraying arm.

Example 9.3.2. A treadle sewing machine uses a rocker/crank mechanism. The operator of a Spong coffee-grinder is part of a rocker/crank mechanism, the user's upper arm providing the input rocker, the forearm the coupler, and the handle being the output crank.

The action of each leg during bicycle riding also is an application of a rocker/crank mechanism. With respect to the bicycle frame as base, the rider's thigh is the rocker, the lower-leg is the coupler and the bicycle crank is the linkage crank.

Problem 9.3.1. *A small child riding an adult's bicycle seems to bob up and down. Would this still be necessary if the saddle could be adjusted to a low enough position?*

Fig. 9.14 A coffee grinder and a bicycle rider

Solution. We have from Grashof's Theorem that, while the child is seated, if either the upper or lower leg is shorter than the bicycle crank then it is physically impossible to completely rotate the bicycle crank. This is one reason why very small children have difficulty riding a bike too large for them - it is not their lack of strength or weight alone that is the problem. A very small child cannot ride in the saddle if the bicycle's crank is longer than either the child's upper or lower leg.

Exercise 9.3.4. Can you identify a rocker/crank mechanism in a child's pedal car? Can you find a 4-bar linkage in an automobile windscreen-wiper.

Example 9.3.3. The lifting mechanism of the plow shown in Figure 9.15 is a double-rocker mechanism.

Rotation is not important in the job for which this linkage is designed. On the contrary, two locking positions of the linkage, where no further rotation is possible, are required. Each of these supports the plow at a fixed height, the lower for plowing and the higher for transporting the plow. The coupler can be fixed to the wheel by a clutch as required enabling the linkage to be moved into one of the two locking positions, each held stable by the weight of the plow pulling two bars straight.

Example 9.3.4. The double-rocker mechanism [32] of Figure 9.16 joins an engine and a massive rotating flywheel used to supply energy for starting

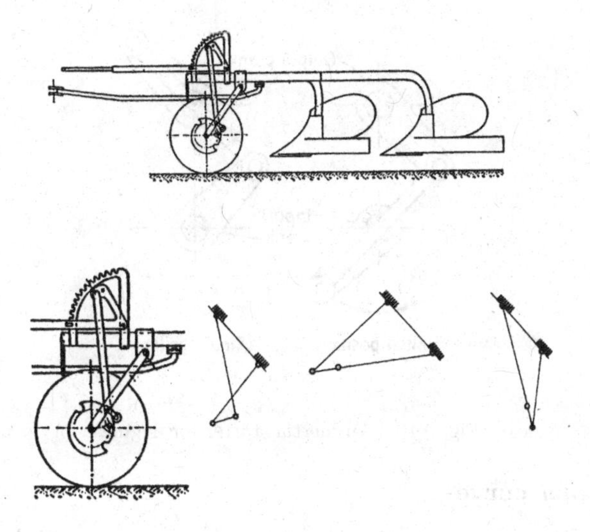

Fig. 9.15 A plow

the engine. The output crank is connected, through step-up gearing, to the starter pinion of the engine. The input crank is connected by a clutch to a shaft turned by the rotating flywheel. To start the engine, the linkage is placed in its "initial" position with the clutch disengaged. The flywheel is brought up to speed, either by hand crank or electric motor. The clutch is then engaged, the linkage driven to its "final" position (with the flywheel momentarily at rest), and then the flywheel is disengaged having transferred all its energy to the engine.

Substantially shockless transfer of kinetic energy from the rotating mass to the engine is accomplished. As the input crank rotates 180° from initial to final position, the output crank also turns 180°. The key feature of the design is that the 4-bar linkage is so proportioned that the angular velocity ratio of input to output crank varies from infinity to zero during the operating period. A consequence of this is the failure of the linkage to satisfy the Grashof condition. In inertia starters that use a friction clutch to directly connect a flywheel to the engine, a large part of the kinetic energy of the flywheel is dissipated in friction at the clutch faces. The above design avoids that energy loss.

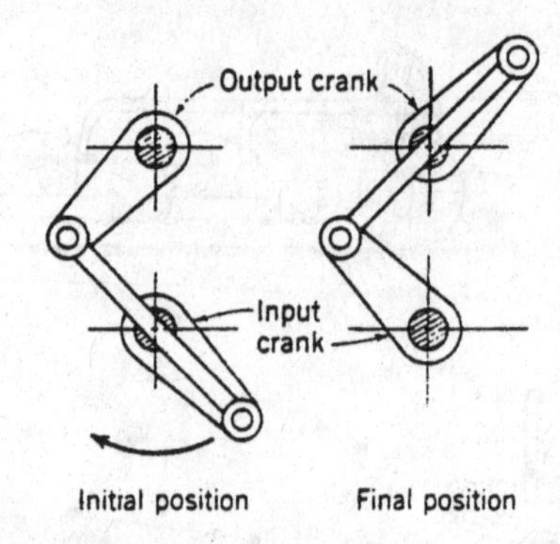

Fig. 9.16 An inertia starter for engines

9.4 Coupler curves

So far we have thought of each planar 4-bar linkage primarily as a convenient method of energy transfer. But as we saw in Example 9.3.3 they have an alternative use, as devices for moving a point along a required path. In this section, we examine this role of linkages.

In some applications it is a significant over-simplification to think of each bar of a 4-bar linkage as a segment. By replacing a bar with a more general rigid body, jointed at two points, we are able to use the linkages more widely. For example, in the case of the automobile-hood hinge of Exercise 9.2.1 we treat a cross-section of the hood as a coupler body. We are interested in the motion of the rear of the hood, in order to ensure that it misses the bulging center of the windscreen.

Definition 9.4.1. *Let a planar rigid body replace the coupler bar of a 4-bar linkage. A* **coupler point** *of the linkage is any point of the body. The set of possible positions of any given coupler point is a* **coupler curve** *of the linkage.*

Construction 9.4.1. Make the model of the mechanism shown in Figure 9.17 and trace the curve followed by the coupler point that is labeled by x. Make the model using bars already to hand, a 13cm bar, and a slightly over-sized cardboard triangle. Punch holes the appropriate distance apart in the triangle. Later we will need models of the two other triangles shown, and you may find it convenient to make all three now. The numbers given are lengths, measured in centimeters.

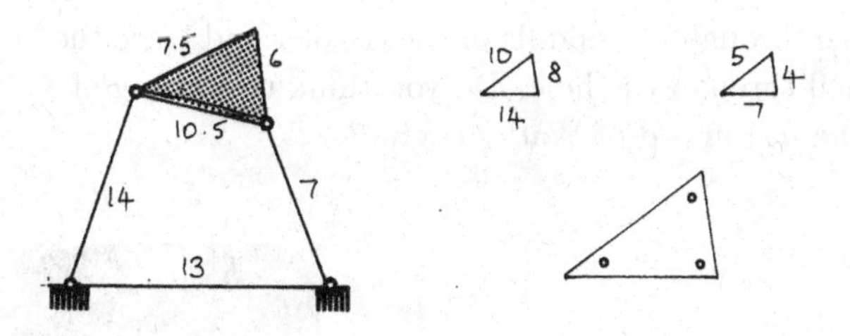

Fig. 9.17 A coupler point of a linkage

In the 18th Century, steam was beginning to power the industrial revolution. James Watt's rotative engine made in 1784 was the first to rotate a shaft directly without using a water wheel fed by a pumping engine. The guiding of the large piston required was difficult - as accurate milling was not available to make reliable straight sliding guides. Watt's solution was to use the linkage shown in Figure 9.18 to guide the piston rod. This seems to have been the first use of a 4-bar linkage as a means of moving a coupler-point along a particular path rather than as an energy transfer device. It created great interest in coupler curves that contain approximately straight sections.

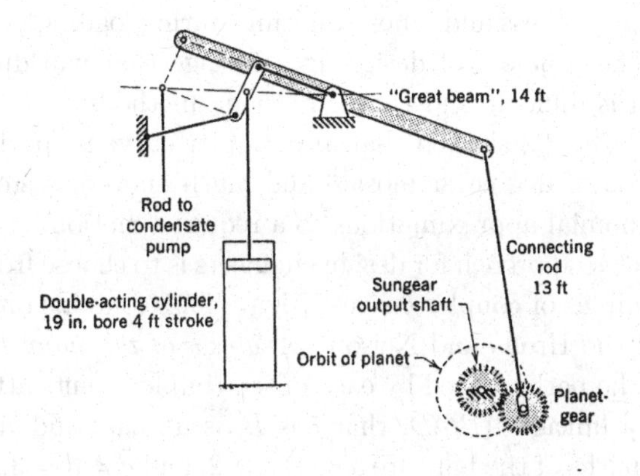

Fig. 9.18 James Watt's steam engine

Construction 9.4.2. Model the guiding 4-bar linkage of Watt's engine shown in Figure 9.19. Use paper and a soft backing board as in Construction 9.3.3 to plot the path of the midpoint of the coupler. It might be helpful to punch a hole through the middle of the coupler and trace the hole's motion with a pencil through the hole. Do you think this linkage is a satisfactory guide for the piston rod of Watt's engine?

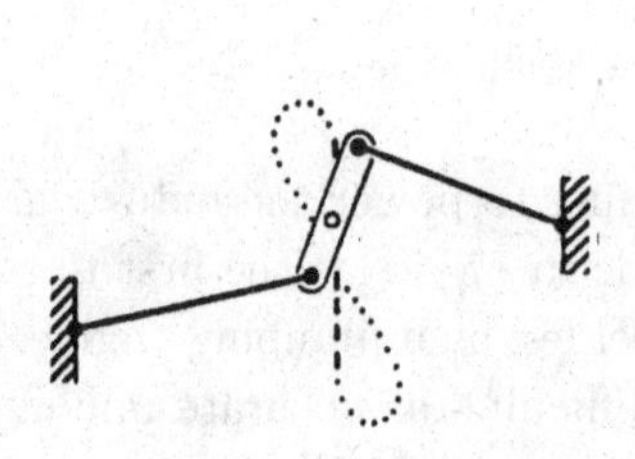

Fig. 9.19 Guiding a piston rod

Fig. 9.20 A level-luffing crane

Some level-luffing cranes use the approximate straight line motion generated by 4-bar linkages in order to load and unload containers efficiently. The load is moved a horizontal distance without un-necessary height change. This saves energy, stress and, above all, time during loading and un-loading operations. The problem of designing a linkage that will duplicate a required motion is difficult and there are many methods of tackling it. For example, R. Beyer [2] showed that any coupler curve is specified by polynomial equation of degree at most 6 and much effort has since gone into choosing polynomial approximations to a required motion.

An alternative approach for design engineers is to choose from a large list of known examples of coupler curves. Figure 9.21 contains one page (from 730 pages) of the Hrons and Nelson "*Analysis of the Four Bar Linkage*" [34], showing the paths traced by each of ten coupler points attached to the coupler D of a linkage $ABCD$, that has B as its base and A as its input crank. The lengths of the bars are $a = 1$, $b = 2$, and $c = d = 3$, respectively. We note that this particular linkage is a crank/rocker mechanism. Two of the ten coupler points shown are the joints of D, their paths being shown as non-dashed arcs. One of these two paths, a full circle, is the path of the end of the crank. The other, an arc of a circle, is that traced out by the end of the rocker. The path of each of the other eight possible coupler points is a dashed curve. Each dash corresponds to the motion of that coupler point during 10 degrees of crank rotation.

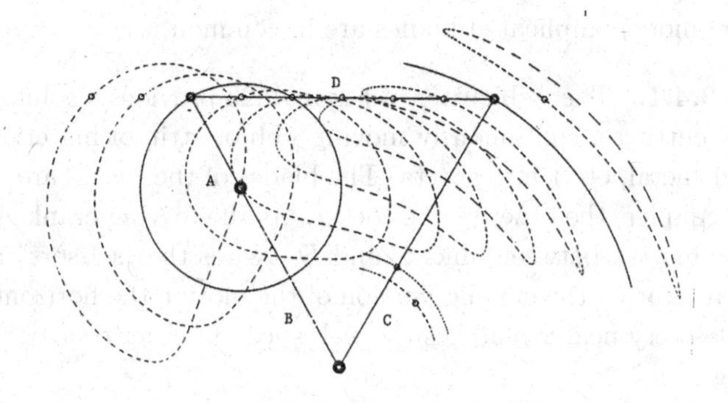

Fig. 9.21 Hrons and Nelson: Analysis of the Four-Bar Linkage

Problem 9.4.1. *Let us try our hand at some engineering design work. Choosing from the selection of coupler curves in Figure 9.21, design a mechanism to slowly slide a box of food through an ionising chamber and quickly return ready for another box.*

Solution. There are many possibilities and a final choice may well depend, among other considerations, on availability and price of standard components. The example illustrated below in Figure 9.22 is based on a choice of coupler point that follows a coupler curve of Figure 9.21 having a reasonably straight slow-moving section and a speedier return path. The mechanism can be arranged to slip the slide below a waiting box, rise and slowly push the box through the chamber, and then quickly drop and return.

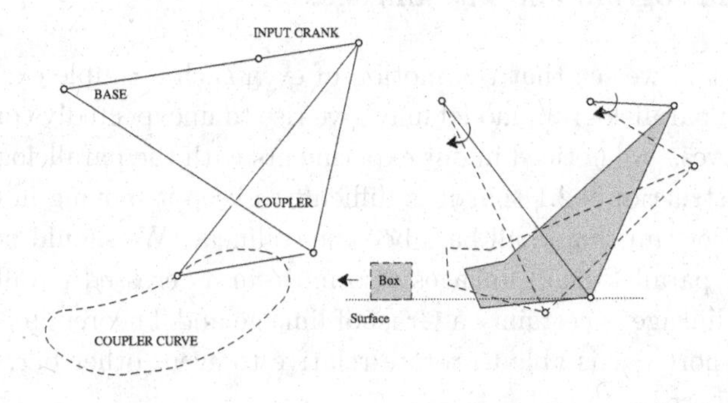

Fig. 9.22 An ionising mechanism

Linkages, such as the following Example 9.4.1, in which several bars are replaced by more complicated bodies are in common use.

Example 9.4.1. The web cutter of Figure 9.23 provides a solution to the problem of cutting a continuously moving web or strip of material (paper, cloth, sheet metal, etc.) into sheets. The blades of the shears are attached, one to the coupler, the other to the rocker. As the driving crank A rotates, the relative motion between links C and D creates the "scissors" action to cut the web. During the cutting portion of the motion the horizontal speed of the blades very nearly matches the web speed in order to avoid buckling the moving strip.

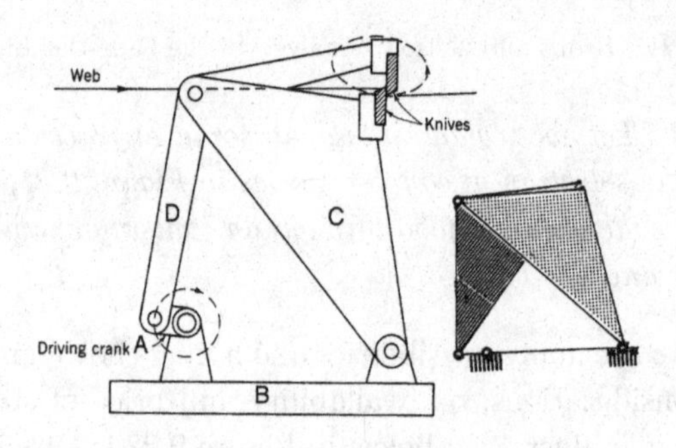

Fig. 9.23 A web cutter

9.5 Parallelogram and kite linkages

In this section, we see that the motion of even such a simple example of a linkage as a parallelogram model may give rise to unexpectedly complicated coupler curves. We noticed in our experiments with the parallelogram linkage of Construction 9.2.1 that it is difficult to keep it moving in the shape of a parallelogram should all bars become collinear. We should not be surprised that parallelogram linkages also move in a "crossed parallelogram" form. The linkage is certainly a Grashof linkage and Theorem 9.3.1 tells us that each short bar is able to rotate relative to every other bar, including the other short bar.

Definition 9.5.1. *A* **linkage** *in which two opposite bars are of equal length* *and the other pair of opposite bars are of equal length is a parallelogram linkage. A linkage in which two adjacent bars are of equal length and the other two bars, also an adjacent pair, are of equal length is a* **kite linkage**.

In fact the following special case of Grashof's theorem characterizes both parallelogram linkages and kite linkages.

Corollary 9.5.1. *Two bars of a Grashof linkage are each able to rotate fully relative to all other bars if and only if they are the shortest bars of a parallelogram linkage or a kite linkage.*

Proof. Let the bars A, B, C, and D of the linkage have lengths a, b, c, and d respectively.

Suppose first that $a = b$, $c = d$ and $a \leq c$. From applying Theorem 9.3.1 in turn to A and B we have that each can rotate fully relative to other bars.

Conversely, suppose that A and B can each rotate fully. Then both $a \leq b$ and $b \leq a$ and each is a shortest bar. Without loss of generality we may assume that D is a longest bar. As A can rotate fully we have that $a + d \leq b + c$, giving $d \leq c$. Therefore $c = d$. □

Exercise 9.5.1. Apply the method of Construction 9.4.2 to the parallelogram linkage of Construction 9.2.1, using a longest side as the base and the midpoint of the opposite side as a coupler point. Trace a coupler curve and compare it with the curve shown in Figure 9.24.

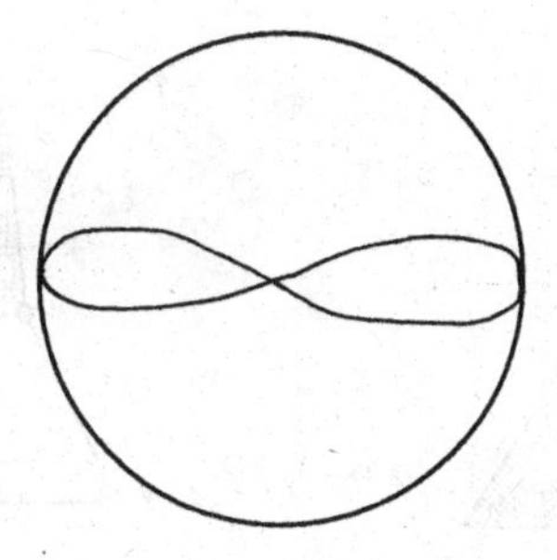

Fig. 9.24 A coupler curve of a parallelogram linkage

Exercise 9.5.2. Locate a parallelogram linkage in a sewing tidy, in an anglepoise lamp, in a child's pop-up book, in a steam locomotive and in a tree pruner.

The uncertainty about whether a parallelogram linkage will switch to "crossed parallelogram" form as it moves through a linear position makes it sometimes a poor choice of mechanism if full rotation of an input bar is required. We also noted the difficulty encountered in the use of a double-crank mechanism in the last experiment of Exercise 9.3.3. Why then does a parallelogram linkage give a successful method of linking driving wheels of steam locomotives?

9.6 Plagiographs

In this section, we examine the behaviour of a parallelogram mechanism, moving in parallelogram form, in which two bars have been replaced by similar triangular panels.

Definition 9.6.1. *A* **plagiograph** *is a parallelogram linkage in which two adjacent bars are replaced by similar triangular panels*

We notice that any plagiograph may be moved into the first position shown in Figure 9.25. Engineers refer to this as the *design position* of the mechanism. We label three points of the mechanism x, y and z respectively, as shown in the Figure, and have the following lemma:

Lemma 9.6.1. *In any parallelogram position of a plagiograph, the triangle xyz is similar to each triangular panel.*

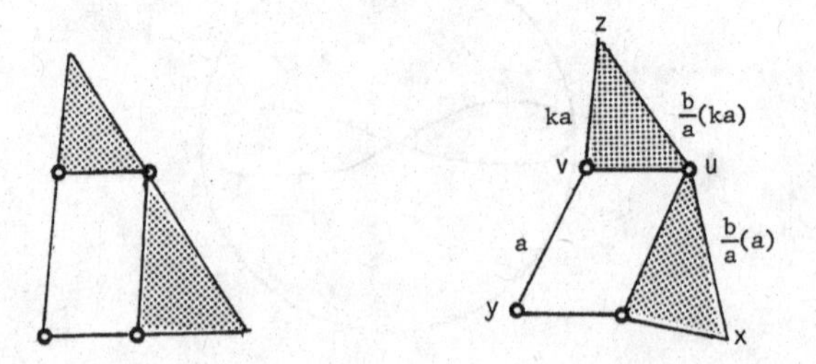

Fig. 9.25 Two positions of a plagiograph

Proof. Any position of the plagiograph is obtained from the design position by turning one of the triangular panels relative to the other, as shown second in Figure 9.25. Applying Theorem 4.7.4 we see that in the second position of the plagiograph the triangles xuz and yvz are similar. Hence the two angles $x\hat{z}u$ and $y\hat{z}v$ are equal. But subtracting $x\hat{z}u$ and adding $y\hat{z}v$ to the angle $y\hat{z}x$ gives an angle equal to $v\hat{z}u$. A similar argument proves that the angle $y\hat{x}z$ in this position of the mechanism and the angle $v\hat{u}z$ are equal. As the sum of the angles of a triangle is constant, the angle $x\hat{y}z$ equals the angle $u\hat{v}z$. From Theorem 4.7.4 we have that the triangle xyz is similar to each triangular panel of the model. □

If we pin a plagiograph to a backing board at y, place a pencil at z, and trace out a path with the point x then the pencil will trace a path, similarly shaped but turned through the constant angle $z\hat{v}u$. Thus any planar figure may be copied on a reduced or enlarged scale by this plagiograph, invented by J.J. Sylvester [42] in 1875. We are perhaps more familiar with a special case, the usual *pantograph* in which the vertices of each triangular panel collapse to three collinear points with $x\hat{v}u = \pi$. This invention by a london barrister A.B. Kempe [42] preceded the more general plagiograph.

Construction 9.6.1. Make the adjustable pantograph illustrated in Figure 9.26. Pinning it to the paper at y, placing a pencil through a hole at z, and tracing the figure to be copied with a stylus through the hole at x enables the pencil to copy the figure. Re-locating the pencil and stylus changes the scale of copying.

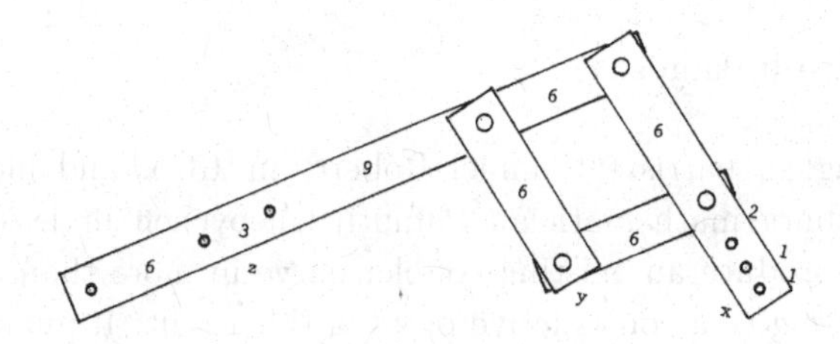

Fig. 9.26 A pantograph

Example 9.6.1. The level-luffing crane of Figure 9.27 is a disguised pantograph.

We earlier met an example of J.J. Sylvester's work, namely Theorem 3.7.1, and we will shortly meet a result of Pafnutij Chebycheff, usually thought of as a mathematical analyst. The stereoscope discussed in Chapter 6 was invented by the scientist George Wheatstone, a pioneer in the

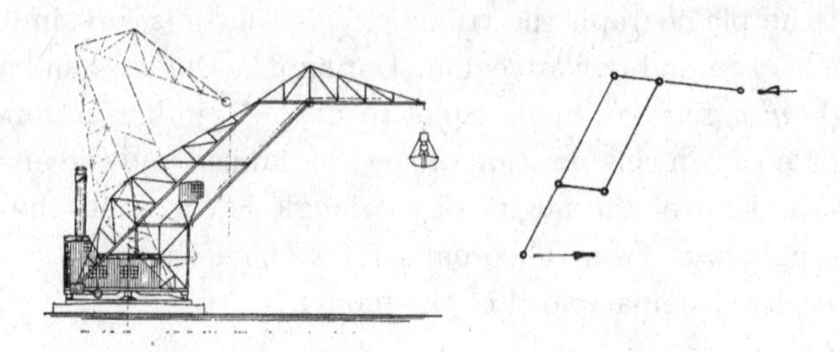

Fig. 9.27 Another level-luffing crane

understanding of electricity. Prior to this century it was not unusual for engineers and mathematicians to work in many fields - extracting inspiration from one for use in others. The era of the "universal scientist", begun by Leonardo da Vinci, was not then finished. Sadly it may now be over, and perhaps the resultant separation of theory and practice in dealing with our world is an undesirable consequence.

Construction 9.6.2. Make a plagiograph, modeling that shown in Figure 9.25, from bars and triangular panels that you already have available. What is the difference in result between using it and using the more familiar pantograph, to copy a figure?

9.7 Cognate linkages

A former English barrister Samuel Roberts in 1875, and independently the St.Petersburg mathematician Pafnutij Chebycheff in 1878, used plagiographs to produce an existing coupler curve in more than one way. In this section, we give a constructive proof of this result. It guarantees engineers a choice from three 4-bar linkage mechanisms that generate the same coupler curve. Moreover if it is possible to use a linkage with a crank to reproduce this curve, it will be one of the three.

Definition 9.7.1. *A linkage is* **cognate** *to another if a coupler point of the first, and a coupler point of the second, have the same coupler curve.*

The Roberts-Chebycheff method combines parallelogram mechanisms in order to construct related 4-bar linkages. We begin in Lemmas 9.7.1 and 9.7.2 by considering the behaviour of two parallelogram linkages that share a bar. Repeated use of Lemma 9.7.2 then enables us to prove the

Roberts-Chebycheff Theorem.

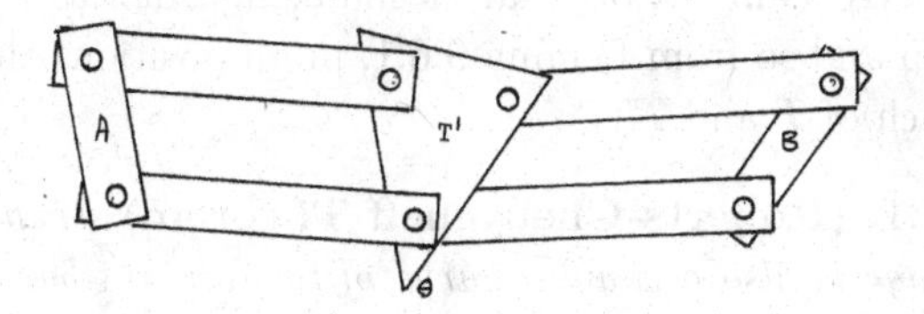

Fig. 9.28 Two linked parallelogram linkages

Lemma 9.7.1. *Let two parallelogram linkages share a triangular panel T' as shown in Figure 9.28. Then in each position of the mechanism the angle between the bars A and B is equal to the angle θ of the triangle T'.*

Proof. The bar A is parallel to one side, and the bar B is parallel to another side, of the triangle T' in each position of the mechanism. Therefore the angle between the bars is equal to θ. □

Lemma 9.7.2. *Let the two triangular panels T and T' of the first mechanism shown in Figure 9.29 be similar. Then in each position of the mechanism the triangle abc is similar to each of T and T'.*

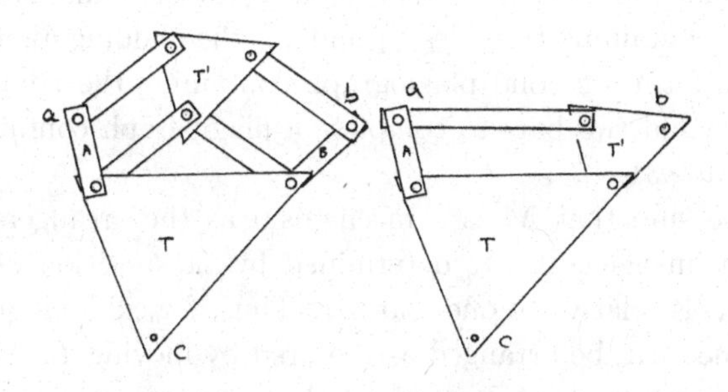

Fig. 9.29 A mechanism

Proof. We replace one of the two parallelogram linkages by the second linkage of Figure 9.29 in which the shortest sides have zero length. From Lemma 9.7.1 we have that the angle between the bars A and B is still θ. The bar B now coincides with a side of the triangular panel T'. This panel is now attached to the panel T. The bar A is able to move in this mechanism exactly as in the original mechanism. But this new mechanism is a plagiograph and so from Lemma 9.6.1, in all positions the triangle xyz is similar to each of T and T'. □

Theorem 9.7.1. (Roberts-Chebycheff Theorem) *Each coupler curve of a 4-bar linkage is also a coupler curve of two other 4-bar linkages.*

Proof. Suppose that the linkage **L** has a given coupler curve. We construct a mechanism as follows: first we remove the base and "flatten out" the linkage **L** as shown in Figure 9.30.

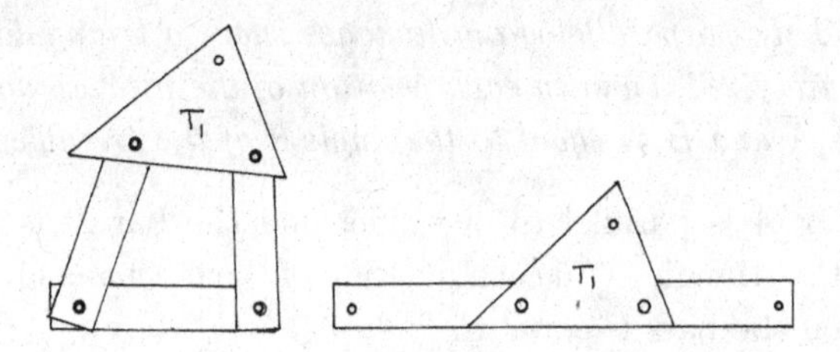

Fig. 9.30 The linkage **L** before and after flattening

Then, by drawing parallel lines, we design the mechanism **M** that can be thought of as adding a bar and triangle T_2 to the remains of **L** to form a plagiograph containing triangles T_1 and T_2, then adding another bar and triangle T_3 to form a second plagiograph containing the triangles T_1 and T_3. Lastly, we add two bars to complete a plagiograph containing T_2 and T_3 as in Figure 9.31.

We can be sure that **M** is a mechanism as the arrangement of each parallelogram linkage is freely determined by the position of two of the triangular panels relative to one another. Thus, for example, the first plagiograph formed can be arranged as required by moving T_2 relative to T_1, and the second plagiograph formed can be independently arranged as required by moving T_3 relative to T_1.

Starting from the initial design position above, and keeping T_1 and T_2 fixed relative to one another, we move **M** until the bar X is in any desired position relative to T_1. From Lemma 9.6.1 we know that the plagiograph with triangular panels T_1 and T_2 is similar to T_1. Applying Lemma 9.7.2, with this plagiograph in place of T and T_3 in place of T', we see that triangle xyz remains similar to T_1 during the motion. During this movement, we

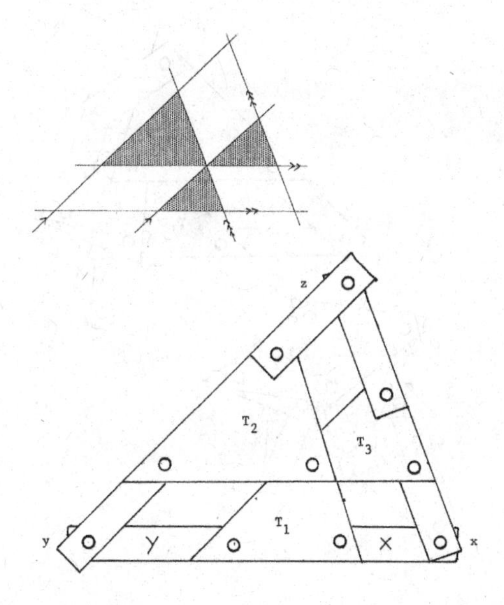

Fig. 9.31 Design for a mechanism and the resulting assembled mechanism

have from Lemma 9.7.2 that the plagiograph that contains the two panels T_1 and T_3 remains similar to T_1, as in the first illustration of Figure 9.32.

From this position, but now keeping T_1 and T_3 fixed relative to one another, we move **M** until the bar Y is in any desired position relative to T_1 as in the second illustration of Figure 9.32. Again from Lemma 9.7.2, this time thinking of the rigid plagiograph of T_1 and T_3 in place of T and T_2 in place of T', we deduce that the shape of triangle xyz remains unaltered. Thus in any position of **M** the triangle xyz is similar to T_1

In this way we "un-flatten" **L**, enabling us to reconnect the removed bar B. This can be done in any position of **L** that is attainable by moving **L** as a 4-bar linkage alone. Thus the coupler point of **L** will trace its complete coupler curve as the new mechanism that contains the bar B, and so contains all of **L**, moves.

During this motion x and y remain fixed relative to B. As the shape of the triangle xyz remains unchanged, z is also fixed relative to B during the motion that moves the coupler point through all its possible positions. We

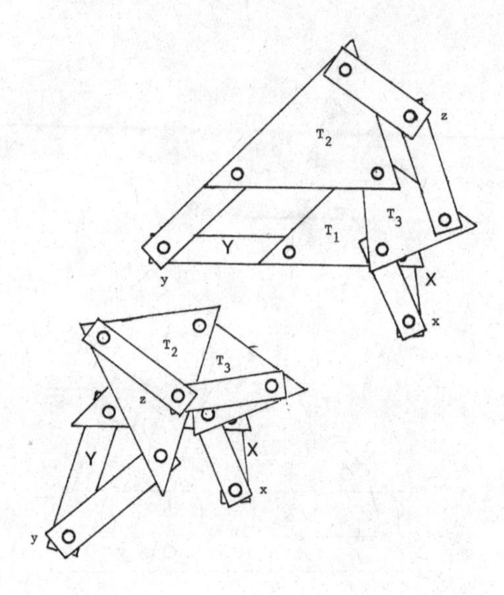

Fig. 9.32 A rotation of X, followed by a rotation of Y

conclude that adding bars $[xz]$ and $[yz]$ to the mechanism will not hinder its motion. The enlarged mechanism contains three linkages: the original **L**, a second 4-bar linkage with base $[yz]$ and coupler triangle T_2, and a third 4-bar linkage with base $[xz]$ and coupler triangle T_3. Each of these linkages share the coupler point z and share its coupler curve. □

Construction 9.7.1. Construct the linkages cognate to that of Exercise 9.4.1 by removing its base and following the design process described in the proof of the Roberts-Chebycheff Theorem. Verify the design position of the mechanism shown in Figure 9.33. Satisfy yourself, by constructing the mechanism and experimenting, that the shape of the triangle xyz remains unaltered during motion of the model. Replace fasteners at the joints x, y and z by thumb-tacks into a soft backing board so that x and y are 13cm apart. The joints y and z should be 9.3cm apart, and x and z 7.4cm apart. Verify that each of the three linkages gives the coupler curve obtained earlier in Construction 9.4.1.

Mechanical motion is most easily available in the form of rotation, by means of electric motors for example. The following Corollary to Theorem 9.7.1 proves that the above design process enables us to capture a given coupler curve via a crank-driven linkage if at all possible.

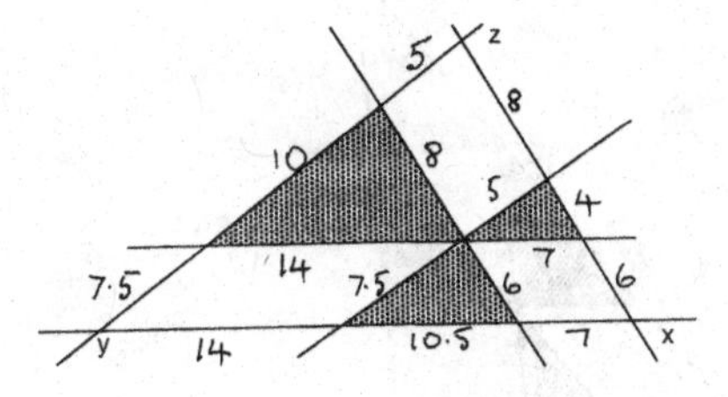
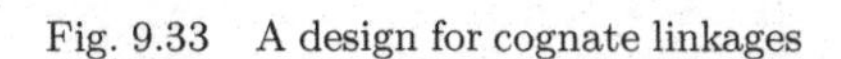

Fig. 9.33 A design for cognate linkages

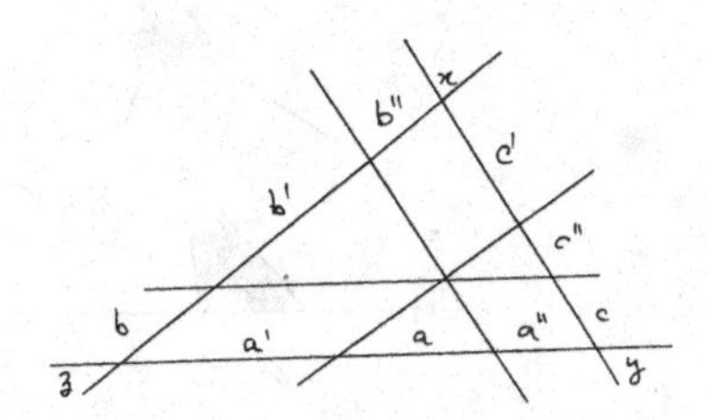

Fig. 9.34 Lengths of cognate linkage components

Corollary 9.7.1. *Each Grashof linkage has a cognate linkage containing a crank. Each cognate of a double-crank linkage is also a double-crank linkage. Each cognate of a non-Grashof linkage is also a non-Grashof linkage.*

Proof. We specify lengths as in Figure 9.34. Without loss of generality, let $a \geq a' \geq a''$. Considering the mechanism in any position, from Theorem 4.7.4 applied to three similar triangles in Figure 9.34 we have that $a/b = a'/b' = a''/b'' = d(y,z)/d(z,x)$.

Suppose that the original linkage is a double crank linkage. Therefore $d(y,z)$ is less than or equal to each of a, a', and a''. Then we have that $d(x,z)/b = d(y,z)/a \leq 1$, $d(x,z)/b' = d(y,z)/a' \leq 1$, and $d(x,z)/b'' = d(y,z)/a'' \leq 1$. Consequently $d(x,z)$ is less than or equal to each of b, b' and b''.

From Theorem 9.3.1 applied to the original linkage we also have that $d(y,z)+a \leq a'+a''$. Therefore $d(x,z)+b = d(y,z)(b/a)+b = (b/a)(d(y,z)+a) \leq (b/a)(a'+a'') = b'+b''$. The second linkage is therefore also a double crank linkage. Similar, but long, arguments prove the other parts of the Corollary. □

Example 9.7.1. In Figure 9.35 we have an example of the application of Theorem 9.7.1 in which a coupler curve is produced by each of three linkages.

9.8 Approximate and exact linear motion

In this section, we give some 4-bar linkages that generate coupler curves with approximately linear sections and some mechanisms that generate

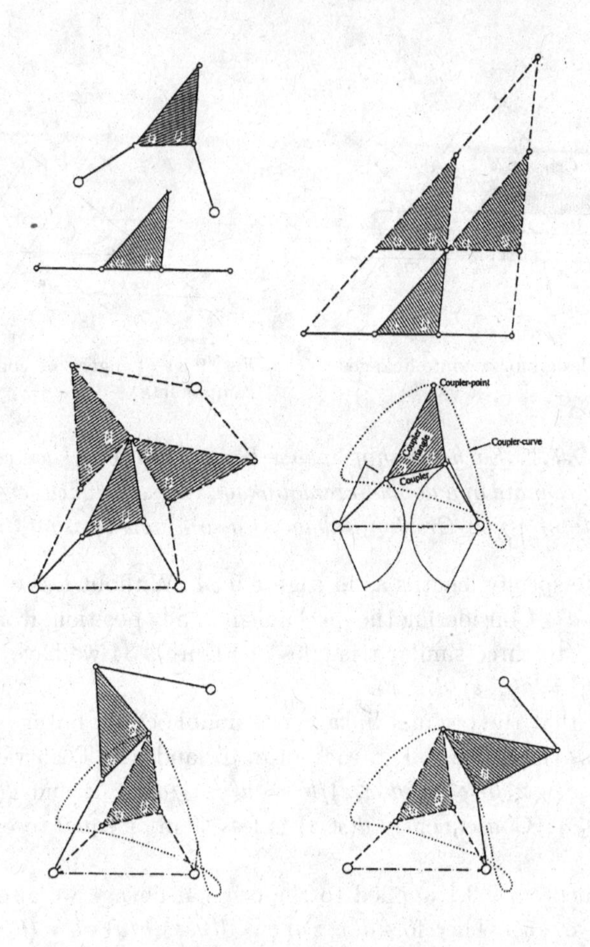

Fig. 9.35 Three cognate linkages

exact linear motion of a point. As we mentioned above, mechanisms that produce linear motion without accurately machined linear guides were in demand from the outset of the Industrial Revolution. James Watt was apparently prouder of the approximately linear motion of his piston guide than of the steam engine *in toto*. Sarrus' non-planar but exact solution to the problem in 1853 is dealt with in Example 10.3.1. Unfortunately it was little-known, and attention concentrated on planar mechanisms. Particular examples of planar 4-bar linkages giving coupler curves with approximately linear sections were invented by Chebycheff in 1850 and, a better example still, by Roberts in 1860.

Example 9.8.1. In Chebycheff's linkage A and C have length 5, B length 4, D length 2 and the coupler point is the midpoint of D. In Roberts' linkage A, C and the sides of the coupler triangle have the same length, and the coupler length is half the length of the base. We may trace the coupler curves of these mechanisms using the method of Construction 9.3.3.

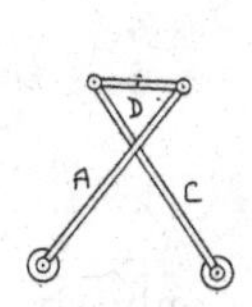

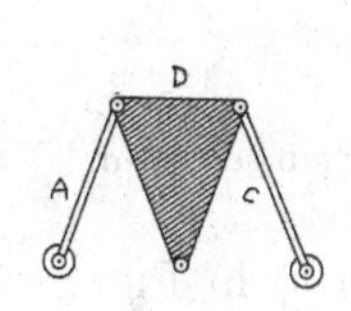

Fig. 9.36 Chebycheff's and Roberts' mechanisms

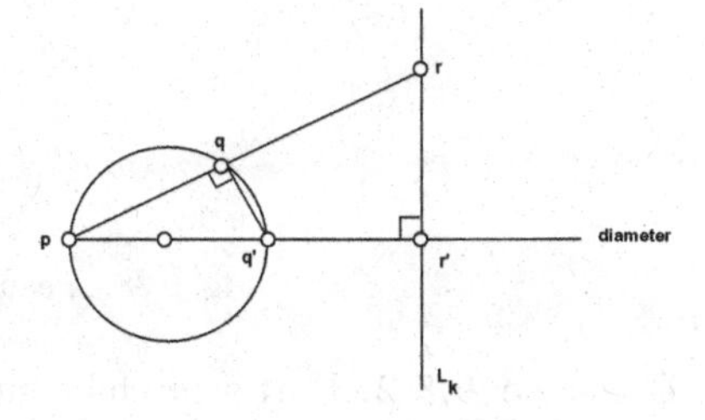

Fig. 9.37 A circle and the line L_k

It is known [42] that no planar bar-and-joint mechanism containing fewer than six bars has a point that traces out a linear path. On the other hand the following result, or a simple variant, is the key to designing mechanisms that do generate linear motion.

Lemma 9.8.1. *Let p be a point on a circle and k be a number. Then there is a line L_k with the following property. Corresponding to each point q, different from p, on the circle, there is a point r on both the line L_k and the line $p \vee q$ that satisfies $d(p,q)d(p,r) = k$.*

Proof. Suppose that the diameter through p meets the circle again in the point q'. Suppose further that r' is the point of this line that satisfies $d(p,q')d(p,r') = k$. We choose L_k to be the line, through r', that is perpendicular to the line $p \vee q'$. The required result follows from the similarity of the two triangles prr' and $pq'q$. □

Exercise 9.8.1. The mechanism shown in Figure 9.38 was invented in 1864 by the French army officer Peaucellier and used to control ventilation pumps for the British Houses of Parliament. The mechanism contains two kite linkages. Figure 9.38 also shows an application to a hand rivetting tool.

Use Pythagoras' Theorem and Lemma 9.8.1 to prove that the point r of Peaucellier's mechanism traces out a line as the mechanism moves.

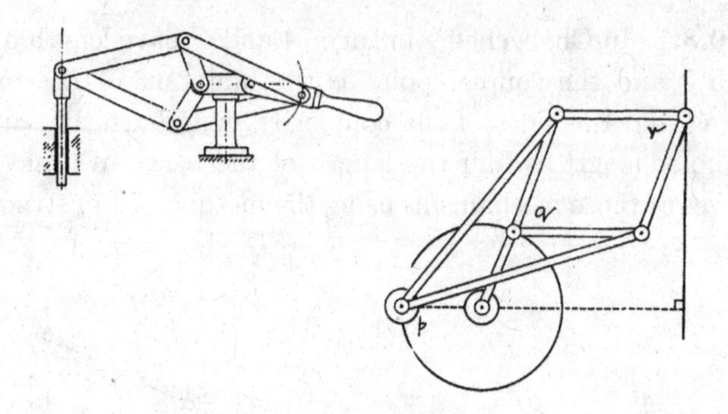

Fig. 9.38 Peaucellier's mechanism

Exercise 9.8.2. Hart's mechanism, shown in Figure 9.39, uses only six bars and contains a kite. It was invented in 1874. Using two similar right-angled triangles, prove that the point r of Hart's mechanism traces out a line as the mechanism moves.

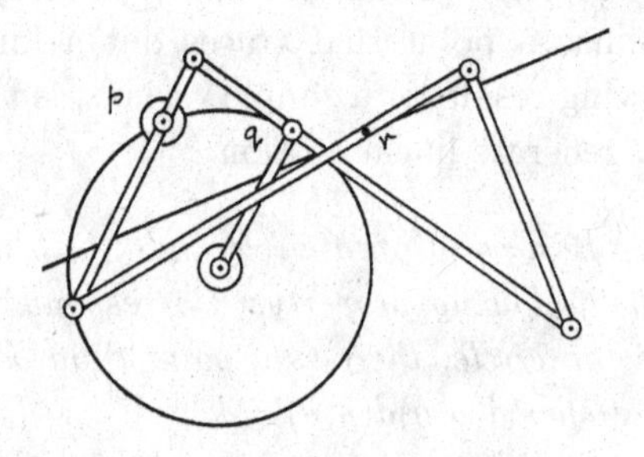

Fig. 9.39 Hart's mechanism

Summary of Chapter 9. We chose the 4-bar linkage as a typical and useful mechanism to investigate in detail. We found examples in many different contexts and learned how to predict their behavior from a simple condition on bar lengths. Experimentation with models led us directly to some practical difficulties of engineering design. We overcame some of these by combining plagiographs to give convenient linkages that generate required motion of a point.

Chapter 10

NON-PLANAR HINGED-PANEL MECHANISMS

We continue the investigation of models that we began in Chapter 9. The practical requirements of model-making and the importance of folding objects such as knock-down cartons, folding doors, and stage scenery lead us to define a class of models that complements the bar-and-joint models of Chapter 9. Those are constructions in which a "skeleton" plays the dominant structural role. In this chapter we examine constructions [15] in which the "skin" is the main structural element. We first define the panels from which they are assembled and then specify rules governing the way in which panels may be hinged together to form a hinged-panel model.

We analyze their behavior in some detail. In doing so we are again entering the realms of the practical engineer and industrial designer. Models that are rigid play a role in structural engineering, while those that are free to change their shape are mechanisms. We concentrate first on hinged-panel cycles, as they are easy to visualize and occur widely in man's activities, often in disguised form. Finally we examine polyhedral models, and see that not all are rigid.

10.1 Hinged-panel models

In this section, we define hinged-panel models and examine some common examples. In Definition 1.2.3 we introduced combinatorial n-gons. If we restrict our discussion [30] to any planar n-gon $a_1a_2\ldots a_n$ with each vertex a Euclidean point, then the union $B[a_1a_2\ldots a_n] = [a_1a_2] \cup [a_2a_3] \cup \ldots \cup [a_{n-1}a_n] \cup [a_na_1]$ of the segments determined by each pair of adjacent vertices is a closed path in the Euclidean plane that contains the n-gon. If this path has no self-intersections then we have, either from intuition or a simple version of the Jordan Curve Theorem [23], that the path divides the

Euclidean plane containing $B[a_1a_2\ldots a_n]$ into an interior and an exterior part.

Definition 10.1.1. *Let $a_1a_2\ldots a_n$ be a planar combinatorial n-gon so that each vertex a_i is Euclidean and $B[a_1a_2\ldots a_n]$ has no self intersections. Then the* **polygonal region** *$[a_1a_2\ldots a_n]$ is the union of the set $B[a_1a_2\ldots a_n]$ and its interior. We call the set $B[a_1a_2\ldots a_n]$ the boundary of the region, each segment $[a_1, a_2]$, ...,$[a_{n-1}, a_n]$, $[a_n, a_1]$ an edge of the region, and each point a_1, a_2, ..., a_n a vertex of the region.*

The first 4-gon in Figure 10.1 is non-planar. The second has an ideal vertex. The remaining examples are planar, with only Euclidean vertices. The third does not have an interior. Only the last has an interior and the union of its boundary and shaded interior constitute a polygonal region.

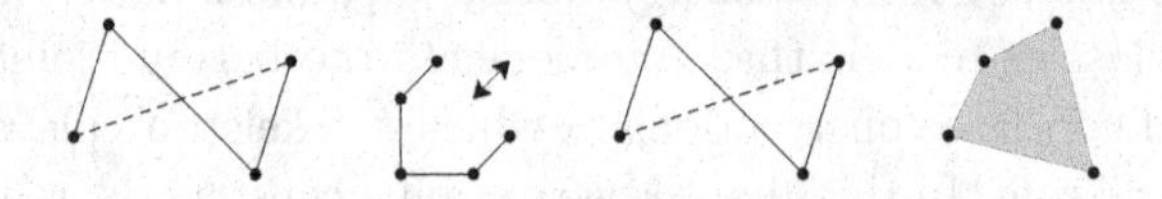

Fig. 10.1 Four 4-gons, only one of which defines a polygonal region

We call any reasonably flat plate [31] that models a polygonal region a *panel.* We meet panels whenever we cut out cardboard shapes, when we cut and fit wall paneling, when we refloor a room with hardboard underlay, and in many other situations. Panels may be combined, the reality of practical construction allowing two panels to be freely hinged along a shared edge. Thus two panels F and G, both containing the edge E, may form a mechanical *hinge*, written $H_{inge}(F, E, G)$ or $H_{inge}(G, E, F)$. We say that the panels and edge are the three components of the hinge. We also say that F and G are *hinged* at E and the line containing E is the *hinge-line* of the hinge.

The way in which several panels are combined into one entity by means of hinges is encapsulated in the following definition:

Definition 10.1.2. *A* **hinged-panel model M** *consists of a list of panels, some of which are hinged and satisfy the following two conditions:*
Condition H1: *For each two panels F, G of* **M** *there is a sequence, $F = F_1$, $H_{inge}(F_1, E_1, F_2)$, F_2, $H_{inge}(F_2, E_2, F_3)$, ..., $H_{inge}(F_{m-1}, E_{m-1}, F_m)$, $F_m = G$, of panels and hinges.*
Condition H2: *Any two hinges share at most one common component.*

A vertex of **M** *is a vertex of any of its panels, an edge of* **M** *is an edge of any of its panels, the shell of* **S** *is the union of its panels, and the skeleton of* **M** *is the union of the edges of its panels.*

Condition **H1** is a convenient way of ensuring that the hinges are the essential "connections" between panels of a hinged-panel model **M**. Condition **H2** reflects the reality of hinge construction. It guarantees that $H_{inge}(F, E, G_1)$ and $H_{inge}(F, E, G_2)$ cannot both be hinges. It also ensures that $H_{inge}(F, E_1, G)$ and $H_{inge}(F, E_2, G)$ are not both hinges, so that two panels are hinged together by at most one edge. When it will not cause confusion, we speak of the *hinge E* rather than the *hinge* $H_{inge}(F, E, G)$.

Around us we see many hinged-panel models. The simplest is undoubtedly the common carpenter's hinge that consists of two panels that hinge at their shared edge. Most road maps are hinged-panel models, cardboard boxes (both before and after gluing), concertina doors, and old-fashioned fans are all hinged-panel models. A book is not a hinged-panel model. An uncreased envelope is not! A folded letter is!

When it is convenient and unambiguous, we use thicker lines in diagrams to distinguish edges that are hinges from those that are not, continuing the use of notation introduced in Figure 2.13 in Chapter 2. We often make hinged-panel models from cardboard panels, taping panels of a hinge together at their shared edge, as for example, in the combinatorial cube models of Examples 2.5.1 and 2.5.2. Each of these two, when opened out, is a hinged-panel model in which all four panels are coplanar, adjacent panels hinging at their common edges. When partially folded each still is a hinged-panel model of four panels and three hinges, but now no two of the four panels are coplanar. Fully clipped into the assembled position each is a model of four panels and four hinges. Thus one mechanism may be used to model different figures.

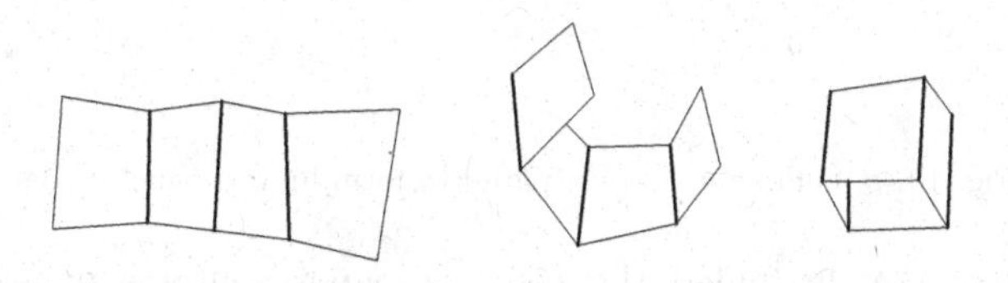

Fig. 10.2 Three 4-panel models

We are able to make one of the tetrahedron models required for Construction 2.7.1, using the method illustrated in Figure 2.12, and taping the valley folds as shown in Figure 10.3. When opened out it is a hinged-panel model, all of whose panels are coplanar and in which each of the three shared edges is a hinge. When partially closed up it is a hinged-panel model having no two panels coplanar. When fully bent up to form the tetrahedron the model has three edges that are not hinges and three that are hinges. If we then tape the remaining edges we have another hinged-panel model - the same panels and shell as before but with each pair of panels forming a hinge with their common edge.

Fig. 10.3 Steps in folding a model of a tetrahedron

Example 10.1.1. Folding a piece of paper into quarters we label the corners as in the first illustration in Figure 10.4. The paper is then a hinged-panel model with panels $F_1 = [0165]$, $F_2 = [1276]$, $F_3 = [2387]$ and $F_4 = [3498]$, and hinges $H_{inge}(F_1, [16], F_2)$, $H_{inge}(F_2, [27], F_3)$ and $H_{inge}(F_3, [38], F_4)$.

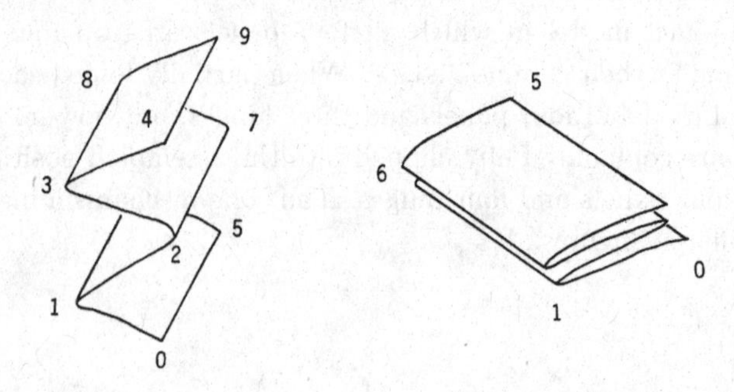

Fig. 10.4 Folding a sheet of paper to form hinged-panel models

When completely folded the piece of paper contains panels labelled $F_1 = [0165]$, $F_2 = [0165]$, $F_3 = [0165]$ and $F_4 = [0165]$, and hinges $H_{inge}(F_1, [16], F_2)$, $H_{inge}(F_2, [05], F_3)$ and $H_{inge}(F_3, [16], F_4)$.

In the first case each panel models a separate polygonal region, but in the second example all four panels model the same polygonal region [0165]. Allowing coplanar panels, and even repetition in the list of panels, in a hinged-panel model allows us to include collapsed models such as the second example above. One of the virtues of hinged-panel mechanisms is the possibility of their folding flat for ease of storage or transport.

Definition 10.1.3. *A hinged-panel model is* **collapsed** *if each panel is in the same plane. A collapsed hinged-panel model, all of whose panels share a common internal point is* **folded**. *A collapsed hinged-panel model, no two of whose panels share a common internal point, is* **unfolded**.

The creased paper model of Example 10.1.1 has two collapsed positions, one folded and one unfolded.

As in Chapter 9 we may think of models as falling naturally into two types; namely those in which parts may move relative to other parts, and those in which the distance between any two parts remains unchanged.

Definition 10.1.4. *A* **hinged-panel mechanism** *is a model in which the panels are able to move relative to one another. A* **model** *in which no panel can move relative to another is rigid.*

Definition 10.1.5. *The first hinged-panel model in Figure 10.5, with F_1 and F_2 hinged, and F_2 and F_3 hinged, is known as a* **Hooke universal joint** *or a* **Cardan joint** [38].

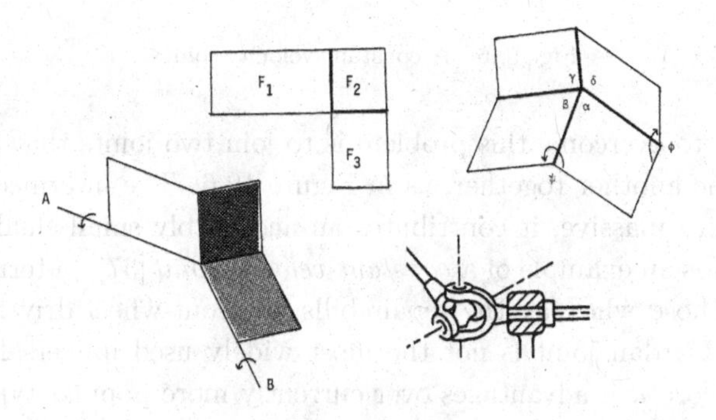

Fig. 10.5 A Hooke universal joint

A Hooke universal joint may be thought of as derived from the 4-panel model of Figure 10.5 by removing one panel, thus allowing the angle α to vary. A Hooke universal joint enables a turning shaft A to turn a shaft B at varying inclination to A as in the third figure in Figure 10.5. Taping a straw along A and a straw along B in a cardboard model enables us to experiment by rotating the first straw and lightly holding the second free to rotate, but inclined at an angle α to the first. Even as the angle α varies, a rotation of the shaft A causes a rotation of the shaft B.

This is the case in the connection of the engine to a front wheel in a front-wheel driven automobile as the suspension flexes. Straightforward trigonometry gives $\tan\phi\tan\psi = -\sec\alpha$. This enables us to quantify a disadvantage of the Hooke universal joint, namely that $\frac{d\psi}{d\phi}$ is not constant. As A rotates with constant speed, B does not, speeding up and slowing down as it rotates. In practice, an automobile engine is designed to turn its axle at constant speed, and the road wheel would also like to turn at constant speed! The resulting shudders would rapidly wear out the mechanism.

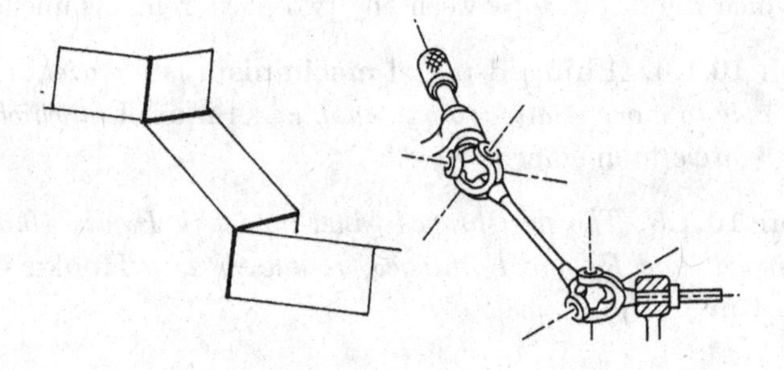

Fig. 10.6 A constant-velocity joint

One way to overcome this problem is to join two joints that are mirror images of one another together, as in Figure 10.6. The intermediate shaft not being very massive, it contributes an acceptably small shudder. This combination is an example of a *constant-velocity joint* [37] - a term only too familiar to those who pay the repair bills for front-wheel driven vehicles. The double-Cardan joint is not the most widely used universal joint but has some theoretical advantages over currently more popular types

The familiar folded letter of Example 10.1.1 is a simple example of a hinged-panel model obtained by folding (and creasing) a sheet of paper. There is a long history of paper folding, much of it full of mathematical significance, [1, 36, 33, 27]. Perhaps the oldest and most carefully recorded examples are the folds of the Japanese art of Origami [48, 53].

Construction 10.1.1. (Some Origami Constructions) The art of Origami [48, 53], or paper folding, provides some delightful examples of hinged-panel models. Let us use two examples of Origami folds to make a bird's foot, an animal's foot, and a human foot. The first fold illustrated in Figure 10.7 gives an animal's foot. The second fold gives a bird's foot, and the first fold carried out twice gives a human foot.

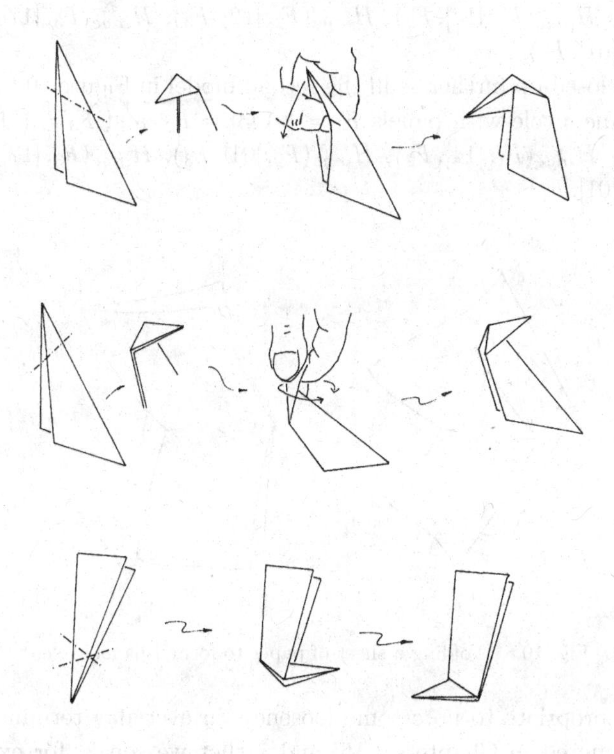

Fig. 10.7 A bird's foot, an animal's foot, and a human's foot

We now turn our attention to panel cycles. They are perhaps the simplest, most common, and most useful hinged-panel models:

Definition 10.1.6. *An n-***panel cycle** *is a hinged-panel model with a list* F_1, F_2, ...,F_n *of at least three panels and with* $\{H_{inge}(F_i, E_{i(i+1)}, F_{i+1}) : i = 1, 2, \ldots, n-1\} \cup \{H_{inge}(F_n, E_{n1}, F_1)\}$ *as its complete set of hinges. A panel cycle is any n-panel cycle. We write* $H_{i(i+1)}$ *to mean the hinge-line of* $H_{inge}(F_i, E_{i(i+1)}, F_{i+1})$.

Each of the two models of cubes in Constructions 2.5.1 and 2.5.2, when assembled, is a 4-panel cycle. It is interesting to note that one is rigid, the other not.

Example 10.1.2. Folding a piece of paper into quarters we label the corners as in the first model in Figure 10.8. The paper is a 4-panel cycle with panels $F_1 = [0145]$, $F_2 = [1234]$, $F_3 = [1287]$, and $F_4 = [0176]$, and hinges $H_{inge}(F_1, [14], F_2)$, $H_{inge}(F_2, [12], F_3)$, $H_{inge}(F_3, [17], F_4)$, and $H_{inge}(F_4, [01], F_1)$.

When closed up further as in the second model in Figure 10.8 the paper is the 4-panel cycle with panels $F_1 = [0143] = F_2$, and $F_3 = [0176] = F_4$, and hinges $H_{inge}(F_1, [14], F_2)$, $H_{inge}(F_2, [01], F_3)$, $H_{inge}(F_3, [17], F_4)$, and $H_{inge}(F_4, [01], F_1)$.

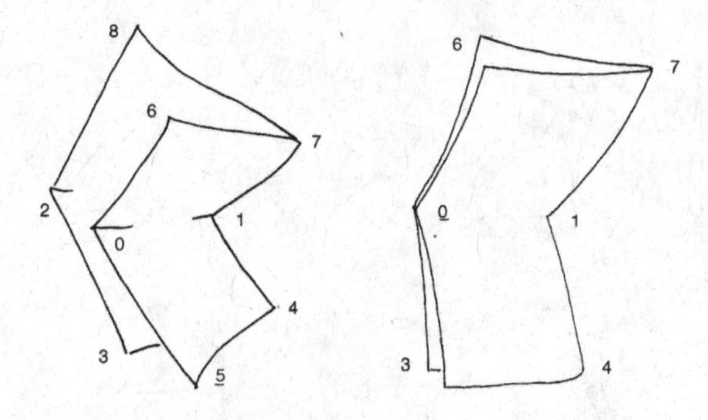

Fig. 10.8 Folding a sheet of paper to form 4-panel cycles

It is appropriate to note some looseness in everyday terminology. We previously noted in Chapters 1, 5, and 7 that we speak, for example, of a "4-gon *abcd*", accompanied by an illustration labeled by a, b, c and d, rather than a "4-gon *abcd* pictured here as $a'b'c'd'$". We use the same label for a point and for its image. We are so familiar with perspective drawings (in fact with diagrams of all kinds) and find them so helpful, that in conversation we almost always identify objects with their images, and this convention is so well understood that we are rarely even aware that we are using it.

In Figure 1.7, Figure 9.32, and elsewhere, we frequently carry this convenient identification a step further. We are representing models, not figures. The labels are intended to pinpoint points of the actual model, not of the structure modeled and not of the image of this structure. Again, we are so used to this convention that we are hardly aware of it. As our concept of a mechanism is applicable only to models, not to figures, the convention should not confuse us.

Exercise 10.1.1. List the panels and hinges of the bird's foot of Construction 10.1.1 in order to prove that it is a panel cycle. Are the animal's foot and the human foot also panel cycles?

Different hinged-panel models may share the same hinge-lines. For example, in Figure 10.9, a 5-panel cycle in which each panel is a 4-gon has the same hinge-lines as another 5-panel cycle that has two triangular panels.

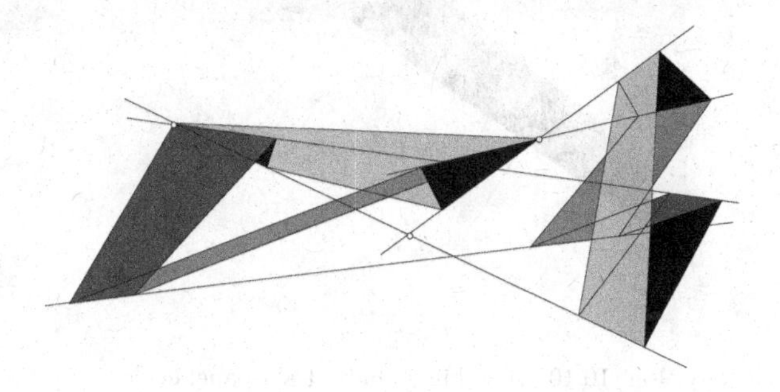

Fig. 10.9 Two 5-panel cycles with the same hinge-lines

Provided that panels do not meet or intersect at points other than hinges, it is only the hinge-lines of each panel, not the actual panel, that determine the motion of panels of a model relative to other panels. Subject to this restriction we have the following theorem:

Theorem 10.1.1. *Let two hinged-panel models have the same hinge-lines and let the list of panels of each model be paired so that corresponding panels share the same hinge-lines. Then the first model is a mechanism if and only if the second model is also a mechanism.*

Proof. Suppose that one model is a mechanism. Then in each position of the model the relative positions of the hinge-lines of one panel are fixed by the rigid panel joining them. So, without restricting the motion of the model, we can attach the corresponding panel of the second model. If we do this for each corresponding pair of panels then both models perform the same motion without hindrance. Thus the second model is also a mechanism. □

Exercise 10.1.2. Draw a 4-panel cycle, that contains exactly two triangular panels, on the hinge-lines of the cycle in Figure 10.10. Draw a 4-panel cycle, that has each panel triangular, on the hinge-lines of the cycle. Label points and list the panels in each case.

Fig. 10.10 The hinge-lines of a 4-panel cycle

10.2 Four-panel cycles

Other than a 3-panel cycle and the 2-panel carpenter's hinge, a 4-panel cycle is the simplest and most useful hinged-panel model. In this section, we determine which 4-panel cycles are mechanisms.

The behavior of 3-panel cycles is central to our arguments and we first establish that each is rigid.

Theorem 10.2.1. *Each 3-panel cycle is rigid.*

Proof. Each two panels are hinged. There is a point of one panel, and a point of a second, both on the remaining panel. These two points are a constant distance apart. Therefore the two hinged panels are fixed relative to one another. Repeating this argument proves each hinge to be rigid, and the model to be rigid. □

We made two 4-panel cycles in Constructions 2.5.1 and 2.5.2, and wondered about the possibility of making a model of a Vámos cube before proving in Theorem 4.4.2 that it is not a figure. With the aid of the following theorem we can better understand the mechanical behavior of the models that we were able to make.

Theorem 10.2.2. *Let a 4-panel cycle have no two of its panels coplanar. Then the cycle is a mechanism if and only if its four hinge-lines are concurrent.*

Proof. If necessary we replace the cycle by a cycle, in which each panel is a 4-gon, that has the same hinge-lines.

Suppose that the cycle has all four hinge-lines concurrent. We may break it into two hinges. Following our discussion of a model cube after Theorem 4.4.2 in Chapter 4 we know that one hinge may be opened to any degree and, provided the other can open wide enough it may be taped to the first completing the cycle, and proving it to be a mechanism. From Theorem 10.1.1 we have that the original 4-panel cycle is also a mechanism.

Suppose that two non-consecutive hinge-lines H_{12} and H_{34} of a 4-panel cycle are concurrent. Then they are coplanar. From Proposition 4.2.1 applied to the model of the cube, the hinge-line H_{23} is concurrent with H_{12} and H_{34}. Similarly H_{14} is concurrent with H_{12} and H_{34}. Thus all four hinge-lines are concurrent.

If no two non-consecutive hinge-lines are concurrent, then we replace the cycle by a cycle containing two triangular opposite panels, but with unchanged hinge-lines, as shown in Figure 10.11.

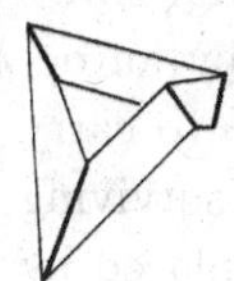
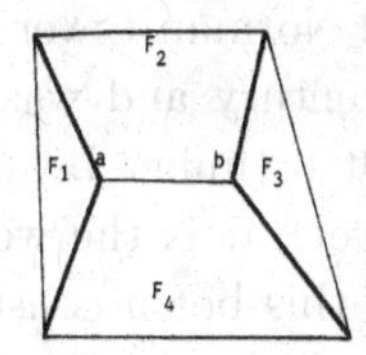

Fig. 10.11 An extra hinge

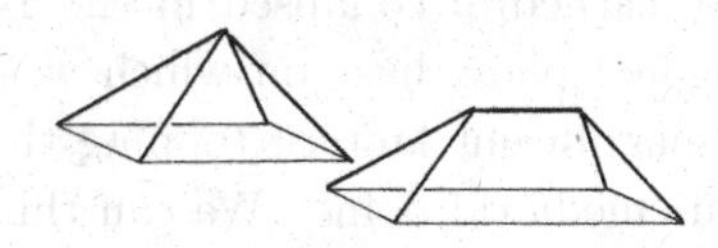

Fig. 10.12 Two 4-panel cycle roof models

In this model the other pair of opposite panels F_2 and F_4 may be hinged without limiting the model's motion, as the edge $[ab]$ is common to both in all positions of the model. From Theorem 10.2.1 we have that that F_1, F_2, and F_4 form a rigid 3-panel cycle. Similarly F_2, F_3 and F_4 form a rigid 3-panel cycle. Thus no panel may move relative to any other panel and the whole model is rigid. From Theorem 10.1.1 we have that the original cycle is also rigid. □

Construction 10.2.1. Make a 4-panel model of each of the two roofs shown in Figure 10.12. The method suggested in Construction 2.7.2 enables us to re-use panels from earlier constructions for the models. The above theorem predicts which of the two cycles is rigid and which is a mechanism. The second example, a hipped roof, is structurally preferable to a peaked roof which relies on extra framing and support to maintain its shape.

Emptied supermarket cartons are usually flattened and stacked as collapsed cycles to be transported for re-cycling or destruction. This is easy because each is basically a 4-panel cycle having all hinge-lines concurrent at an ideal point, and is consequently a mechanism. The origami bird and animal feet of Construction 10.1.1 are each 4-panel cycle mechanisms. This is quite a relief! It means that we are unlikely to tear the paper as we make the creases and fix the final shape. The piece of paper creased in Construction 10.1.2 is a mechanism, as the four creases are concurrent. This is one reason why it is possible to easily unfold a letter in order to read it.

As we saw in the template that we devised in Figure 4.9, the cube model of Construction 2.5.2 has concurrent hinge-lines and so is a mechanism. But the cube of Construction 2.5.1 has only four 4-point circuits and so opposite hinge-lines of any model of it are not coplanar and therefore do not meet. Thus the hinge-lines cannot be concurrent, forcing the model to be rigid. At last we are able to understand the reason for the difference in mechanical behavior of these two cube models.

Example 10.2.1. Part of the central Norman tower of Cambridgeshire's Ely cathedral collapsed in the 14th century and was replaced by an octagonal plane base on which was built a timber beam structure to carry a magnificent lantern topping the tower. It is the world's only surviving true medieval dome. We can think of this beam construction replaced by an equivalent hinged-panel model, in which each triangular framework of beams is replaced by a triangular panel. We then have a model containing a cycle of sixteen panels, each panel being also hinged to a base panel.

The model has a vertical plane of symmetry and thinking of it as halved by this vertical plane, we see in Figure 10.14 that the base, F_1, F_2, and F_3 are the panels of a 4-panel cycle mechanism. Thus, relative to the base, the position of F_1 determines the position of F_3. Again the position of F_3 determines the position of F_5, and so on. We have a mechanism with the

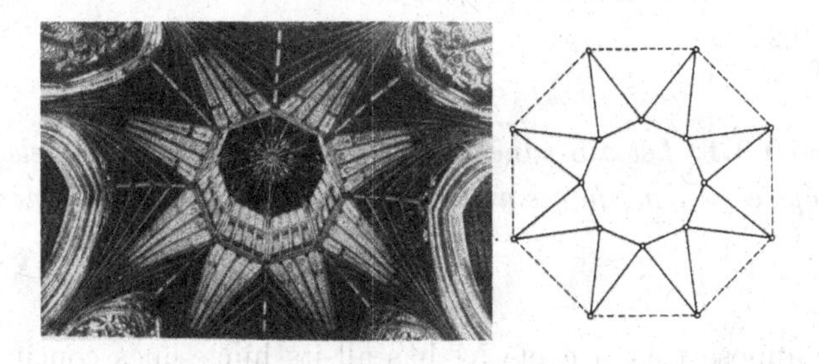

Fig. 10.13 The lantern at Ely cathedral

motion of F_1 determining that of F_9. Any such motion can be mirrored in the vertical plane of symmetry, giving a symmetric motion to the whole seventeen-panel model.

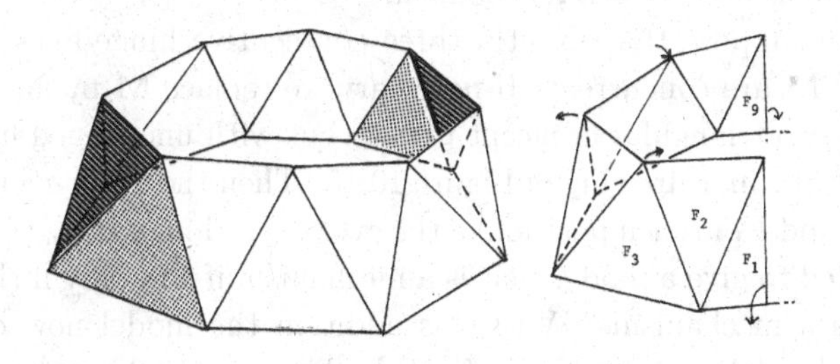

Fig. 10.14 A hinged-panel model of the Ely lantern

The 14th century builders made a mechanism, not a framework. The lantern is beautiful, but perhaps an engineering error! It was strengthened in the 18th century, damaged by storm in 1990, and subsequently repaired [57]. An interesting model in the cathedral explains how the 400 ton structure was put in place.

10.3 Panel cycles of at least five panels

In this section, we prove that the likelihood of a cycle being rigid decreases as the number of its panels increases. We begin with a result similar to

Theorem 10.2.2.

Theorem 10.3.1. *Let a 5-panel cycle have no two of its panels coplanar. Then the cycle is a mechanism if and only if its five hinge-lines are concurrent.*

Proof. Suppose that a cycle **M** has all its hinge-lines concurrent. We replace F_1 and F_2 by a panel in the plane of H_{51} and H_{12}, and from Theorem 10.2.2 we have that this 4-panel cycle, and consequently the original 5-panel cycle, is a mechanism.

Suppose that four, without loss of generality, necessarily consecutive hinge-lines H_{12}, H_{23}, H_{34} and H_{45} of the cycle **M** are concurrent. Then in each position of **M** this point of concurrence is in the plane of the panel F_1 and in the plane of the panel F_2. But any point in the planes of adjacent panels belongs to their hinge-line. Thus H_{15} is concurrent with the other four hinge-lines and **M** is a mechanism.

Next we suppose that exactly three consecutive hinge-lines H_{12}, H_{23}, and H_{34} of **M** are concurrent. If necessary we replace **M** by another cycle containing two triangular adjacent panels, but with unchanged hinge-lines, as shown in the first drawing of Figure 10.15. Then the edge $[ab]$ is common to both F_1 and F_4 in each position of the cycle. The hinge $H_{inge}(F_1, [ab], F_4)$ can be added to give a model that is a mechanism if and only if the original cycle **M** is a mechanism. With this addition the model now contains a 3-panel cycle with panel set $\{F_1, F_4, F_5\}$. Theorem 10.2.1 guarantees that F_1 and F_4 remain fixed relative to one another in any motion of the model. So the model can be replaced by another, in which a panel hinged to F_1 with hinge-line H_{12} and hinged to F_4 with hinge-line H_{34} as shown in the second drawing in Figure 10.15 is added. The new model is a mechanism if and only if the original is a mechanism. But the new model contains a 4-panel cycle with non-concurrent hinge-lines, and so by Theorem 10.2.2, is rigid. We therefore know that **M** is rigid.

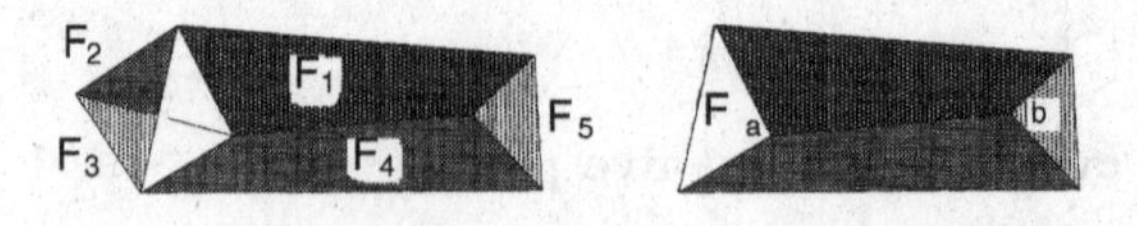

Fig. 10.15 A model with three concurrent hinge-lines

If no two non-consecutive hinge-lines of the original 5-panel cycle **M** are concurrent, then we are able to obtain, if necessary, a second model shown in Figure 10.16. This has the same hinge-lines as **M** and has two triangular panels. The freedom of this model is unchanged by adding a triangular panel $[abc]$ that hinges to each of F_1, F_3 and F_5. Theorem 10.2.1 guarantees the rigidity of the 3-panel cycle containing the panels F_1, $[abc]$, and $F_5\}$. So the lines H_{12} and $a \vee b$ are fixed relative to one another, and F_1 and F_5 are each fixed relative to these two lines.

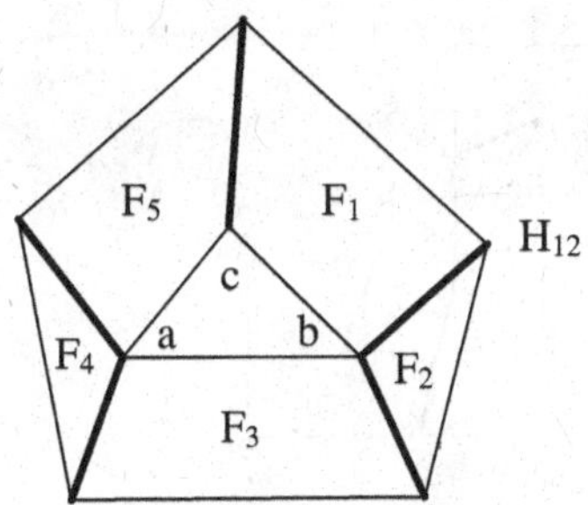

Fig. 10.16 A model having no two non-consecutive hinge-lines concurrent

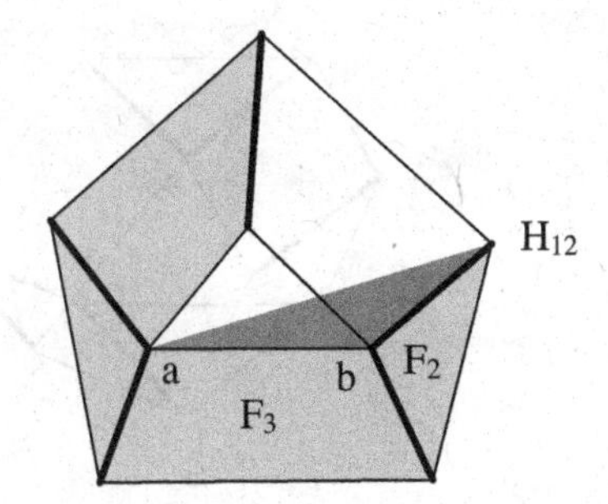

Fig. 10.17 A model obtained after removing and adding panels

We may include a panel F hinged to F_2 and F_3 with hinge-lines H_{12} and $a \vee b$ respectively, provided that we first remove F_1 and $[abc]$. In this new model the panels F_2 and F_3 are allowed all motion permitted in the second model and thus also all motion allowed in **M**. But the 3-panel cycle containing F, F_2 and F_3 ensures that F_2 and F_3 are fixed relative to one another and to the lines $a \vee b$ and H_{12}. From a similar argument we have that F_3 and F_4 are also each fixed relative to the lines $a \vee b$ and H_{12}. Thus all five panels of **M** are fixed relative to the lines H_{12} and $a \vee b$, and **M** is rigid. □

We may investigate the rigidity of some 6-panel cycles using methods similar to those in the proofs of Theorems 10.2.2 and 10.3.1. In this way we are able to deduce that a 6-panel cycle model, no two of whose panels are coplanar, is a mechanism if all six hinge-lines are concurrent or if at least four consecutive hinge-lines are concurrent. It is also a mechanism if three consecutive hinge-lines are concurrent and the remaining three are also concurrent.

Example 10.3.1. A 6-panel cycle mechanism that has two groups of three consecutive hinges concurrent, each at an ideal concurrence point, was described by P.F. Sarrus [18] in 1853 as a solution to the problem of providing reliable straight-line motion. This was a difficult problem at the time as no milling machines were available to make long-wearing and accurate sliding guides - so pistons and slides frequently jammed and bent. Sarrus's mechanism was the first realistic solution, using functional mechanical hinges that were available at the time.

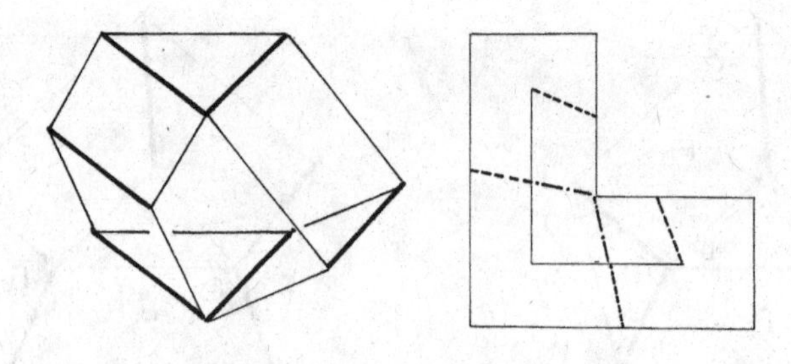

Fig. 10.18 Sarrus' hinge and Goldberg's hinge

Four of the panels of Sarrus' mechanism are rectangular, ensuring that the motion of a point of one triangular panel is along a line perpendicular to the plane of the second triangular panel. The American 20th century geometer Michael Goldberg [28] devised a similar mechanism, also illustrated in Figure 10.18, that flattens neatly into an unfolded planar model. It has Euclidean concurrence points.

Exercise 10.3.1. Design a version of Sarrus' mechanism that flattens into an unfolded model.

The method of Theorem 10.3.1 enables us to prove that any 6-panel cycle, in which only three consecutive hinge-lines are concurrent, is rigid. We now make some models that have no three hinge-lines concurrent in order to understand the variety of mechanical behavior possible in 6-panel cycles.

Construction 10.3.1. Make a 6-panel cycle in which all panels $[afb]$, $[fbd]$, $[bdc]$, $[dce]$, $[cea]$, and $[eaf]$ are equilateral triangles. A good way to make it is by the technique of Construction 2.7.2. We can even re-use the panels of that construction. We need, in total, six panels and six rubber bands.

Construction 10.3.2. We make another 6-panel cycle as follows. Cut out the panels $[afb]$, $[fbd]$, $[bdc]$, $[dce]$, $[cea]$, and $[eaf]$, using the drawing in Figure 10.19 as a template. It may be easiest to hinge the faces as shown, then bend the hinges appropriately until the two panels of the remaining hinge are touching, and tape them together. If you wish to make a larger and more accurate model, a template may be made by drawing two concentric circles and choosing points a and d outside both. The tangents from a and d to the larger circle meet at c and f. The tangents to the smaller circle meet at b and e, as in Figure 10.19.

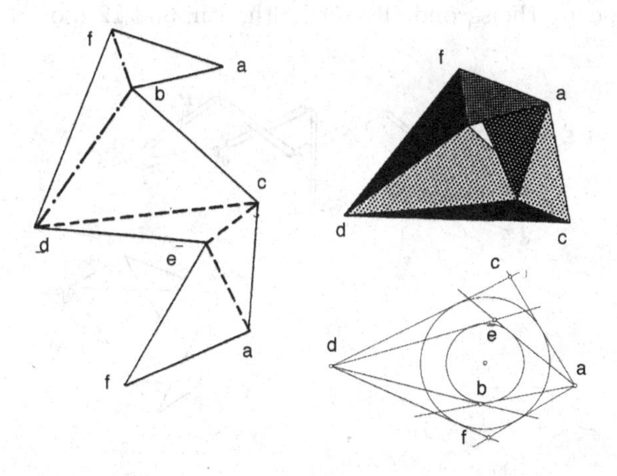

Fig. 10.19 A Bricard 6-panel cycle model, and template

Details of this 6-panel cycle were published by the French mathematician Rauol Bricard [7] in 1897. Our model seems to be far from rigid. Do you see a 4-bar linkage contained within it? Perhaps the fact that the four edges of this "crossed parallelogram" linkage remain coplanar, and its four joints belong to a circle, during motion of the model provides a clue to the mobility of the model.

In the proof of Theorem 10.3.1 we used the fact that three panel edges joined in the form of a triangular bar-and-joint model form a rigid framework in order to insert an extra triangular panel. A triangular panel may be interchanged with its skeleton in a model without affecting the behavior of the model. We made use of this observation in Example 10.2.1. Thus a bar-and-joint model of the skeleton of a hinged-panel model, that contains only triangular panels, is rigid if and only if the hinged-panel model itself is rigid. These bar-and-joint models often are easier to make and examine than are the panel models. We start with a model of the skeleton of Construction 10.3.1.

Construction 10.3.3. Thread twelve 6 centimeter lengths of drinking straw in a necklace and tie the string without slack. Tying together joints with a common label gives the required model. To avoid confusion during construction it is helpful to label each end of each straw indelibly by the appropriate letter as shown in Figure 10.20. If you have different colored straws available, then it might be helpful to use a second color for the hinges. These are the second, 4th, 5th, 6th, 9th, and 12th on the list.

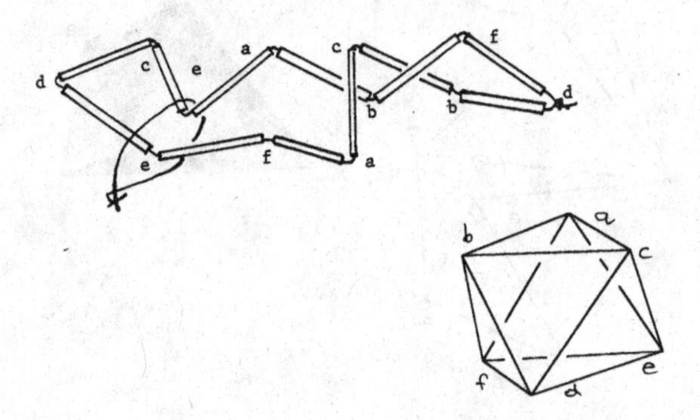

Fig. 10.20 Making a model skeleton

This technique is useful even when edges differ in length. Even though the appearance and mechanical behavior of the models of Constructions 10.3.1 and 10.3.2 are quite different we observe that the labeling of their skeletons is identical. Corresponding edges seem to differ only in length, so we vary the lengths of straws appropriately and join them as in Construction 10.3.3 to give a skeleton of the Bricard cycle.

Construction 10.3.4. Thread lengths (measured in centimeters) 12, 7, 9, 9, 7, 12, 12, 10, 9, 9, 10, and 12, *in this order*, of drinking straw in a necklace and tie the string without slack. To avoid confusion during construction, label each end of each straw indelibly by the appropriate letter. Tying together joints with a common label gives the required model. Again, if you have different colored straws available it might be helpful to use a second color for the hinges. Again these are the second, 4th, 5th, 6th, 9th and 12th straws on the list.

It is no accident that the last two models have triangular panels. If no three consecutive hinge-lines of a 6-panel cycle **M** are concurrent and no two successive hinge-lines are parallel, then from Theorem 10.1.1 we may replace each panel of **M**, if necessary, by a triangular panel as shown in Figure 10.21 to give a model that is a mechanism if **M** is a mechanism. We can consequently also add a panel $[abc]$ hinged to each of F_1, F_3 and F_5 as shown in Figure 10.21, again without hindering any motion of the model.

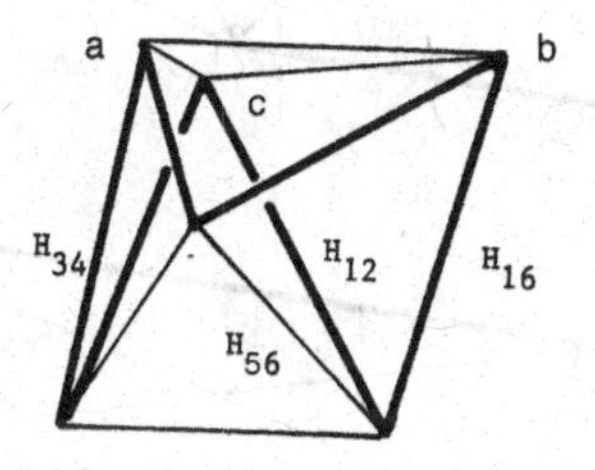

Fig. 10.21 A six-panel cycle and an additional panel

We can think of the 4-panel cycle $\{F_1, F_2, F_3, [abc]\}$ in which the angle ϕ between $[abc]$ and F_1 determines the angle between $[abc]$ and F_3, the 4-panel cycle $\{F_3, F_4, F_5, [abc]\}$, then determining the angle between $[abc]$ and F_5, and the 4-panel cycle $\{F_5, F_6, F_1, [abc]\}$ determining the angle ψ between $[abc]$ and F_1. If $\phi = \psi$ for a range of values of ϕ then the model, and consequently also **M**, would be a mechanism. Bricard showed in a long analysis that this occurs in exactly three ways. In the first possibility, made in Construction 10.3.4, the model has two collapsed positions. The second possible mechanism has an axis of symmetry. All three mechanisms are shown in Figure 10.22.

We will understand Bricard's cycles much better if we make examples of each type. We use the technique of Construction 10.3.4 to make two more skeletons, in each case threading twelve lengths of drinking straw in a necklace, then tying together those ends with a common label.

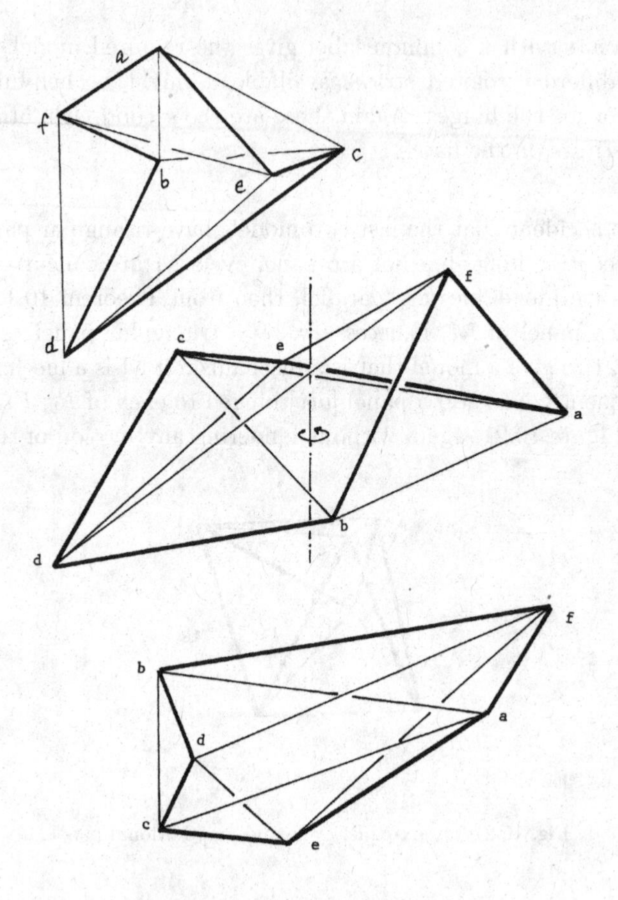

Fig. 10.22 Bricard's three 6-panel cycles

Construction 10.3.5. We again use the technique of Construction 10.3.3. To make Bricard's second model we thread on the following lengths (measured in centimeters) 11, 6, 10, 12, 6, 8, 10, 11.2, 8, 11, 11.2 and 12 in the given order. To make Bricard's third model, the lengths, in correct order, are 12, 13, 11.2, 6, 10.2, 2.4, 10, 9.3, 3.8, 10.6, 6.5 and 4.7. If you have different colored straws available, the hinges of each model are the second, 4th, 5th, 6th, 9th and 12th on the respective list of straws.

Bricard's results complete the proof of the following theorem:

Theorem 10.3.2. *Let a 6-panel cycle* **M** *have no two of its panels coplanar. Then* **M** *is a mechanism if and only if one of the following is true:* **M**

has at least four consecutive concurrent hinge-lines, **M** *has three consecutive hinge-lines concurrent and the other three consecutive hinge-lines also concurrent, or* **M** *has the hinge-lines of one of the three Bricard cycles.*

Panel cycles containing many panels are not of great interest as mechanisms. We might suspect this if we have had many dealings with bicycle chains or tank tracks. Experience suggests that they tangle easily. Why is this? Cycles having many panels are very "floppy" - which is why bicycle chains cannot easily be held rigid in one's hands. This excessive freedom of movement would be a handicap in, for example, hinging a door. We outline a proof for the following theorem:

Theorem 10.3.3. *Each panel cycle that has no two panels coplanar and contains more than six panels is a mechanism.*

Proof. Suppose that **M** is an n-panel cycle, $n \geq 7$. If no three consecutive hinge-lines are concurrent, then we replace **M** by a model with the same hinge-lines and the panels $F_1, F_2, \ldots, F_{n-1}, F_n$ as in Figure 10.23.

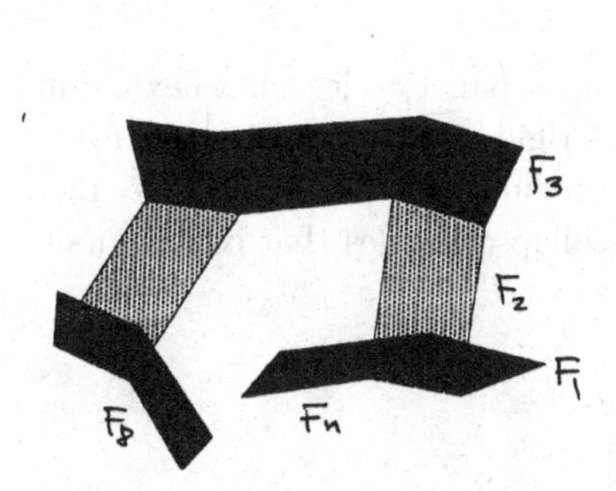

Fig. 10.23 A choice of panels of an n-panel cycle

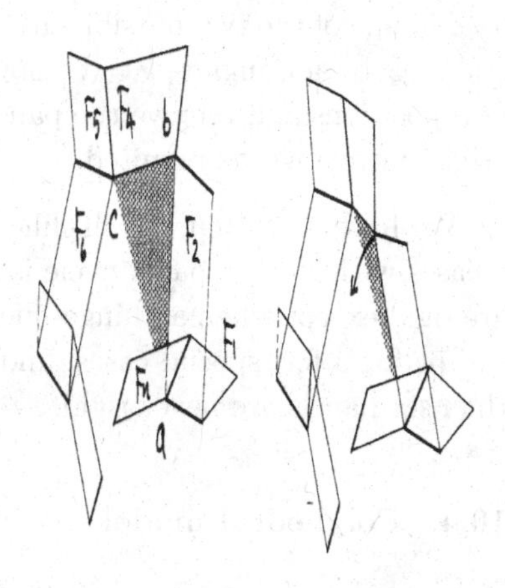

Fig. 10.24 Before and after folding the model about a line

Omitting F_3 from the cycle model leaves a sequence $F_1, F_2, F_4 \ldots, F_{n-1}, F_n$ of panels, and $n-2$ hinges. We keep $F_6 \cup \ldots \cup F_n$ rigid, add a panel $[abc]$, and fold the new model about the line $a \vee c$ as shown in Figure 10.24.

As we fold, the Hooke universal joint of panels F_n, F_2, and F_1 determines the position of F_1, and the Hooke universal joint of panels F_6, F_5, and F_4

determines the position of F_4. Thus we change the angle between H_{23} and H_{34} by a small angle θ.

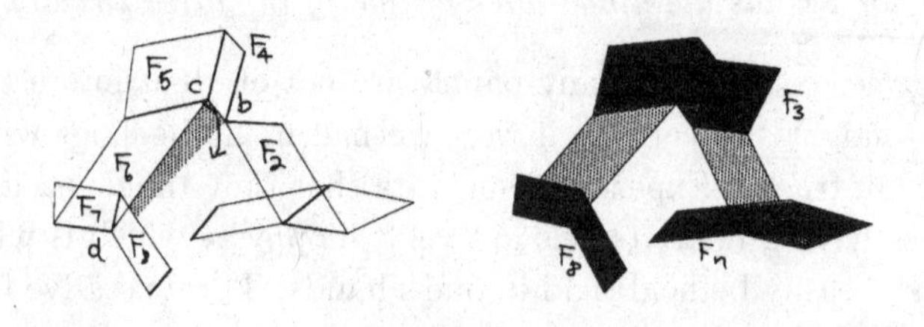

Fig. 10.25 Replacing F_3 after a second folding

Next we remove the panel $[abc]$, keep $F_8 \cup \ldots \cup F_n \cup F_1 \cup F_2$ rigid, and add a panel $[bcd]$. If $\mathbf{M}$ is a 7-panel cycle, then we keep $F_1 \cup F_2$ rigid. As we fold the Hooke universal joint of panels F_8, F_7, and F_6 determines the position of F_6 and the Hooke universal joint of panels F_6, F_5, and F_4 determines the position of F_4. By folding the new model about the line $b \vee d$ in one of the two possible directions, we change the angle between H_{23} and H_{34} by an angle $-\theta$. We now re-attach F_3 to H_{23} and H_{34} in their new positions and remove the panel $[bcd]$ to give a new arrangement of the original cycle $\mathbf{M}$, as required. □

We have seen that the likelihood of an n-panel cycle being flexible increases with n. A 3-panel cycle is always rigid, 4-panel and 5-panel cycles are rigid except when all hinge-lines are concurrent, 6-panel cycles are rigid except for a few special cases, and non-collapsed cycles that contain more than six panels are never rigid.

10.4 Polyhedral models

Panel cycles contain few hinges. Polyhedral models, on the other hand, are commonly occurring hinged-panel models that have as many hinges as possible. In this section, we examine some examples, paying particular attention to their skeletons.

Definition 10.4.1. *A* **polyhedral model** *is a hinged-panel model in which each panel F and edge $E \subseteq F$, together belong to some hinge.*

Definition 10.4.1 tells us that in a polyhedral model each edge of a panel binds it to another panel. A *tetrahedral model* is the simplest example of a polyhedral model. Its vertices are those of a combinatorial tetrahedron, its panels are the triangular panels defined by any three of the four vertices of the combinatorial tetrahedron, and each pair of panels together with their common edge make up its six hinges. The ubiquitous liquid container made by Sweden's big export earner, TetraPak, is a commonly seen tetrahedral model.

Example 10.4.1. If $[abc]$ is a triangular region, then the rigid polyhedral model that contains four panels F_1, F_2, F_3, and F_4, each equal to $[abc]$, and that contains hinges $H_{inge}(F_1, [ac], F_2)$, $H_{inge}(F_3, [ac], F_4)$, $H_{inge}(F_2, [ab], F_3)$, $H_{inge}(F_1, [ab], F_4)$, $H_{inge}(F_1, [bc], F_3)$, and $H_{inge}(F_2, [bc], F_4)$ is a folded polyhedral model.

Polyhedra are often classified according to the number of their panels: a tetrahedral model has four panels, a *pentahedral model* has five, a *hexahedral model* six, a *heptahedral model* seven, an *octahedral model* eight, and so on.

Construction 10.4.1. The octahedral model shown in Figure 10.26 is an example of a polyhedral model. Each panel is an equilateral triangle. It may be modeled by adding two triangular panels to the 6-panel cycle model of Construction 10.3.1. As well we need six more rubber bands to ensure that all shared edges model hinges.

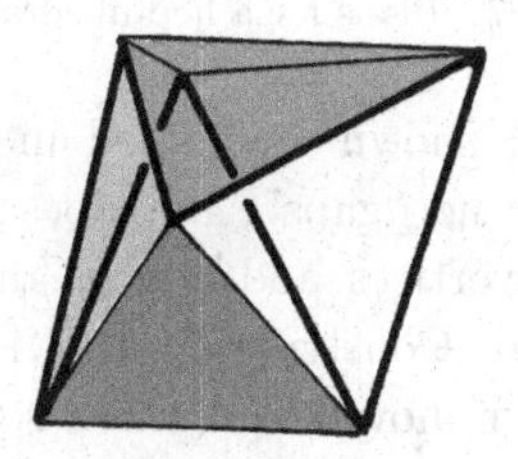

Fig. 10.26 An octahedral model

Construction 10.4.2. Following the instructions of R. Hughes Jones [35], a heptahedral model may be made from six copies of the tabbed right-angled triangular panel and eight copies of the equilateral triangular panel in Figure 10.27, using four colors 1, 2, 3, and 4, by gluing the tabs, shown dotted, as indicated. Each of the three triangles colored 1 then have a second, same-colored triangle glued to their back, leaving unglued pockets for the three unglued tabs de, ef and fd. The model can then be folded as indicated by the lettering, slipping the tabs into appropriate pockets. The two triangles colored 2 then form a square panel intersected by each of the two square panels colored 3 and 4 respectively.

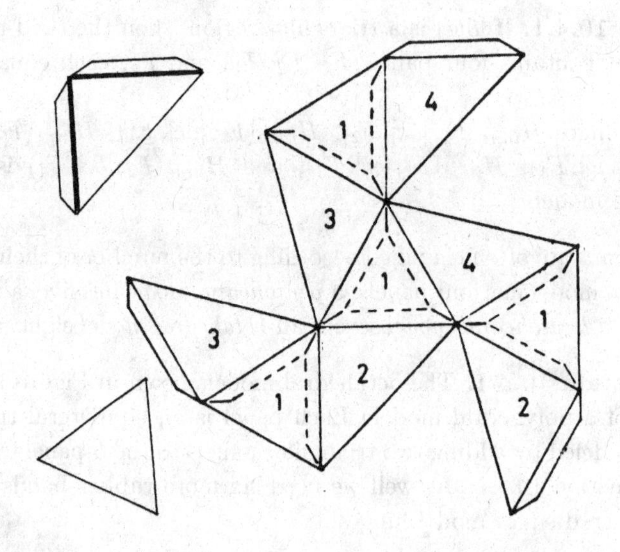

Fig. 10.27 Plans for a heptahedral model

This particular model, shown assembled in Figure 10.28, has all its edges hinged, so there are no "gaps", but does it have an "outside" and an "inside"? It has the vertices and edges, and therefore the skeleton, of the octahedral model of Construction 10.4.1, but alternate panels of the octahedral model are removed and three squares through the center introduced in their place, and so it has four triangular panels and three square panels.

Exercise 10.4.1. List the panels and hinges of this heptahedral model and compare its skeleton with that of the octahedral model of Construction 10.4.1.

Exercise 10.4.2. We derive another hinged-panel model from this heptahedral model as shown in Figure 10.29. It also has the vertices of the octahedral model modeled in Construction 10.3.1. List its six panels and

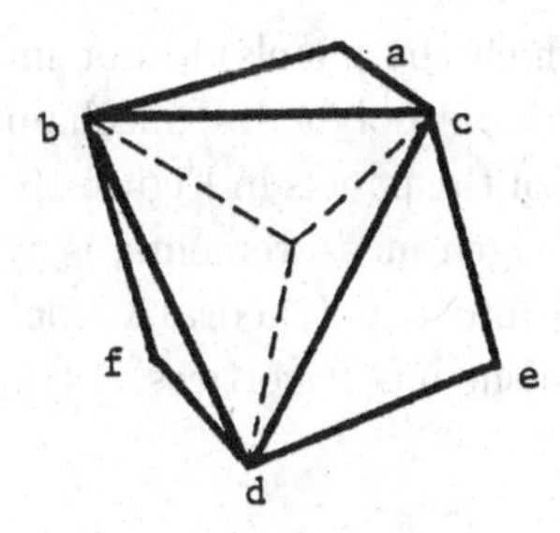

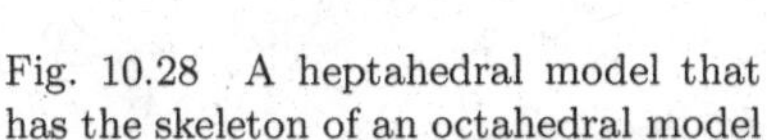
Fig. 10.28 A heptahedral model that has the skeleton of an octahedral model

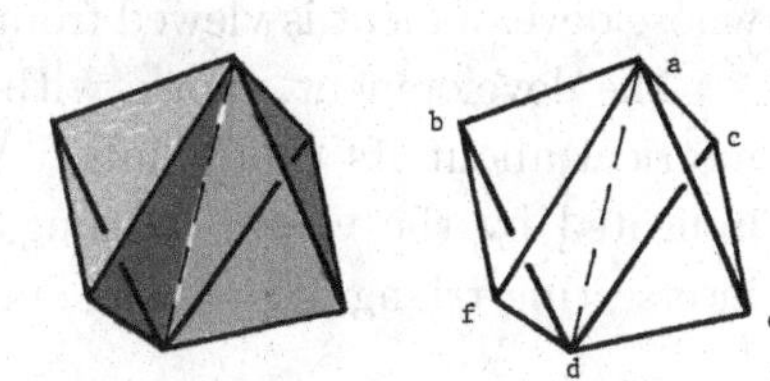

Fig. 10.29 A hexahedral model

its hinges. Is it a polyhedral model? If you can devise a satisfactory method of constructing this hexahedral model, do so.

In a similar fashion to the construction of an octahedral model from a 6-panel cycle in Construction 10.3.1, we may add two triangular panels to each Bricard 6-panel cycle pictured in Figure 10.22 in order to obtain an octahedral model. Thus the skeletons we made in Constructions 10.3.4 and 10.3.5 are also skeletons of octahedral models. The panels of each Bricard octahedral model intersect and in practice we could only add part of each of the seventh and eighth panels in order to allow the mechanism to move. This is no accident, as each octahedral mechanism (which must be made in this way) has bad panel intersections. Just to hint at the beauty and intricacy of Bricard's octahedral models, we note that each gives rise to a pair of combinatorial tetrahedra, each vertex of the first being coplanar with a panel of the second, and *vice versa.*

Just as the boundary of a panel separates the Euclidean plane containing it into an interior and an exterior part, so a polyhedral model with no intersecting panels appears to separate Euclidean space into a part interior to the shell and a part exterior to the shell. The French mathematician A.L. Cauchy considered *convex models.* In a convex model no two panels are coplanar, and for each pair of interior points a and b, the segment $[ab]$ is completely contained in the interior of the model. In 1813 he proved the following theorem:

Theorem 10.4.1. (Cauchy's Theorem) *Each model of a convex polyhedron is rigid.*

It was long thought that this result might also hold for any model without panel intersections. But in 1977 Robert Connelly[12] constructed a variation of a Bricard octahedral model in which the panels do not intersect. In 1978 this led Klaus Steffen to the simple polyhedral mechanism whose development is viewed from the outside of the panels in Figure 10.30.

The development, shown with some side-length measurements, is symmetrical about its center line. After folding into a polyhedral model as indicated by the vertex labeling, the mechanism has 9 vertices, 14 non-intersecting triangular panels, and 21 edges.

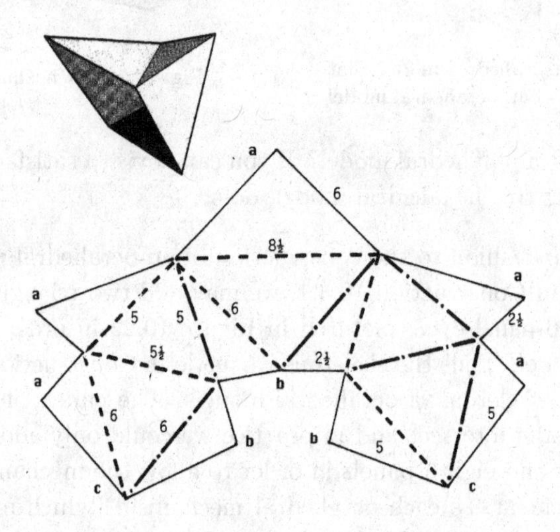

Fig. 10.30 A development for a polyhedral mechanism

Construction 10.4.3. Construct Steffen's mechanism.

Summary of Chapter 10. We defined hinged-panel models and observed the wide occurrence of these models in everyday life, looking at examples, and some of their more common applications. We distinguished between models that are mechanisms and those that are rigid, proving that panel cycles and polyhedral models are of both types. We characterized those panel cycles that are mechanisms.

Chapter 11

GRAPHS, MODELS AND GEOMETRIES

In Chapter 9 we became familiar with the practical uses of simple bar-and-joint models. But many engineering works are quite complicated bar-and-joint models, containing many hundreds of bars. In order to analyze the rigidity of such a construction we define an underlying mathematical structure, namely a graph, that is modeled by the bar-and-joint model A graph is formally a listing of some pairs of objects that belong to a specified set. It is an important mathematical structure with an associated large body of theory and we lay the groundwork for this theory. In particular we introduce the notions of a connected graph and an acyclic graph. These ideas enable us to determine the rigidity of a framed building, and define the developments of hinged-panel models. A development of a hinged-panel model is an intermediate step of efficient construction for the model itself. Graph theory also plays a fundamental role in our investigation of spherical polyhedra in Chapter 12. But perhaps of most interest in the overall context of geometries is the last section of the chapter. There we prove certain graphs to be geometries in their own right.

11.1 The definition of a graph

A graph formalizes a selection of pairs of objects. Thus, for example, in a bar-and-joint model a pair of joints could be singled out for mention if the joints belong to a common bar of the model. In this section, we introduce some basic notation and fundamental properties of graphs.

Definition 11.1.1. *A* **graph** G *consists of a finite set* $V(G)$, *of vertices, and a finite list* $E(G)$ *of edges. Each edge is an unordered pair of (not necessarily distinct) vertices. Two identical edges are called multiple, or* *parallel, edges. An edge* $\{u, u\}$ *is called a loop.*

Example 11.1.1. The set $V(G) = \{1, 2, \ldots, 5\}$ and the list $E(G) = \{\{1,2\}, \{2,3\}, \{3,4\}, \{1,4\}, \{1,4\}, \{1,5\}, \{2,5\}, \{3,5\}, \{4,5\}, \{3,3\}\}$ define a graph G. It has two parallel edges and one loop.

In a more restrictive alternative definition of graph repetition is not allowed, neither in $E(G)$ nor within each edge. However we use the adjective *simple* to distinguish those graphs that have neither parallel edges nor loops.

Definition 11.1.2. *A* **simple graph** *is a graph that contains neither parallel edges nor loops.*

Example 11.1.2. For each natural number n the set $V(K_n) = \{1, 2, \ldots, n\}$ of vertices, and the set $E(K_n) = \{\{i, j\} : 1 \leq i < j \leq n\}$ of edges, define a simple graph, denoted by K_n, and called the complete graph on n vertices.

The bar-and-joint models of Chapter 9 provided the motivation for Definition 11.1.1, and each can be thought of as a physical model of a graph in which each joint represents a vertex and each bar models an edge.

Definition 11.1.3. *Let* $\mathbf{M}$ *be a bar-and-joint model. Then the set* $V(G) = \{u : u \text{ is the end of a bar}\}$ *and the set* $E(G) = \{\{u, v\} : u \text{ and } v \text{ are the ends of a common bar}\}$ *define a simple graph* G*. We say that* $\mathbf{M}$ **models** G *and write* $\mathbf{M} = \mathbf{M}(G)$ *and* $G = G(\mathbf{M})$.

Exercise 11.1.1. Specify each graph that is modeled by a flail, a triangular framework, or a 4-bar linkage. Are K_3 and K_4 among these graphs?

Graphs may be presented visually in a way reminiscent of sketches of planar figures.

Definition 11.1.4. *Let* G *be a graph. A* **drawing** *of* G *is a collection of points in a Euclidean plane, each labeled by a different vertex of* G*, and a collection of arcs in the plane, one arc being drawn between* u *and* v *for each edge* $\{u, v\}$ *of* G.

As each drawing of a graph G contains all the information required for the graph's definition, we may blur the distinction between G and any drawing of it without ill-effect. There is no requirement that an arc representing an edge of G be straight but often we use the segment $[uv]$ as the arc that represents an edge $\{u, v\}$. This is possible for each edge of any simple

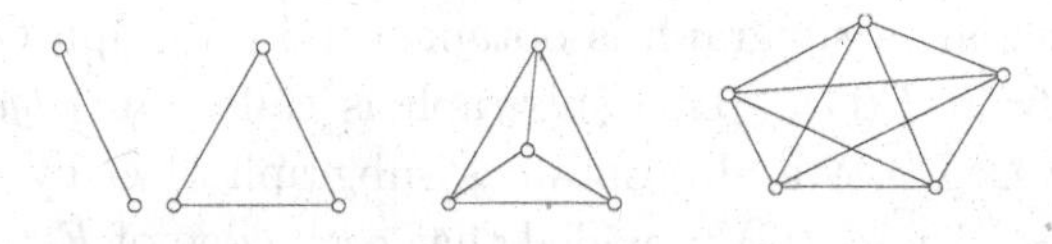

Fig. 11.1 A drawing of each K_n, for $n = 1, 2, 3, 4$, and 5

graph. Figure 11.1 contains such a drawing of each K_n, for $n = 1, 2, 3, 4$, and 5.

Road maps, airline route diagrams, and electrical wiring diagrams are some, among many, documents that have the characteristics of a drawing of a graph, suggesting that graph theory should find wide application.

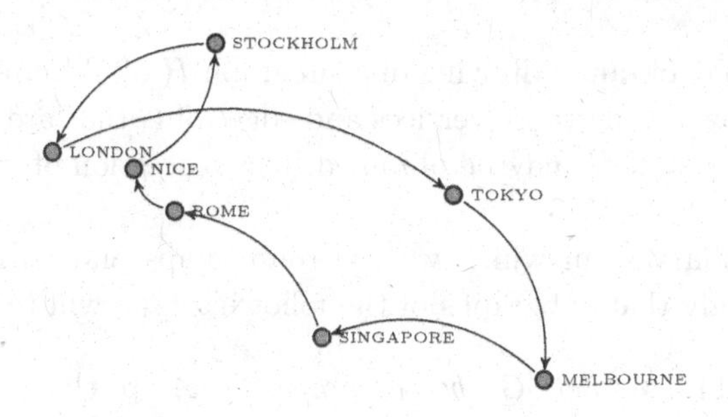

EUROPE VIA SOUTHEAST ASIA & JAPAN
Fly with British Airways or Qantas from Melbourne to Singapore and then to Rome. Stop over in Nice, Stockholm, and then London before flying to Tokyo with British Airways. Return home with Qantas.
MEL - SIN - ROM - NCE - STO - LON - TYO - MEL
6 stops - 23,668 miles.

Fig. 11.2 An airline route map

We now build on Definition 11.1.1, deriving various properties of graphs that we later use for our investigation of bar-and-joint models and hinged-panel structures. We start by examining structures induced on some subsets of vertices and edges of a graph. The result of erasing some parts of a drawing of a graph is a drawing of a graph provided that any edge ending in an erased vertex is also erased. More formally we say:

Definition 11.1.5. *Let G be a graph. Then a graph H is a* **subgraph** *of G if $V(H)$ and $E(H)$ are subsets of, respectively, $V(G)$ and $E(G)$.*

In particular, a subgraph is obtained from a graph G by deleting any subset E' from $E(G)$. This subgraph is called an *edge deletion* and is denoted by $G \backslash E'$. We also obtain a subgraph of G by deleting a subset V' of vertices from $V(G)$ provided that each edge of $E(G)$ that contains a member of V' is deleted from $E(G)$. This subgraph is a *vertex deletion* and is denoted by $G \backslash V'$.

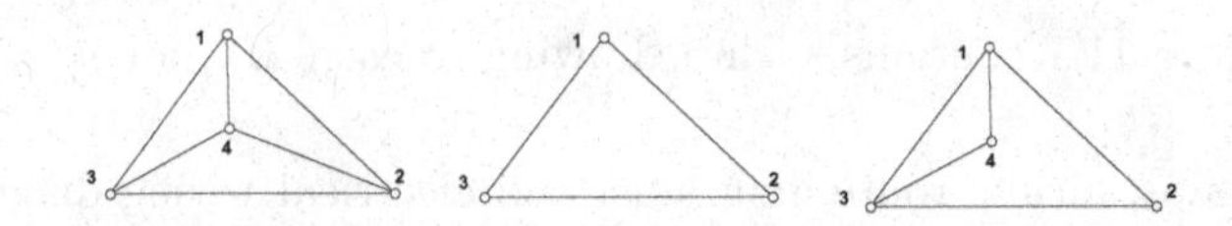

Fig. 11.3 A drawing of each of K_4, $K_4 \backslash \{4\}$, and $K_4 \backslash \{\{2,4\}\}$

Thus we may obtain a drawing of a subgraph H of G from any drawing of a graph G by erasing those vertices and edges of G that are not elements of H. Any subgraph H may be obtained by a succession of single-element deletions.

The common way in which we use route maps and wiring diagrams suggests strongly that subgraphs of the following type will repay study.

Definition 11.1.6. *Let G be a graph. A* **path** *of G is any subgraph H of G with $V(H) = \{v_1, v_2, \ldots, v_n\}$ and $E(H) = \{\{v_1, v_2\}, \{v_2, v_3\}, \ldots, \{v_{n-1}, v_n\}\}$, the vertices v_i being pairwise distinct. We say that the path joins v_1 and v_n in G.*

Exercise 11.1.2. Write down the vertices of a path, that joins Stockholm and Singapore, of the airline route map in Figure 11.2.

11.2 Connected graphs and acyclic graphs

In this section, we analyze a graph in terms of its paths.

Definition 11.2.1. *A graph is* **connected** *if each pair of distinct vertices has a path joining them. A graph that is not connected is* **disconnected**.

It is useful to study subgraphs of a graph G that are connected, but minimally so. These subgraphs give us insight into hinged-panel constructions and they have a more general importance [43]. The existence of the following graph as a subgraph ensures a redundancy among the paths of G.

Definition 11.2.2. *Let G be a graph. A* **cycle** *of G is any subgraph H of G with $V(H) = \{v_1, v_2, \ldots, v_n\}$ and $E(H) = \{\{v_1, v_2\}, \ldots, \{v_{n-1}, v_n\}, \{v_n, v_1\}\}$, the vertices v_i being pairwise distinct. A graph is acyclic if it contains no cycles. A forest is an acyclic graph, and a tree is a connected acyclic graph.*

The airline route map in Figure 11.2 is connected, the wiring diagram is disconnected. Neither is acyclic. A flail models a tree, and a pair of flails model a forest. A 4-bar linkage models a graph that is a cycle.

Exercise 11.2.1. List the vertices of each cycle of K_3. List the vertices of each cycle of K_5 that contains four or five vertices.

Lemma 11.2.1. *Let G be a simple graph. If H is a cycle of G, then $|V(H)| = |E(H)| \geq 3$.*

Proof. The inequality is a consequence of the impossibility of repetition within $E(H)$ and of repetition within any edge of H. □

We call a subgraph of a graph G a *maximal tree* of G if it is a tree, and is not a subgraph of any other tree in G. The graph modeled by a flail is the only example of a tree that we have seen so far. But it is straightforward to find further examples. If we delete an edge that belongs to a cycle of a graph, and repeat this process, eventually we obtain an acyclic subgraph of the original graph. This experiment suggests the following property of trees:

Theorem 11.2.1. *Let G be a connected graph. Then G contains a maximal tree. Any tree of G is a subgraph of a maximal tree of G and it is a maximal tree if $V(T) = V(G)$.*

Proof. If T is a tree of G and $u \in V(G) - V(T)$, then we choose a path joining u and a vertex v of T. There is an edge $\{u', v'\}$ of the path so that u' is in $E(T)$ and v' is not in $E(T)$. Adding this edge to $E(T)$ and adding the vertex v' to $V(T)$ gives a graph that is a tree and has T as a subgraph. We continue in this way, until we obtain a tree containing all the vertices of G. Starting this process from any single vertex proves that each connected graph contains a subgraph that is a maximal tree.

Suppose that a subgraph T of G is a tree, and further suppose that $V(T) = V(G)$. If T is a subgraph of a tree T' in G, then $V(T) \subseteq V(T') \subseteq V(G) = V(T)$. Thus $V(T) = V(T')$. Also any edge e in $E(T') - E(T)$ is $e = \{u, v\}$ for some u and v in $V(T)$. There is a path joining u and v in T, and e is not an edge of this path. The path is also a path in T'. Adding e to the edge set of this path gives a cycle in T', contradicting the definition of a tree. So T' and T have the same vertex sets and the same edge sets, and are therefore the same graph and T is a maximal tree in G. $\square$

Exercise 11.2.2. By deleting edges of the complete graph K_5 obtain a maximal tree of K_5. How many maximal trees does K_5 contain?

It is an important task to find all the maximal trees in a graph [6]. The problem is of interest in particular to chemists [43]. For example, each of the many possible carbon-based molecules with a particular formula can be identified with a tree, each carbon bond representing an edge of the tree. It is important for a chemist to know whether a molecule can exist before trying to create it. The problem is also of increasing interest in communication networks. The links between n nodes of a proposed network would model edges of a graph, and a cheap means of connecting each pair of nodes requires a minimal sets of links. These are the maximal trees of the complete graph K_n.

Lemma 11.2.2. *Let T be a tree and P a path of T so that $E(T)$ is not equal to $E(P)$. Then there is a vertex of $V(T) - V(P)$ that belongs to exactly one edge of $E(T) - E(P)$.*

Proof. Let Q be a maximal path of T from a vertex u to a vertex v such that u is the only vertex of Q that lies on P. Then v belongs to an edge of Q that is in $E(T) - E(P)$. Moreover, v belongs to no other edge of $E(T) - E(P)$ as T is acyclic and Q is maximal. $\square$

Lemma 11.2.3. *If T is a tree, then $|V(T)| = |E(T)| + 1$.*

Proof. We argue by induction on $|V(T)|$. The result is true for $|V(T)| = 1$. Suppose that the result holds for any tree T satisfying $|V(T)| \leq n$. Let T be a tree with $n + 1$ vertices. From Lemma 11.2.2 applied to the empty path P, we have the existence of a vertex x of T that belongs to only one edge of T. Then $T \backslash \{x\}$ is a tree T' that satisfies $|V(T')| = |E(T')| + 1$. Observe that $|V(T)| = |V(T')| + 1$ and $|E(T)| = |E(T')| + 1$. Thus $|V(T)| = |V(T')| + 1 = |E(T')| + 1 + 1 = |E(T)| + 1$. Thus the result holds by the Principle of Mathematical Induction. $\square$

11.3 The rigidity of a one-story building

In Definition 11.1.3 we usefully think of any bar-and-joint model **M** as a model $G(\mathbf{M})$ of a simple graph G in which each universal joint models a vertex, two joints being connected by a rigid bar if the set of two vertices that they model is an edge. Thus two bars of the model are joined together at an end whenever they model edges that contain a common vertex. We examined flails, triangular frameworks, and 4-bar linkages in Chapter 9. In this section, we apply the connectivity of the graph $G(\mathbf{M})$ in order to determine the physical behavior of the bar-and-joint model **M**.

The difficulty and expense of constructing freely movable universal joints has meant that non-planar models are primarily of interest as rigid structures. We are all familiar with space frames, building scaffolding, bridge trusses and geodesic domes. In these, and other, applications rigidity is the desired quality, not mobility. We look at one application [55], the framing of a one-story building. An example is shown in Figure 11.4.

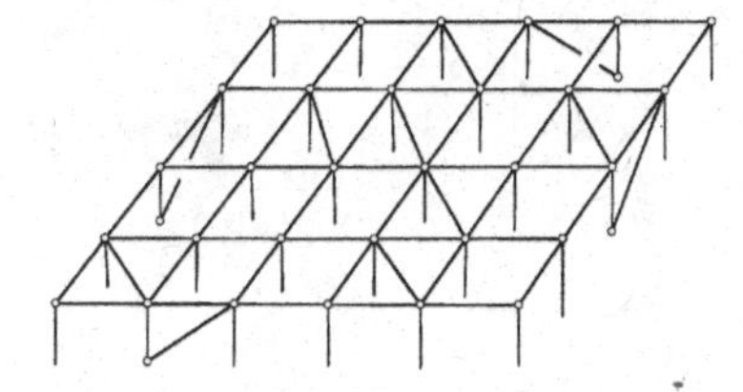

Fig. 11.4 A typical warehouse frame

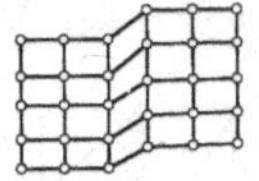

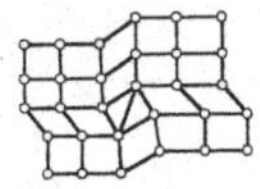

Fig. 11.5 Planar movement of a roof

Problem 11.3.1. *We are designing a post-and-beam framed warehouse so that machinery is able to move through it as freely as possible. Consequently we brace it only with diagonal braces within roof squares and external wall squares. Where should the braces be placed in order to use fewest possible?*

Solution. First we examine possible motion of the roof beams within the horizontal plane of the roof. We try to rotate a beam through an angle θ. If we model the warehouse frame as a bar-and-joint structure, each square being a parallelogram linkage, then each cross-bar in the same line of squares of the roof grid also turns through θ. If there is a braced square in this line, then all bars of this square turn through θ. Consequently, all cross-bars of the other line containing the square turn through θ also. Thus any braced square in a moved line forces a second line to move. If each bar is forced to turn through θ, then no bar has turned relative to any other and the roof is rigid. As each bar is a cross-bar of some line, this is equivalent to each line being forced to move.

We may summarize this information concisely in terms of the *bracing graph* of the roof. Each complete line of squares of the roof-grid is a vertex of the bracing graph. Each two lines form an edge of the bracing graph if the two meet in a braced square. In the light of Definition 11.2.1 and Lemma 11.2.1 we have the following result:

Lemma 11.3.1. *Let the roof be a k by ℓ rectangle. Then the roof is rigid within its horizontal plane if and only if the bracing graph is connected. At least $k + \ell - 1$ braces are required to make the roof rigid. A set of $k + \ell - 1$ braces suffices if and only if the bracing graph is a tree.*

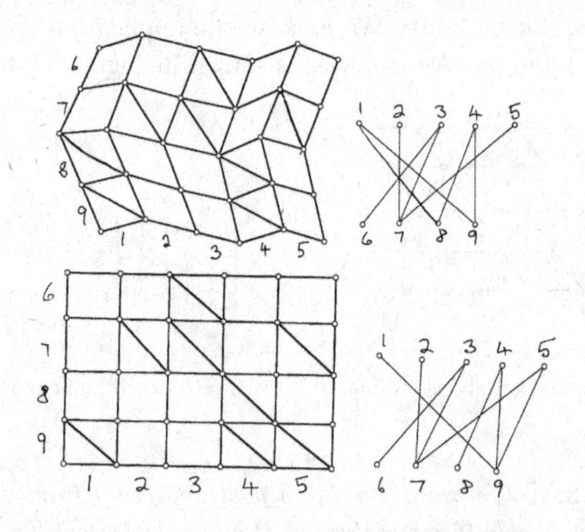

Fig. 11.6 A roof that is not rigid and one that is rigid

Proof. First we suppose that the bracing graph is connected. We consider any two bars of the roof grid, one is a cross-bar of the line i, and the other is a cross-bar of the line j. If $i = j$, then any rotation through an angle θ of one bar forces a rotation through θ of the second bar as we outlined above. If $i \neq j$, then there is a path in the bracing graph from i to j. For any edge $\{u, v\}$ in this path the lines u and v of the roof grid meet in a braced square and so each cross-bar of the line u turns through the same angle as does each cross-bar of the line v. Applying this argument sequentially to the edges of the path we deduce that any two bars of the roof grid turn through the same angle. In other words, the grid is rigid.

Conversely, we suppose that the roof grid is rigid. If there is no path from the line i to the line j of the roof grid, then a rotation of any cross-bar of the line i gives rise to a motion of the roof grid that does not rotate the cross-bars of the line j, and the grid deforms.

Therefore the roof is rigid if and only if the bracing graph is connected. We apply Lemmas 11.2.3 and 11.2.1 to complete the proof. □

The roof of the warehouse shown in Figure 11.4 is not rigid. In Figure 11.6 we have added a brace in two ways. In the first the resulting grid is still not rigid, in the second the resulting grid is rigid. Next we suppose that each wall is only permitted to move within its vertical plane. Then a diagonal brace in a wall prevents movement of the wall within its plane. Two such braces, one in each of two intersecting walls as in Figure 11.7, keep both walls rigid. Consequently, a brace in each of the four external walls will keep the four corners of the roof in place, if the roof is rigid within its horizontal plane. These four braces, together with bracing in the roof sufficient to keep it rigid within its plane, will suffice to keep the building frame rigid. So we know that $k + \ell + 3$ braces, properly placed, will keep the building in shape. The bracing shown in Figure 11.4 is not adequate to make the frame of the building rigid. We can make it rigid by adding one more horizontal brace in, for example, the lower right-hand square of the roof grid.

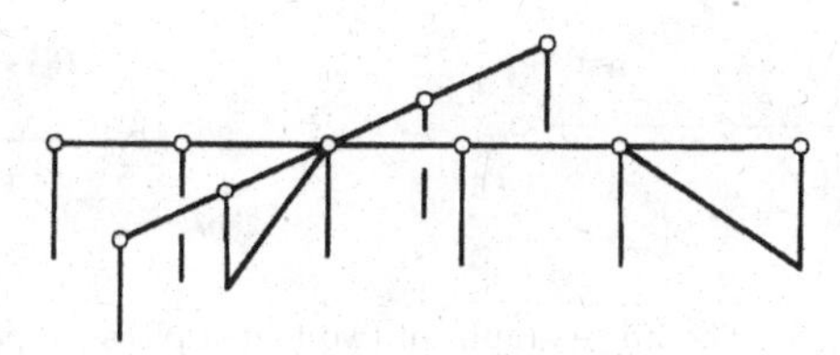

Fig. 11.7 Bracing intersecting walls

11.4 Hinge graphs and developments

In this section, we associate a graph with each hinged-panel model **M**. Consequently the existence and properties of maximal trees enable us to rigorously expand our intuitive notion of developments of a hinged-panel model.

Definition 11.4.1. *Let* $\mathbf{M}$ *be a hinged-panel model. The* **vertices** *of the hinge graph* $H(\mathbf{M})$ *of* $\mathbf{M}$ *are the panels of* $\mathbf{M}$. *A pair* $\{F, G\}$ *of panels* F *and* G *is an* **edge** *of* $H(\mathbf{M})$ *if* F *and* G *are hinged in* $\mathbf{M}$.

From Condition **H1** of Definition 10.1.2 we also have that each hinge graph is a connected graph. Condition **H2** tells us that each hinge graph is simple. We further note that a hinged-panel model is a panel cycle if and only if its hinge graph is a cycle.

Example 11.4.1. The hinge graph of the folded paper of Example 10.1.1 is drawn first in Figure 11.8. The hinge graph of the folded paper of Example 10.1.2 is drawn second. In each case the graph does not depend on how tightly the paper is folded, but on the number and interrelation of the creases alone.

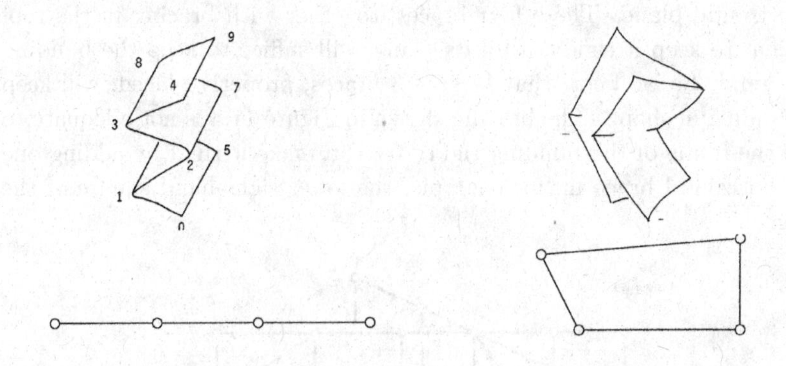

Fig. 11.8 The hinge graphs of two folded pieces of paper

Example 11.4.2. Figure 10.2 shows the three models obtained in the process of assembling a cube. Figure 10.3 gives four models obtained during construction of a tetrahedral model. We show, in Figure 11.9, the hinge graph of each of these seven hinged-panel models.

Exercise 11.4.1. Draw the hinge graph of each of the three polyhedral models of Constructions 10.4.1, 10.4.2, and 10.4.2.

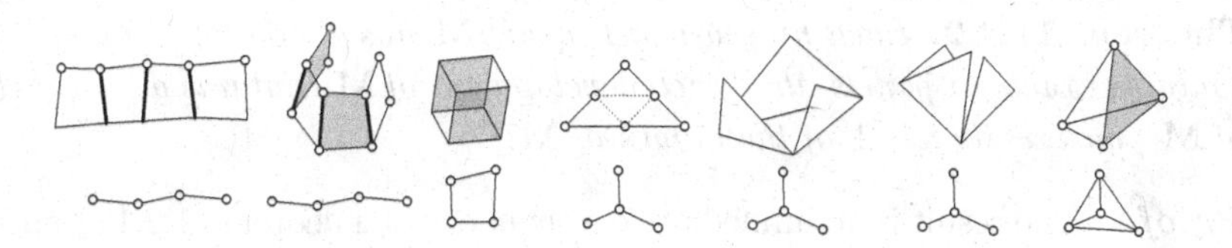

Fig. 11.9 The hinge graphs of seven models

Our proofs of the conditions that govern the mechanical freedom of cycles use models obtained by removing some panels. We now see exactly when this can be done. Let **M** be a hinged-panel model. Then any hinged-panel model **M**′, with its list of panels a sublist of the panels of **M** and with each hinge also a hinge of **M**, is *contained in* **M**.

Theorem 11.4.1. *Let* **M** *be a hinged-panel model. Then any connected subgraph of* $H(\mathbf{M})$ *is the hinge graph of a model contained in* **M**. *Conversely, the hinge graph of any model contained in* **M** *is a connected subgraph of* $H(\mathbf{M})$.

Proof. First we suppose that K is a connected subgraph of $H(\mathbf{S})$. We let $\mathbf{M}'$ consist of those panels of **M** that are vertices of K, hinged only if they are joined by an edge of the subgraph K. Then, for each pair of vertices u and v of K, the existence of a path joining u and v in K guarantees that the vertices of this path and the hinges that give rise to the edges of the path satisfy Condition **H1** of Definition 10.1.2. As each edge of K is also an edge of $H(\mathbf{M})$, and so represents a hinge of **M**, Condition **H2** is satisfied for $\mathbf{M}'$ and it is a hinged-panel model of K, $\mathbf{M}' = \mathbf{M}(K)$.

Conversely, if $\mathbf{M}'$ is contained in **M**, then $H(\mathbf{M}')$ is a subgraph of $H(\mathbf{M})$, and, from Condition **H1** applied to $\mathbf{M}'$, we have that $H(\mathbf{M}')$ is connected. □

Definition 11.4.2. *A hinged-panel model* **D** *is a* **development** *if the hinge graph* $H(\mathbf{D})$ *is a tree. A development of a hinged-panel model* **M** *is any development* **D** *so that* $H(\mathbf{D})$ *is a maximal tree in* $H(\mathbf{S})$.

We have already used the word "development" in the above sense, but without formal definition, in Exercise 2.5.1, Construction 10.3.4, Construction 10.4.3, and elsewhere. The most common example of a development is the usual carpenter's hinge consisting of two panels hinged together. Some graph theory gives us useful properties of the developments of a hinged-panel model.

Theorem 11.4.2. *Each hinged-panel model* $\mathbf{M}$ *has a development. If* $\mathbf{M}$ *contains exactly* n *panels, then each development of* $\mathbf{M}$ *contains all* n *panels of* $\mathbf{M}$ *and exactly* $n-1$ *of the hinges of* $\mathbf{M}$.

Proof. The result is an immediate consequence of Theorem 11.2.1 applied to the trees of $H(\mathbf{M})$. □

Exercise 11.4.2. A set of unfolded developments of the sealed box of Exercise 2.5.2, and the hinge graph of the box, are shown in Figure 11.10. Comment on these developments and their hinge graphs in the light of Theorems 11.2.1 and 11.4.2.

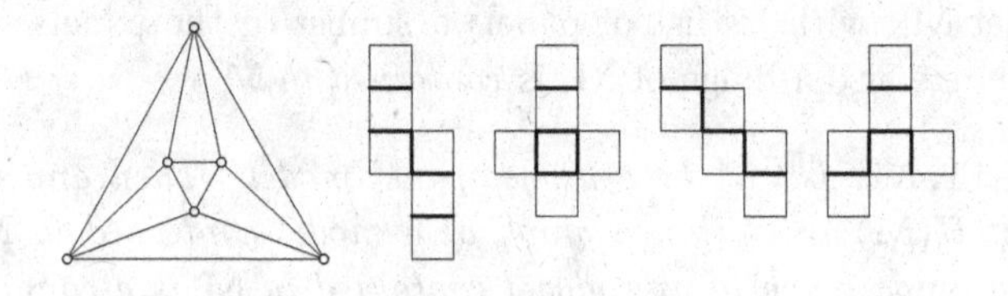

Fig. 11.10 Some unfolded developments of a box

The Hooke universal joint and associated constant-velocity joint of Chapter 10 are developments, and the mobility they display is characteristic of every development.

Definition 11.4.3. *A* **flap** *of a hinged-panel model* $\mathbf{M}$ *is a panel of* $\mathbf{M}$ *that belongs to exactly one hinge.*

Practical experience with the lids of cardboard boxes suggests that a flap is free to move about its one hinge. The freedom of a flap is basic to the extreme flexibility of developments, as we now prove.

Theorem 11.4.3. *Each hinge of a development is free to move. Consequently any development of a hinged-panel model* $\mathbf{M}$ *is a mechanism that has a collapsed position as well as each position of* $\mathbf{M}$ *itself.*

Proof. We see directly that any development of $\mathbf{M}$ can take a position of $\mathbf{M}$ itself by removing those hinges of $\mathbf{M}$ that are not hinges of the development. This gives the development in the required position.

We complete the proof by arguing inductively on the number of hinges of the development. The result is true for the carpenters hinge. Suppose it is true that both each hinge of a development containing at most n hinges moves freely and the development has a collapsed position. Let $\mathbf{D}$ be any development containing $n+1$ hinges. From Lemma 11.2.2 applied to the empty path in $H(\mathbf{D})$, we have that the hinge graph of $\mathbf{D}$ contains a vertex belonging to one edge of the graph. This vertex is a flap of $\mathbf{D}$. Removing it and its hinge from $\mathbf{D}$ gives a development in which each hinge moves freely. Clearly a flap of a model may hinge freely relative to the remainder of the model. Therefore each hinge of $\mathbf{D}$ moves freely as required. Attaching the flap to the remainder in a collapsed position gives a collapsed position for $\mathbf{D}$. This completes the proof that each hinge of a development is free to move and the development has a collapsed position. □

In fact the way that a development moves through its range of positions mimics the way in which we usually assemble a model $\mathbf{M}$, starting with a collapsed development, that is often unfolded, and adding hinges as the panels are folded into their final positions. Theorem 11.4.2 guarantees that each hinged-panel model can be made in this way. An unfolded development is a convenient starting point, as it may be cut from one sheet of paper or cardboard that is then creased to form the hinges.

Construction 11.4.1. A convenient model dodecahedron can be made by combining two copies of the development in Figure 11.11. Twisting a rubber band about both as shown causes the model to pop up to an assembled position. But it can be flattened easily by gentle pressure for storage.

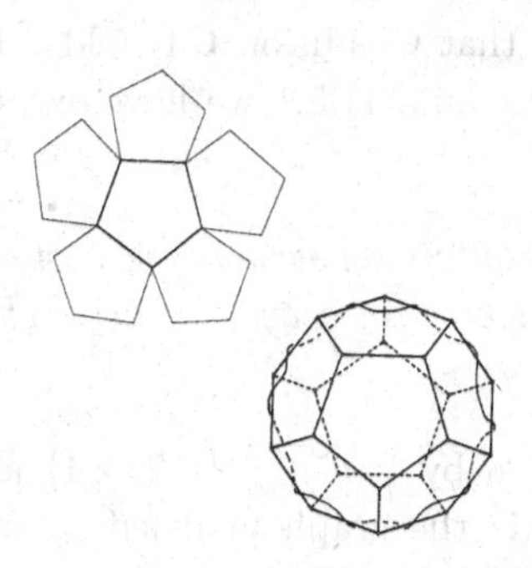

Fig. 11.11 A model dodecahedron

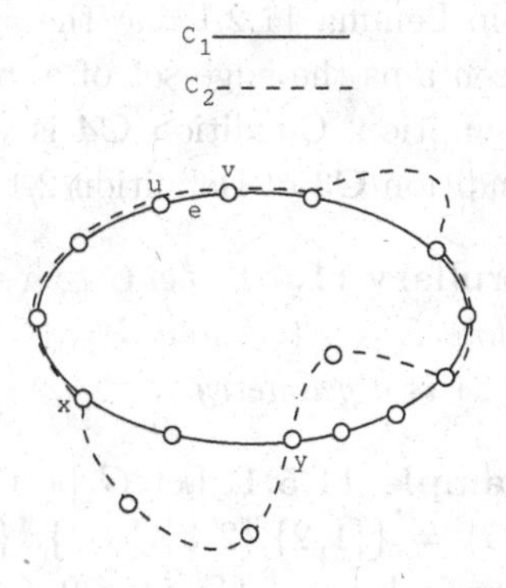

Fig. 11.12 A cycle avoiding e

11.5 Graphs that are also geometries

In this section, we highlight a common structure that underlies both geometries and graphs. The key to this structure is the following observation about the union of the edge-sets of two cycles of a graph.

Lemma 11.5.1. *Let G be a graph. Suppose further that E_1 and E_2 are the edge sets of distinct cycles C_1 and C_2, respectively, of G, and that e is an edge in both E_1 and E_2. Then there is a subset of $(E_1 \cup E_2) - \{e\}$ that is the edge set of a cycle of G.*

We omit the proof of this lemma and note that Figure 11.12 suggests that it is not difficult to find such a cycle avoiding the edge e mentioned in the statement of the lemma.

We recall that when testing a collection $\mathcal{C}$ of subsets of a set E as the collection of circuits of a geometry $(E, \mathbf{C})$, the requirement that is often the most troublesome is the Elimination Condition, **C3**, of Definition 2.1.1. The form of Lemma 11.5.1 suggests the edge set of a graph as a possible example of the set of points of a geometry, and we see that this is sometimes the case:

Theorem 11.5.1. *Let G be a simple graph in which each five-element subset of $E(G)$ contains the edge-set of a cycle. Define $E = E(G)$ and $\mathcal{C} = \{C : C$ is the edge set of a cycle of $G\}$. Then $(E, \mathcal{C})$ is a geometry.*

Proof. Removing any edge from a cycle C leaves the edge set of an acyclic graph. So no proper subset of a cycle is a cycle. Thus Condition **C2** is valid. From Lemma 11.2.1 and the assumption that each five-element subset of E contains the edge set of a cycle we have that Condition **C1** holds. By assumption, Condition **C4** is valid. From Lemma 11.5.1 we have exactly Condition **C3** of Definition 2.1.1. □

Corollary 11.5.1. *Let G be a simple graph containing at most five vertices. Define $E = E(G)$ and $\mathcal{C} = \{C : C$ is the edge set of a cycle of $G\}$. Then $(E, \mathcal{C})$ is a geometry.* □

Example 11.5.1. Let G be the graph given by $V(G) = \{1, 2, 3, 4\}$ and $E(G) = \{\{1, 2\}, \{2, 3\}, \{3, 4\}, \{1, 4\}\}$. This is the graph modeled by any 4-bar linkage of Chapter 9. Applying Corollary 11.5.1, we obtain that $E = \{\{1, 2\}, \{2, 3\}, \{3, 4\}, \{1, 4\}\}$ and $\mathcal{C}$ contains just one member, namely the set $\{\{1, 2\}, \{2, 3\}, \{3, 4\}, \{1, 4\}\}$. For convenience we relabel each member of E by a single symbol, $E = \{a, b, c, d\}$ and $\mathcal{C} = \{\{a, b, c, d\}\}$. Now we recognize this geometry. It is the figure of four coplanar points in general position shown in Figure 11.13.

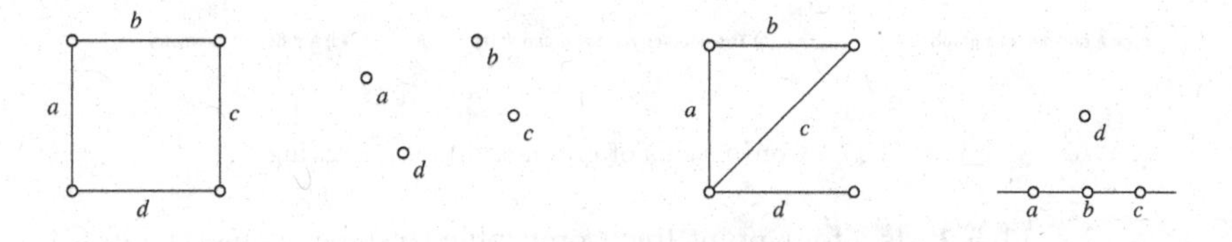

Fig. 11.13 A graph and four points in a plane

Fig. 11.14 Another graph and its associated figure

Example 11.5.2. Let G be the graph given by $V(G) = \{1, 2, 3, 4\}$ and $E(G) = \{\{1, 2\}, \{2, 3\}, \{1, 3\}, \{1, 4\}\}$. Applying Corollary 11.5.1, we obtain the set $E = \{\{1, 2\}, \{2, 3\}, \{1, 3\}, \{1, 4\}\}$ of points, the set $\mathcal{C}$ of circuits containing just one member, namely the set $\{\{1, 2\}, \{2, 3\}, \{1, 3\}\}$. Again relabeling, we have $E = \{a, b, c, d\}$ and $\mathcal{C} = \{\{a, b, c\}\}$. This is the 4-point figure containing one 3-point line drawn in Figure 11.14.

Example 11.5.3. We have already met a geometry isomorphic to that obtained from K_3. It is the three-point line. We have also met a geometry obtained from K_4, it is drawn in Figure 11.15.

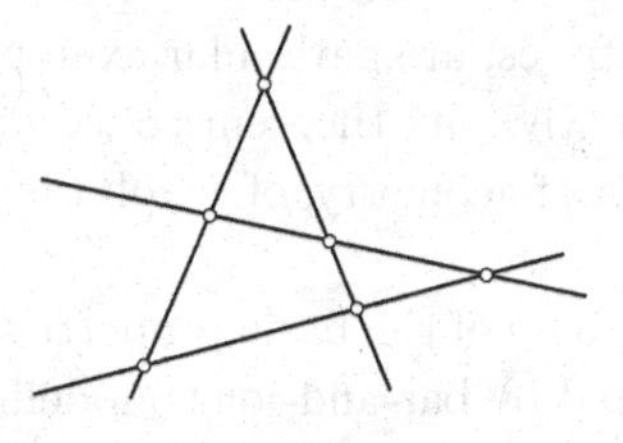

Fig. 11.15 A figure that is isomorphic to the geometry of K_4

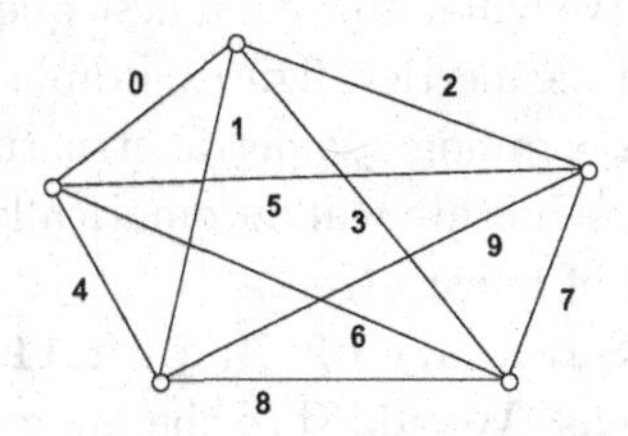

Fig. 11.16 A complete graph with labeled edges

We need to be a little careful when drawing graphs and their associated geometries. For example, the first illustration in Figure 11.17 is a drawing of a graph. The graph defines the figure in the second illustration. Our usual notation for a triangle, in the third illustration in Figure 11.17, distinguishes it from a drawing of the graph.

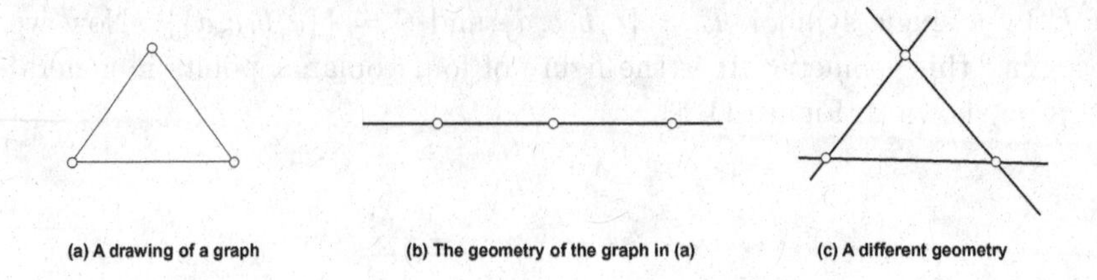

Fig. 11.17 Conventions of graph and figure drawing

Exercise 11.5.1. Is a four-point line isomorphic to some geometry $(E, \mathbf{C})$, where G is a graph, $E = E(G)$ and $\mathcal{C} = \{C : C$ is the edge set of a cycle of $G\}$?

Problem 11.5.1. *Here is a beautiful question! Have we previously met the geometry $(E, \mathcal{C})$ obtained from the complete graph on five vertices, where $E = E(K_5)$ and $\mathcal{C} = \{C : C$ is the edge set of a cycle of $K_5\}$? If so what is it?*

Solution. An answer springs to mind if we compare the three-point circuits listed in Definition 2.3.3 with the list of members of $\mathcal{C}$ above, labeled according to the labeling of edges of K_5 shown in Figure 11.16. But caution! We should verify that the lists of all circuits are identical, not just the lists of three-point members. We conclude that we have a non-planar Desargues figure.

We will not pursue these questions further now, but in Chapter 14 we see that geometries, figures, graphs and even matrices, are particular examples of a common geometric structure. Consequently, any theorem concerning this structure will automatically be a theorem of geometry, of graph theory and of linear algebra.

Summary of Chapter 11. We derived some of the basic properties of graphs. We added to the understanding offered by bar-and-joint modeling, with which we were already familiar, by an application to decide the rigidity of one-story building frames. We drew graphs in a Euclidean plane, and used a graph naturally associated with each hinged-panel model, namely its hinge graph, to define developments of the model. Finally, we found an intriguing glimpse of an underlying common structure of graphs and combinatorial geometries.

Chapter 12

SPHERICAL POLYHEDRA

We begin by examining figures that have polyhedral hinged-panel models. We introduce the notion of a drawing of a polyhedron by using selected drawings of an associated graph. We examine spherical polyhedra in detail, obtaining information about them by subjecting some of their drawings, rather than the polyhedra themselves, to scrutiny. These drawings are planar embeddings of three-connected planar graphs, and enable us to count the vertices, edges, and faces of any spherical polyhedron. Finally we apply our results in some detective work on a stone block pictured in Albrecht Dürer's "*Melencolia*".

12.1 The definition of a polyhedron

We introduced polygonal regions in Definition 10.1.1. We may think of each hinged-panel model of Chapter 10 as modeling a figure that consists of a union of polygonal regions. In this section, we single out for attention figures that have polyhedral models. Our main weapon in this attack is our knowledge of the structure of a graph that we associate with each figure.

Definition 12.1.1. *A* **polyhedron F** *is a union of pairwise distinct polygonal regions, each called a face of the polyhedron, satisfying the requirement that each edge of a face is an edge of exactly two faces.*

We consider only polyhedra **F** that are connected in the sense that for each two faces F, G of **F** there is a sequence, $F = F_1, E_1, F_2, E_2, F_3, \ldots, F_{m-1}, E_{m-1}, F_m = G$, of faces and hinges so that E_i is the edge common to F_i and F_{i+1}, for each i in $\{1, 2, \ldots, m-1\}$. A *vertex* of a polyhedron **F** is a vertex of any of its faces, and an *edge* of **F** is an edge of any of its faces. The *skeleton* of $\mathbf{F}$ is the union of the edges of $\mathbf{F}$.

Clearly polyhedra are figures that are modeled by the polyhedral models of Chapter 10, and we write $\mathbf{M}(\mathbf{F})$ to mean any model of $\mathbf{F}$. Panels F and G of $\mathbf{M}(\mathbf{F})$ stand for the faces that are the polygonal regions modeled by F and G respectively, and a hinge $H_{inge}(F, E, G)$ corresponds to the edge contained in both these faces. Thus, for example, the results of Constructions 10.4.1, 10.4.2, and Exercise 10.4.2 model polyhedra. Every polyhedron has a model but a polyhedral mechanism in a position, perhaps collapsed, such that two hinges coincide, or two panels coincide, is not modeling a polyhedron at that moment of coincidence. Moreover, there are other examples of polyhedral models that do not model polyhedra as we see in Example 12.1.1.

Example 12.1.1. The polyhedral model defined in Example 10.4.1 does not model a polyhedron.

Let two tetrahedral models touch along a common edge E. Suppose that $H_{inge}(F_1, E, F_2)$ is a hinge of the first and $H_{inge}(F_1', E, F_2')$ a hinge of the second model. Then the collection of the panels of both, together with the hinges of both - with the exception that the two above are replaced by $H_{inge}(F_1, E, F_1')$ and $H_{inge}(F_2', E, F_2)$ - is a polyhedral model as shown in Figure 12.1. This polyhedral mechanism does not model a polyhedron.

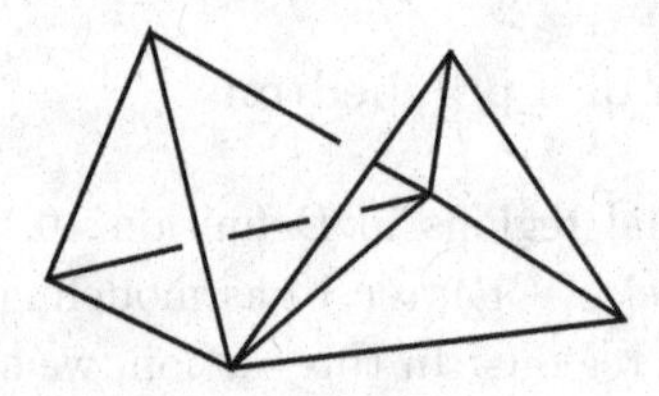

Fig. 12.1 A model of the union of two tetrahedra

Our discussion of perspective drawings and scene analysis has shown us both the advantages and difficulties of using planar information to analyze non-planar figures. We use the previously defined notion of a drawing of a graph in order to pictorially represent polyhedra by associating the following simple connected graph with a polyhedron $\mathbf{F}$.

Definition 12.1.2. *Let* $\mathbf{F}$ *be a polyhedron. The* **vertices** *of the graph of* $\mathbf{F}$*, denoted by* $G(\mathbf{F})$*, are the vertices of* $\mathbf{F}$*. A pair* $\{u, v\}$ *of vertices is an* **edge** *of* $G(\mathbf{F})$ *if* $[uv]$ *is an edge of* $\mathbf{F}$.

The skeleton of any hinged-panel model $\mathbf{M}(\mathbf{F})$ may be thought of as a bar-and-joint model of the graph $G(\mathbf{F})$. The result of applying Constructions 10.3.3, 10.3.4, and 10.3.5 is in each case a skeleton of an octahedral model of an octahedron. The common labeling of the vertices in Figures 10.20 and Figure 10.22 demonstrates that in each case the graph modeled is the same.

We already conveniently represent any graph by a drawing as specified in Definition 11.1.4. We could use a drawing of $G(\mathbf{F})$ in order to represent the polyhedron $\mathbf{F}$, bearing in mind that ideally we would like the representation to be a perspective drawing of the skeleton of $\mathbf{F}$. We explore the possibilities of this in Chapter 13. However perspective rendition is neither quick nor easy in practice. Let p be a general viewpoint of the set of vertices of a polyhedron $\mathbf{F}$. Then, in any perspective drawing of $\mathbf{F}$ that is drawn from p, the image of each face is a polygonal region. This suggests that we only allow drawings of $G(\mathbf{F})$ in which the boundary of a face appears as the boundary of a polygonal region.

Definition 12.1.3. *Let* $\mathbf{F}$ *be a polyhedron. Then any drawing of* $G(\mathbf{F})$*, in which each arc is a segment and in which the boundary of each face of* $\mathbf{F}$ *is drawn as a boundary of a polygonal region, is a* **drawing** *of the polyhedron. We say that the face has the polygonal region as its image.*

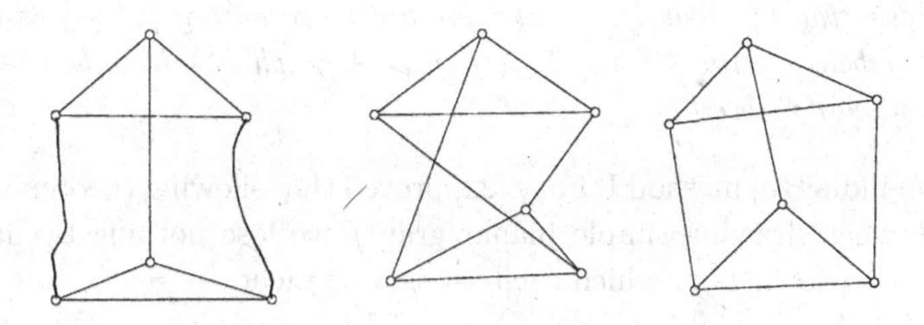

Fig. 12.2 Three attempts, one successful, at drawing a polyhedron

Even a drawing that fails to be a perspective drawing of the skeleton of a polyhedron $\mathbf{F}$ may efficiently convey information about $\mathbf{F}$ to the viewer. However we must be cautious. As we discovered in examining the Penrose "triangle" in Problem 7.2.1, a convincing planar figure may not be the scene it purports to be. To be sure of a polyhedron we need to verify its existence, perhaps by identifying it as the subject of a perspective drawing, perhaps in some other way.

Unfortunately a drawing may be of two "different" polyhedra. For example, we saw in Exercise 10.4.1 that the octahedron and the heptahedron modeled in Constructions 10.4.1 and 10.4.2, respectively, have the same graph. Therefore a drawing of one may be a drawing of the other. But no matter how the vertices of the octahedron and heptahedron are labeled the two cannot have identical lists of faces, as one contains eight faces and the other contains only seven. This raises the question of the meaning of "different" in the context of polyhedra and we examine it further during our later investigation of the existence of polyhedra.

12.2 Planar graphs and three-connected graphs

In this section, we characterize those graphs that have drawings appropriate to an elegant class of polyhedra that has long engaged the attention of geometers. The drawings are traditionally called planar embeddings, their arcs meeting only at ends, thereby giving a convenient partition of the Euclidean plane into regions that correspond to faces of polyhedra.

Definition 12.2.1. *A* **planar embedding** *of a graph G is a drawing of G so that, for each pair $\{\{u,v\},\{u',v'\}\}$ of edges, the intersection of the arc representing the edge $\{u,v\}$ and the arc representing $\{u',v'\}$ is the set of points labeled by the set $\{u,v\}\cap\{u',v'\}$. A graph is said to be* **planar** *if it has a planar embedding*

By an inductive method I. Fáry [22] proved the following theorem, showing that when drawing simple planar graphs we lose nothing by limiting ourselves to drawings in which each arc is a segment.

Theorem 12.2.1. *Let G be a simple planar graph. Then G has a planar embedding in which each arc is straight.*

Lemma 12.2.1. *In a planar embedding of a simple graph G each cycle is drawn as the boundary of a polygonal region.*

Proof. Suppose that v_1, v_2, ..., v_n are the vertices, in cyclic order, of a cycle of G. From the definition of planar embedding we have that no two segments touch or cross, and so $[v_1v_2]\cup[v_2v_3]\cup\ldots\cup[v_{n-1}v_n]\cup[v_nv_1]$ is the boundary of one polygonal region $[v_1v_2\ldots v_n]$. □

Consequently any planar embedding of the graph $G(\mathbf{F})$ of a polyhedron $\mathbf{F}$ is also a drawing of the polyhedron, as each cycle has a polygonal region boundary as its image.

We see from Figure 11.1 that each graph K_n is planar for $n = 1, 2, 3$, and 4. Each subgraph of a planar graph is planar, and it is straightforward to prove inductively that each tree is planar. Each planar embedding of the graph of a polyhedron $\mathbf{F}$ is automatically a drawing of the polyhedron. But not every graph is planar, as we now prove:

Lemma 12.2.2. *The graph K_5 is not planar.*

Proof. Suppose that K_5 had a planar embedding. Then, in this embedding, one of the triangular regions [123], [124], and [125] would have a vertex in the interior and a vertex in the exterior. The arc joining these two vertices would contradict the requirements of a planar embedding. So no such planar embedding of K_5 exists. □

Exercise 12.2.1. Make two drawings of the graph of a tetrahedron, one a planar embedding and the other not. Make three drawings of the graph of a cube, one not a drawing of the cube, one a drawing of the cube but not a planar embedding of the graph, and the last a planar embedding of the graph. Verify, by experimenting, that the graph of the nonahedron of Figure 12.3 has a planar embedding.

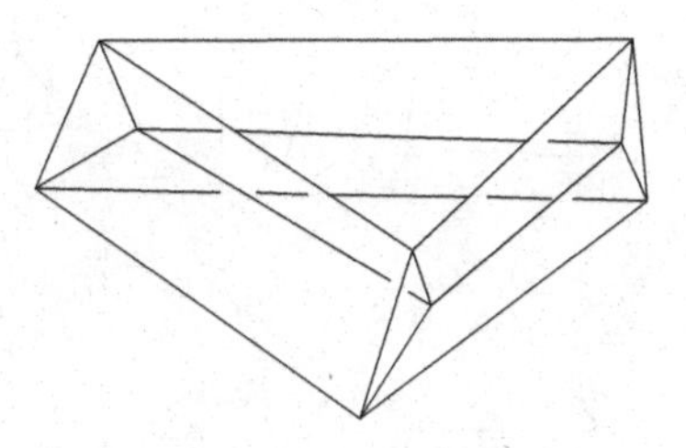

Fig. 12.3 A nine-faced polyhedron

Any planar embedding of a graph G partitions the points, not belonging to any arc, of the Euclidean plane of the embedding as follows. We call two points of the plane that are not on an arc equivalent if they are the same point or if there is an arc joining one to the other, in the Euclidean plane, that does not intersect any arc of the embedding. It is straightforward to prove that this is an equivalence relation, and therefore partitions the points that are not on any arc into disjoint subsets of equivalent points. Exactly one of these subsets is unbounded. By a slight misuse of terminology we call the union of any bounded subset and its boundary a *face* of G. We also call the set of all those Euclidean points not in the unbounded subset a *face* of G.

Definition 12.2.2. *Let a simple graph G have at least $n+1$ vertices. Then G is n-***connected** *if each deletion of at most $n-1$ vertices from G leaves a connected subgraph.*

A graph is 1-connected if and only if it is connected. An n-connected graph is also $(n-1)$-connected. Each tree is connected but not 2-connected. If G is a cycle and $|V(G)| \geq 3$, then G is 2-connected. Each complete graph K_n, with $n > 1$, is $(n-1)$-connected.

Example 12.2.1. The graph G of a polyhedron is drawn twice in Figure 12.4. The first drawing is not a planar embedding of G, but the second is. We see, from either drawing, that deleting a single vertex leaves a connected graph. Deleting a pair of vertices leaves one of the four graphs shown, and as these are also connected we deduce that G is 3-connected.

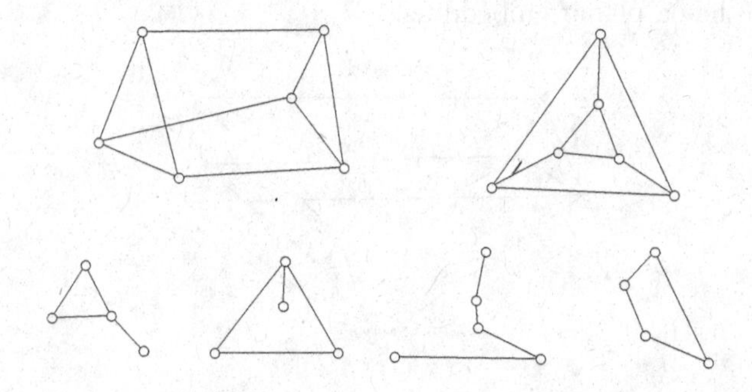

Fig. 12.4 A graph and some subgraphs

Exercise 12.2.2. Use a common vertex labeling for each of the four drawings in Figure 12.5 and so verify that each is a drawing of the same graph G. Prove that this graph is both planar and 3-connected.

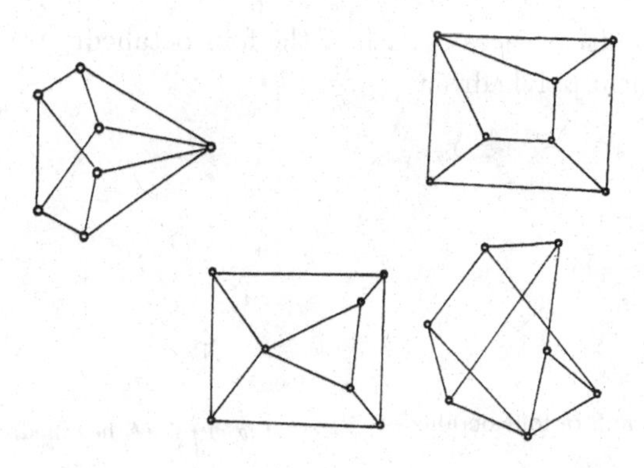

Fig. 12.5 Drawings of a graph

12.3 The definition of a spherical polyhedron

In this section, we specify spherical polyhedra by means of their graphs. The definition uses only graphs that are three-connected in order to ensure that the class contains only polyhedra, exactly those that have traditionally been the subject of study by geometers.

Definition 12.3.1. *Let the graph of a polyhedron* **F** *be 3-connected. Suppose that* $G(\mathbf{F})$ *has a planar embedding in which the faces of* $G(\mathbf{F})$ *are exactly the images of the faces of* **F**. *Then we call* **F** *a* **spherical polyhedron.**

From Definition 11.1.4, the vertices of the image of any cycle of $G(\mathbf{F})$ in an embedding of the graph have the same labels as the vertices of the cycle. Therefore the boundary of any face $[v_1v_2\ldots v_n]$ is drawn as the boundary of a face of the embedding if and only if $[v_1v_2\ldots v_n]$ is a face of the embedding. In testing that a polyhedron **F** is a spherical polyhedron we therefore check a suitable planar embedding of $G(\mathbf{F})$ in order to verify that a list of the faces of $G(\mathbf{F})$ is also a list of the faces of **F**.

We examined four octahedral models in Chapter 10, a convex model and three Bricard models. From their skeletons, given by Constructions 10.3.3, 10.3.4, and 10.3.5, and labeled in Figures 10.20 and 10.22, we saw that each of the four octahedra modeled has the same graph G. A planar embedding of G is drawn in Figure 12.6. An exhaustive test by deleting pairs of vertices reveals only connected subgraphs, proving that G is 3-connected. A list of faces of the embedding is $[ace]$, $[abc]$, $[edf]$, $[aef]$, $[abf]$, $[bdf]$, $[bcd]$ and $[cde]$.

As this is also a list of faces of each of the four octahedra we conclude that each is a spherical polyhedron.

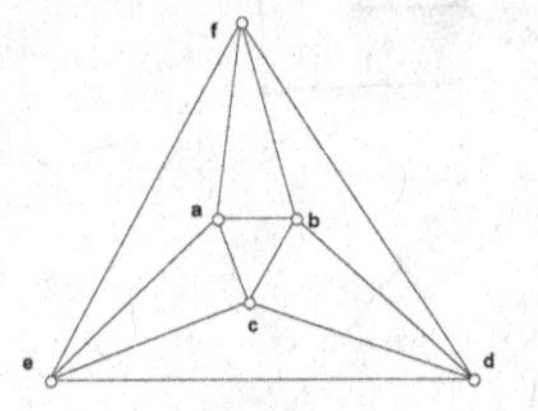

Fig. 12.6 The graph of four octahedra

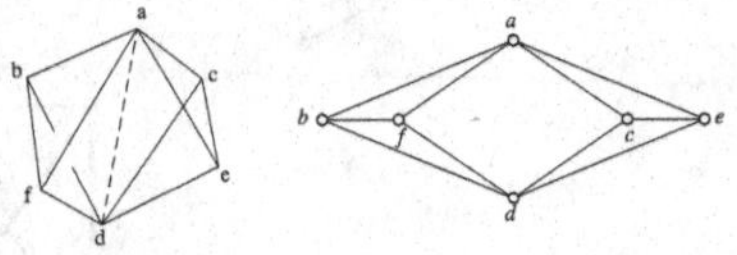

Fig. 12.7 A hexahedron and its graph

In Exercise 10.4.1 we proved that the heptahedron modeled in Exercise 10.4.2, and pictured in Figure 10.28, has the same graph as the octahedra above. Therefore it has a 3-connected planar graph. But the faces of the heptahedron are $[def]$, $[bcd]$, $[abf]$, $[ace]$, $[abde]$, $[acdf]$, and $[bcef]$. Not all the faces in this list are in the list of eight faces that arise from the embedding of G in Figure 12.6. Therefore we are unable to deduce from this embedding that the heptahedron is a spherical polyhedron.

Figure 12.7 contains a planar embedding of the graph G_1 of the hexahedron modeled in Exercise 10.4.2. The list of faces of the embedding of G is exactly the list of faces of the hexahedron, not forgetting the face $[abde]$. But deleting vertices a and d produces a disconnected subgraph of G_1, proving that G_1 is not 3-connected. We conclude that this hexahedron is not a spherical polyhedron.

12.4 Vertices, edges, and faces of a spherical polyhedron

In this section, we obtain relations between the numbers of vertices, edges and faces of each spherical polyhedron by counting the vertices, arcs and faces of a planar embedding of its graph. The 18th Century Prussian mathematician, Leonhard Euler, first proved the following result:

Theorem 12.4.1. (Euler's Formula) *Let* **F** *be a spherical polyhedron that contains exactly v vertices, e edges, and f faces. Then $v + f = e + 2$.*

Proof. Consider any planar embedding of the graph of the spherical polyhedron. This drawing of a planar 3-connected graph has v vertices, e arcs and divides its plane into f regions. If any bounded face is non-triangular we add an arc joining two non-adjacent vertices as shown in Figure 12.8.

Fig. 12.8 Changes to the embedding of the graph of a spherical polyhedron

This increases the number of regions by one, increases the number of arcs by one, and does not change the number of vertices. Repeating this process if necessary we obtain a drawing in which each bounded region is bounded by three arcs. The value of $v + f - e + 2$ is the same for the new drawing as it was for the original.

If the drawing contains more than one triangle, then we delete an arc bounding the unbounded region, decreasing f by one and decreasing e by one. If the drawing contains a vertex that belongs to exactly one arc, then we delete it and the arc, reducing each of v and e by one. We continue with a sequence of these two processes, giving precedence to the second, until we are left with a triangular drawing. Each deletion leaves $v + f - e - 2$ unchanged. Thus for the original drawing, and also for the spherical polyhedron, $v + f - e - 2 = 3 + 2 - 3 - 2 = 0$. □

Example 12.4.1. There is no spherical polyhedron with 11 vertices, 16 edges, and 8 faces as $11 - 16 + 8 \neq 2$. However there may be one having 11 vertices, 17 edges, and 8 faces, but we cannot yet be sure.

The numbers of vertices, edges, and faces of the octahedron modeled by Construction 10.4.1 satisfy Euler's Formula, but those of the heptahedron modeled in Construction 10.4.2 do not. Consequently it is not a spherical polyhedron - resolving a question that we were unable to answer with certainty above. Interestingly, the hexahedron modeled in Exercise 10.4.2, although not a spherical polyhedron, does satisfy Euler's Formula.

Definition 12.4.1. *The number of edges containing a vertex of a graph is the* **degree** *of the vertex in the graph and in any polyhedron having that graph.*

Clearly each vertex of a 3-connected graph has degree at least three. We call a vertex of degree three *trivalent*.

Proposition 12.4.1. *Let* **F** *be a spherical polyhedron. Suppose that* **F** *has* v *vertices,* e *edges, and* f *faces. Then* $2e \geq 3v$*, and* $2e \geq 3f$.

Suppose also that **F** *has* f_i i*-gonal faces and* v_i *vertices of degree* i*. Then* $\sum_i (6-i)f_i \geq 12$, $\sum_i (6-i)v_i \geq 12$*, and* $\sum_i (4-i)(v_i + f_i) = 8$.

Proof. We mark each edge twice, once at each end. We will count the $2e$ marks in another way. Each vertex has degree at least 3, giving at least $3v$ marks. Next we again mark each edge twice, but in a different way, near its center and once in each bordering face. Again there are $2e$ marks. As each face is bordered by at least 3 edges, there are at least $3f$ marks.

As each edge borders two faces, $\sum_i if_i = 2e$. Also $\sum_i f_i = f$. Hence $\sum_i (6-i)f_i = 6(\sum_i f_i) - \sum_i if_i = 6f - 2e = 12 - 6v + 4e \geq 12$. A similar argument gives the second inequality. Now $\sum_i (4-i)f_i = 4f - 2e$ and $\sum_i (4-i)v_i = 4v - 2e$, and adding we obtain $\sum_i (4-i)(f_i + v_i) = 4v - 4e + 4f = 8$. □

Corollary 12.4.1. *Each spherical polyhedron contains at least one face that has at most five sides. If it contains neither a triangular nor a four-sided face, then it contains at least twelve pentagonal faces. Each spherical polyhedron contains either a triangular face or a trivalent vertex.*

Proof. The first two claims follow immediately from the observations that $\sum_i (6-i)f_i$ contains at least one positive term and totals 12. The last claim follows from the requirement that $\sum_i (4-i)(f_i + v_i)$ has a positive term. □

Example 12.4.2. Any spherical polyhedron that contains exactly seven edges. would satisfy the conditions $3f \leq 14$ and $3v \leq 14$, giving $f \leq 4$ and $v \leq 4$, and $v - e + f \leq 1$. This contradicts Euler's Formula, forcing the conclusion that there is no spherical polyhedron containing exactly seven edges.

Exercise 12.4.1. Let m and n be positive integers. We call a spherical polyhedron in which each face has an m-gonal boundary and each vertex has degree n combinatorially regular. Prove that there are at most five possible pairs of m and n for which combinatorially regular spherical polyhedra exist.

The five Platonic polyhedra [18], namely the *regular tetrahedron, cube, regular octahedron, regular dodecahedron,* and *regular icosahedron* are shown in affine projection, together with planar embeddings of their respective graphs, in Figure 12.9. With some difficulty, and the possibility of experimental error, we could verify their existence by applying the techniques of Chapter 7 to these affine projections.

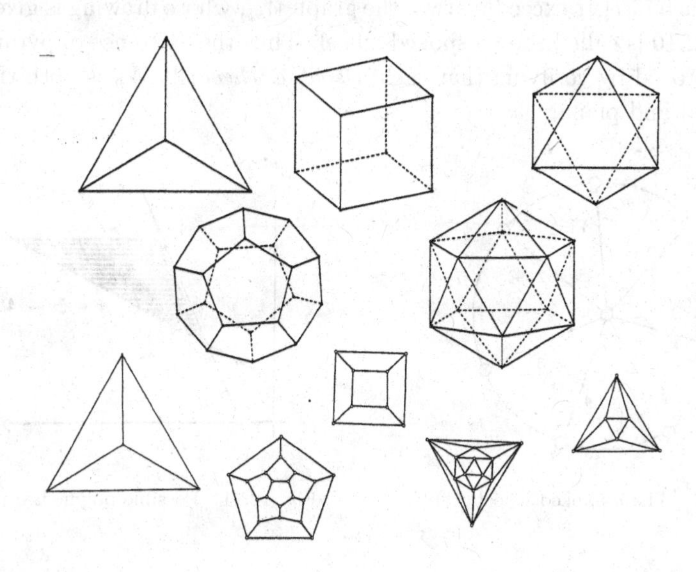

Fig. 12.9 The platonic polyhedra

12.5 The existence of spherical polyhedra

The question of the existence of the Platonic polyhedra suggests that the question of the existence of certain polyhedra, in general, may be complex.

It is easy to see that *pyramidal* polyhedra exist. Let $[a_1 a_2 \ldots a_n]$ be any polygonal region and p be a Euclidean point not in the plane of the region. Then the union of the regions $[a_1 a_2 \ldots a_n]$, $[a_1 p a_2]$, $[a_2 p a_3]$, …, $[a_{n-1} p a_n]$, and $[a_n p a_1]$ is a polyhedron.

However it is not so easy to be sure that more complicated proposals do exist. We meet again the problem of the existence of a proposed non-planar figure. Fortunately in 1922 Ernst Steinitz proved the following remarkable theorem. The proof is long and difficult [66] and we do not give it here.

Theorem 12.5.1. (Steinitz' Theorem) *Let G be a simple three-connected planar graph. Then a spherical polyhedron* **F** *exists such that*

$G = G(\mathbf{F})$, and the faces in each planar embedding of G are exactly the images of the faces of $\mathbf{F}$.

For each integer exceeding two, the graph W_n, whose drawing is given in Figure 12.10 is called the $n-$spoked wheel. Thus the existence of pyramids can be proved by verifying that each *n-spoked wheel* [49] W_n is both three-connected and planar.

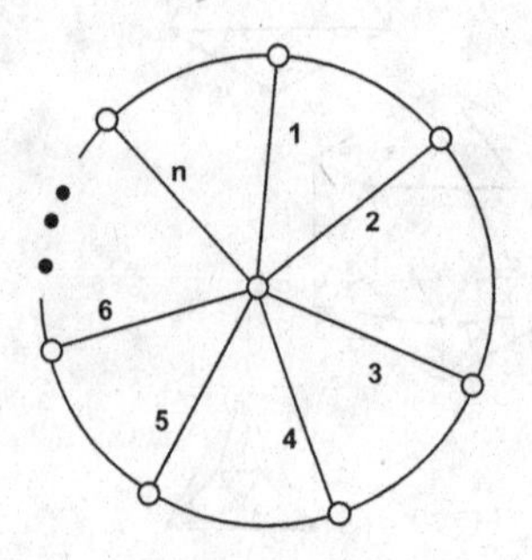

Fig. 12.10 The n-spoked wheel graph

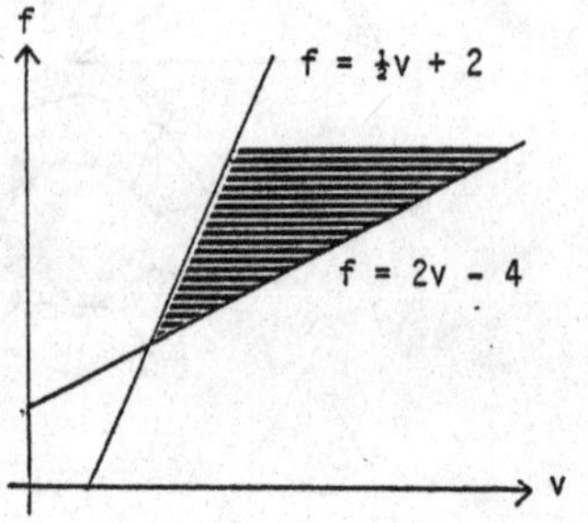

Fig. 12.11 Possible polyhedral pairs

Exercise 12.5.1. Prove, by examining the drawings of Exercise 12.2.1, that the cube and tetrahedron exist and are spherical polyhedra.

From Euler's Formula we have that any two of the numbers of vertices, edges, and faces of a polyhdedron determine the third. If a spherical polyhedron containing exactly v vertices and f faces exists then we call the ordered pair (v, f) a *polyhedral pair.* A polyhedral pair satisfies the following inequality.

Lemma 12.5.1. *If (v, f) is a polyhedral pair, then $\frac{1}{2}v + 2 \leq f \leq 2v - 4$.*

Proof. The lemma follows from Euler's Formula and the inequalities of Proposition 12.4.1. □

Consequently, points of a Euclidean plane representing polyhedral pairs must lie in the shaded portion shown in Figure 12.11. But which points in the shaded area that have integral coordinates do represent polyhedral pairs? The existence of pyramids with three-sided, four-sided, and five-sided bases guarantee that $(4, 4)$, $(5, 5)$, and $(6, 6)$ are polyhedral pairs.

There are two simple ways of obtaining new spherical polyhedra from old. The first is by capping a triangular face, that is, adding a vertex and three edges to give a spherical polyhedron with the original triangular face replaced by three new triangular faces. The new vertex is trivalent, and v is increased by 1, e by 3, and f by 2. A second way of obtaining a new spherical polyhedron from an old is by truncating at a trivalent vertex, that is, replacing a trivalent vertex by a triangular face, giving a spherical polyhedron in which the original trivalent vertex is replaced by three new trivalent vertices. The number v is increased by 2, e by 3, and f by 1. The following proposition in combination with Steinitz' Theorem proves that these two processes do give spherical polyhedra.

Proposition 12.5.1. *Let G be a three-connected graph. Suppose that u, v, and w are the vertices of a cycle $\mathcal{C}$ of G, and further that G has a planar embedding in which the image of the cycle bounds a triangular region. Then the graph obtained by adding a vertex x to $V(G)$ and edges $\{u,x\}$, $\{v,x\}$, and $\{w,x\}$ to $E(G)$ is three-connected and planar.*

Suppose that x' is a vertex of G. Then the graph obtained by replacing each edge $\{x',u_i\}$ of G by an edge $\{x_i,u_i\}$, removing x' from $V(G)$, and adding u_i to $V(G)$ is three-connected and planar.

Proof. Consider the embedding of $\mathcal{C}$. Choosing any point in the region bounded by the image of $\mathcal{C}$ to be labeled by x, we draw the required three arcs, giving a planar embedding of the new graph. In order to verify that the new graph is three-connected we need only note that the deletion of two vertices cannot disconnect x from G.

Next we consider the image of the vertex x' in any planar embedding of G. We may choose a small enough circle, centered at this image, that cuts each arc representing an edge containing x once, and no other edge. These three points of intersection will suffice as images of the points u_i, u_2, and u_3. We erase that part of the drawing inside the circle and draw three segments to complete the required embedding. Again the question of the three-connectedness of this newly drawn graph is answered by noting that deletion of any subset of the added vertices leaves a path between two vertices if there is a path between the two in $G/\{x'\}$. □

Capping a triangular face gives a polyhedral pair $(v+1, f+2)$ from a pair (v,f). Truncating at a trivalent vertex gives the pair $(v+2, f+1)$ from the pair (v,f). By combinations of truncating and capping, as in Figure 12.12, starting with each of the three pyramids above we see that each pair (u,v) of integers represented in the shaded area of Figure 12.11 is a polyhedral pair. We have proved the following theorem:

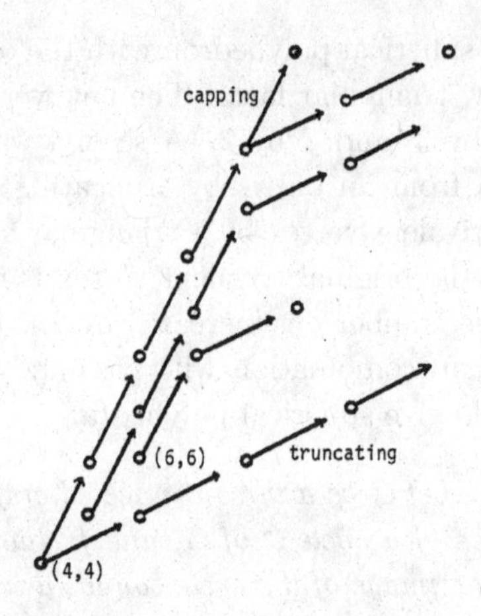

Fig. 12.12 Sequences of spherical polyhedra

Theorem 12.5.2. *Let v and f be positive integers, each at least four and satisfying $\frac{1}{2}v + 2 \leq f \leq 2v - 4$. Then there is a spherical polyhedron containing exactly v vertices, f faces, and $v + f - 2$ edges. These are the only spherical polyhedra.*

Exercise 12.5.2. Obtain the first spherical hexahedron drawn in Figure 12.13 from a tetrahedron by some combination of capping triangular faces and truncating at trivalent vertices.

Prove that the cube cannot be obtained from a pyramid by truncating at trivalent vertices and capping triangular faces.

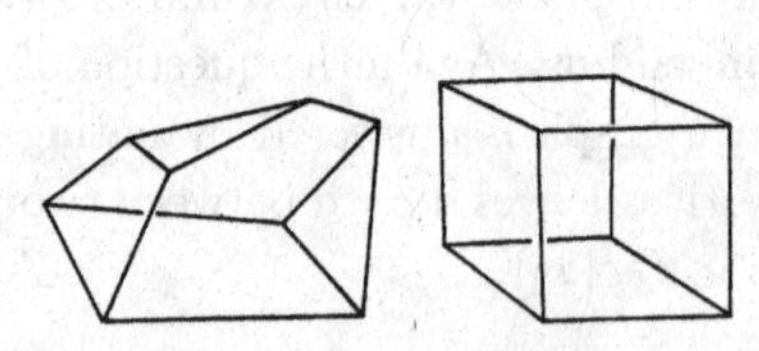

Fig. 12.13 Two spherical hexahedra

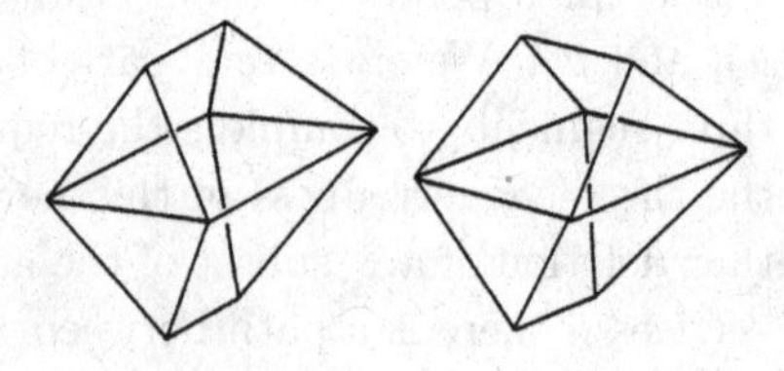

Fig. 12.14 Two non-isomorphic spherical octahedra

We described two polyhedra above as "different" because one contained only seven faces and the other contained eight faces. Even though the two polyhedra in Figure 12.13 each contain the same number of vertices, of edges, and of faces we probably feel that they are not "the same". We clarify this question, by first asking what we mean by saying that two graphs are isomorphic.

Definition 12.5.1. *Two graphs are* **isomorphic** *if the vertices of the first can be paired with the vertices of the second, and paired vertices given the same label, so that a list of edges of the first graph is identical to a list of edges of the second graph.*

Lemma 12.5.2. *Two graphs are isomorphic if and only if the two share a common drawing.*

Proof. If the two are isomorphic, then each vertex set can use the same set of labels so that a drawing of one is automatically a drawing of the other. Conversely, any common drawing induces a pairing between vertex sets that is the required isomorphism. □

Definition 12.5.2. *Two spherical polyhedra that have isomorphic graphs are* **combinatorially isomorphic**.

We note, in passing, that the name "spherical polyhedron" is used because each spherical polyhedron is isomorphic to a spherical polyhedron that is homeomorphic to a sphere. Thus the four octahedra modeled in Chapter 10 are pairwise isomorphic, but the two hexahedra of Figure 12.13 are not. The notion encompasses a qualitative similarity, as for example, models of the octahedra differ markedly in appearance and mechanical behavior but the same order of connecting edges gave skeletons of all four

Exercise 12.5.3. Prove that the two spherical octahedra drawn in Figure 12.14 are not isomorphic.

Figure 12.15 contains a list of planar embeddings of pairwise non-isomorphic spherical polyhedra, each having at most seven faces.

Exercise 12.5.4. Locate the spherical polyhedra of Figure 12.13 and those modeled by a polyhedral model of each of the roofs of Figure 10.12 in the above list.

Exercise 12.5.5. In Exercise 12.5.2 we proved that not every spherical polyhedron can be obtained from a pyramid by truncating at trivalent vertices and capping triangular faces. You might like to see how many of the ten spherical polyhedra that each have at most six faces are obtainable from pyramids by truncating at trivalent vertices and capping triangular faces.

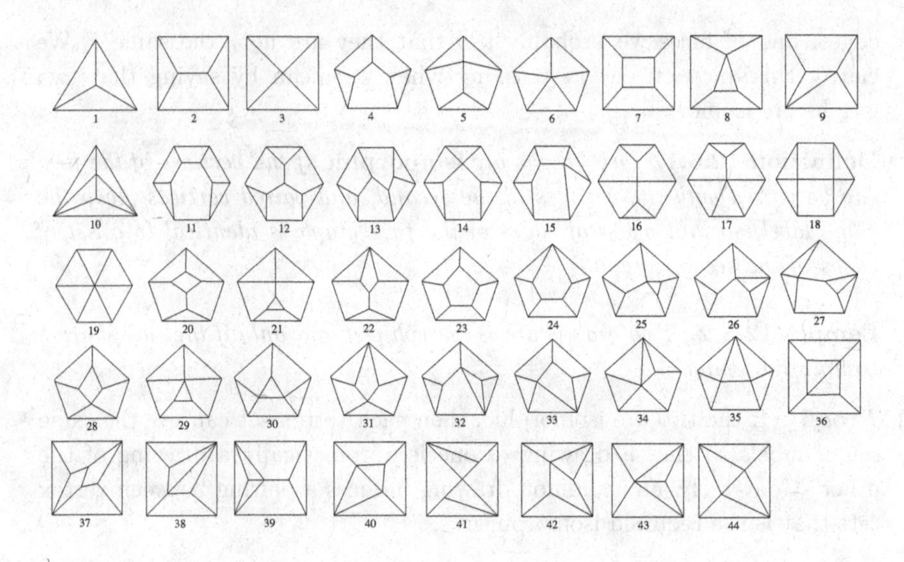

Fig. 12.15 All spherical polyhedra that each contain at most seven faces

Euler's Formula expresses the number of edges of a spherical polyhedron as a function of the number v of its vertices and the number f of its faces. Below is a list of the numbers of pairwise non-isomorphic spherical polyhedra with small values of v and f. The symmetry of the numbers on diagonals of this list cannot be accidental! There must surely be some interesting reason for it. The right hand column contains information about the number of non-isomorphic spherical polyhedra having the same number of faces. Hermes calculated the number of spherical heptahedra and octahedra in 1899, and Frederico obtained the number of spherical nonahedra [24, 25] in 1969.

12.6 Dürer's melancholy octahedron

In this section, we examine the large stone block pictured by Dürer in his etching "*Melencolia*", and reproduced below in Figure 12.16. Let us assume that the style of the part that we see is representative of the whole block and investigate the simplest possibilities for the remainder.

Problem 12.6.1. *Is the surface of the stone block that Dürer drew in "Melencolia" a spherical polyhedral model?*

f	Number of vertices									
	4	5	6	7	8	9	10	11	12	Total
14						50				
13						219				
12					14	558				$>> 10^6$
11					38	768	6134			$> 425,000$
10				5	76	633	2635	6134		$> 31,000$
9				8	74	296	633	768	558	2606
8			2	11	42	74	76	38	14	257
7			2	8	11	8	5			34
6		1	2	2	2					7
5		1	1							2
4	1									1

Fig. 12.16 Part of a stone block in "*Melencolia*"

Solution. We can see ten vertices, thirteen edges, and four faces. Let us suppose that the surface of the block is a model of a spherical polyhedron that contains f faces, e edges, and v vertices. We label the visible vertices of Figure 12.16, and from Lemma 12.5.1 it follows that $f \geq \frac{1}{2}v + 2 \geq 7$.

Thus the simplest possibility would be for the block to be a heptahedral model. Then the unique possible value for v is ten, and from Euler's Formula the polyhedron contains exactly fifteen edges. As the vertices 4, 6, 7, and 9 are at least trivalent the only possibility for the two hidden edges would be the first in Figure 12.17. On practical grounds we reject this possibility as, resting on an edge, the block would be unstable!

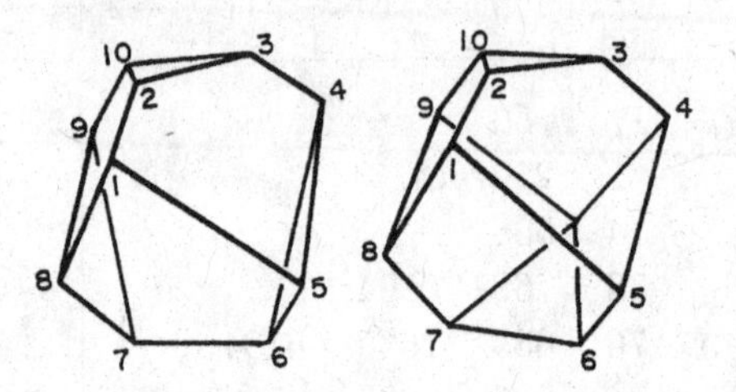

Fig. 12.17 An unstable heptahedral, and a stable octahedral, stone block

The next simplest possibility would be for the block to be octahedral. Then, from Lemma 12.5.1, we have that $8 \geq \frac{1}{2}v + 2$. It follows from this inequality that $v = 10, 11$, or 12. But if the polyhedron had only ten vertices, as in the heptahedral case, there would be no hidden vertices and the block, resting on an edge, would again be unstable.

If the model has eleven vertices, then, from Euler's Formula, we deduce that it contains seventeen edges. One vertex is hidden, and four edges are hidden, from our view. As each vertex of a spherical polyhedron is at least trivalent, then at least three of these hidden edges meet at the hidden vertex. The same reasoning leads us to the conclusion that a hidden edge meets each of 4, 6, 7, and 9 ensuring that the eleventh vertex has degree at most four. If it has degree four, then the only possibility is the second shown in Figure 12.17. If the eleventh vertex is trivalent then there are edges from the hidden vertex to three of the vertices 4, 6, 7, and 9. In the interests of stability, the base on which the octahedral block rests must be triangular containing 6, 7, and the hidden vertex. There are six possibilities, but each is, to within combinatorial isomorphism, one of the three possibilities shown in Figure 12.18.

If the model has twelve vertices, then it contains eighteen edges, giving two hidden vertices and five hidden edges. The necessary inequality in Proposition 12.4.1 is satisfied only if each vertex is trivalent. If the base is triangular we have only the first possibility shown in Figure 12.19. If the base is 4-gonal, then Proposition 12.4.1 rules out all other possibilities than the second in Figure 12.19.

The second possibility in Figure 12.17 and the two in Figure 12.19 can be thought of as being cut from a cube, as shown in Figure 12.20.

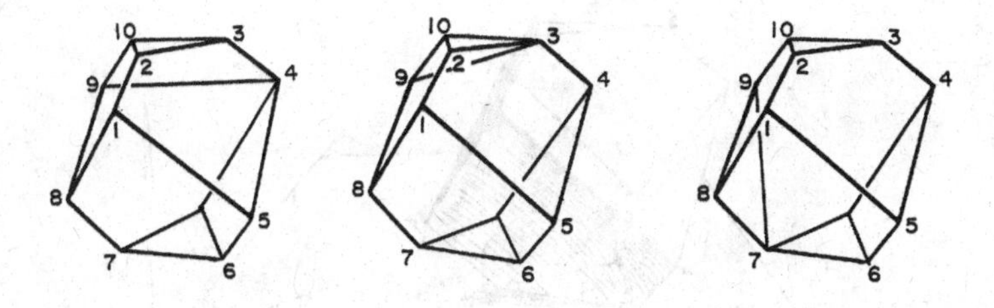

Fig. 12.18 Three possible octahedral blocks

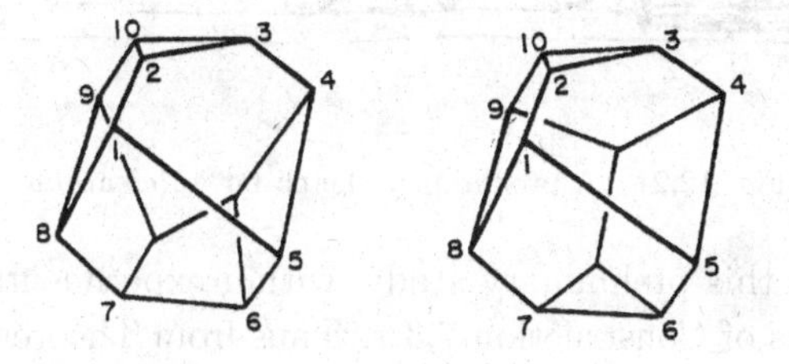

Fig. 12.19 Two more possibilities

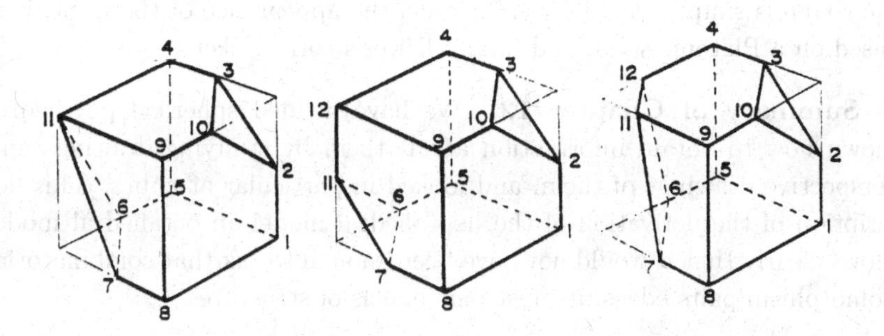

Fig. 12.20 Truncations of three cubes

The second of these has some nice features, having two triangular and six pentagonal panels symmetrically arranged. It can be obtained simply, from a cube, by truncating at two of the vertices. In fact the preliminary sketch by Dürer in Figure 12.21 suggests that this is exactly what he had in mind for the stone.

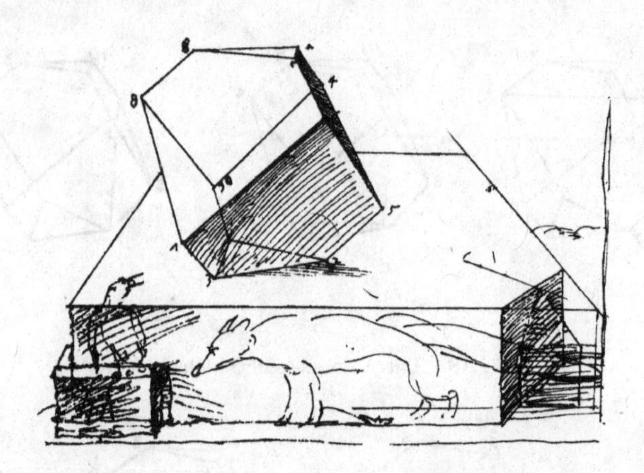

Fig. 12.21 A preliminary sketch for "*Melencolia*"

The lines in this preliminary study, within experimental error, satisfy the requirements of Construction 7.3.1. Thus from Theorem 7.3.1 we have that Dürer's preliminary sketch contains a perspective drawing of a box. From Theorem 7.1.1, or more easily from Theorem 13.1.1, we may verify that the sketch is a correct perspective drawing of an octahedral spherical polyhedron. The evidence is overwhelmingly in favor of the truncated cube: the shape is simple, it is in keeping with the appearance of the front, it is based on a Platonic solid, and finally Dürer's correct sketch.

Summary of Chapter 12. We have defined spherical polyhedra, shown how to obtain information about them by studying drawings and perspective drawings of them, and looked in particular at cubes. This description of the derivation of the heptahedral model an octahedral model shows clearly that it would not have been enough to ask that combinatorial isomorphism pairs edges, it must pair panels of structures.

Chapter 13

SCENE ANALYSIS AND SPHERICAL POLYHEDRA

Spherical polyhedra, apart from being traditional objects of study, repay attention as their models are among the most common "shapes" built in man's various civilizations. Here we apply the scene analysis methods of Chapter 7 to any drawing of a spherical polyhedron, and characterize those drawings [61] that are perspective drawings of the polyhedron.

13.1 Perspective drawings of a spherical polyhedron

Two major difficulties in applying the scene analysis techniques of Chapter 7 in general are that, first, the appropriate labeling of the lines of the supposed scene is not always apparent, and, second, an application of Theorem 7.1.1 requires many verifications. The application to spherical polyhedra both makes the labeling automatic and significantly reduces the number of checks that need to be performed.

We consider the following problem: Is a given drawing of a simple graph a perspective drawing of a spherical polyhedron?

We can think of the problem in four parts. In the first part of the problem we decide if the graph G is planar, most easily by constructing a planar embedding of G. In the second part of the problem we verify that deleting each set of at most two vertices leaves a connected subgraph of G. If both these tests are successful, then, from Steinitz' Theorem, we have that there is a spherical polyhedron $\mathbf{F}$ so that $G = G(\mathbf{F})$ and a planar embedding of G in which the faces of $\mathbf{F}$ have faces of G as images. In the third part of the problem we check that each face of $\mathbf{F}$ also has a polygonal region as image in the original drawing. If this is so, then we are certain that the drawing is the drawing of a spherical polyhedron. With this information in hand we use Theorem 13.1.1 in order to complete the investigation and determine whether the drawing is a perspective drawing.

Theorem 13.1.1. *Let* $\mathbf{F}'$ *be a drawing of a spherical polyhedron. Then the following two conditions are equivalent:*

1. *The drawing* $\mathbf{F}'$ *is a perspective drawing, drawn from any given point* p *as viewpoint, of the skeleton of a spherical polyhedron* $\mathbf{F}$. *The point* p *is a general viewpoint of the set of vertices of* $\mathbf{F}$.

2. *Let a listing of the face images in* $\mathbf{F}'$ *be given. Then, for each* $\{i, j\} \subseteq \{1, 2, \ldots, f\}$ *with* $i \neq 1$, *if the* i*th and* j*th images share an edge, then the line containing the edge is labeled by both* ij *and* ji. *If the first and* i*th images do not share an edge, then a line labeled by* $1i$ *may be drawn in the plane of the drawing so that, for each* $\{i, j\} \subseteq \{2, 3, \ldots, f\}$, *the three lines labeled* $1i$, ij, *and* $1j$ *are concurrent.*

Proof. First suppose that Condition 1 is true. We write the faces of the spherical polyhedron $\mathbf{F}$ as $F_1, F_2, \ldots, F_f$. For each i and j we label the image of the intersection of the plane that contains F_i and the plane that contains F_j by ij and by ji. Lemma 5.3.2 then gives the required concurrences in the drawing.

Conversely, suppose that labeled lines as required by Condition 2 exist. We associate a graph $H(\mathbf{F}')$ with the drawing $\mathbf{F}'$. The definition is analogous to that of the hinge graphs of Chapter 11. The vertices of $H(\mathbf{F}')$ are the face images in the drawing. A pair $\{F', G'\}$ of face images F' and G' is an edge of $H(\mathbf{F}')$ if F' and G' share a common edge. It follows from the comment immediately after Definition 12.1.1 that this simple graph $H(\mathbf{F}')$ is connected.

The remainder of the proof depends on the following two lemmas .

Lemma 13.1.1. *Let* T *be a tree. Suppose that* $v_1, v_2, \ldots, v_r$ *are the vertices, in order along the path, of a path* P *of* T. *Then all the vertices of* T *can be listed so that* v_i *is the* i*th element of the list, for each* i *in* $\{1, 2, \ldots, r\}$, *and so that each vertex of the list, after the first, shares an edge with exactly one earlier vertex in the list.*

Proof. We prove the lemma by induction on the number of vertices of T. The result is true for the tree consisting of an isolated vertex. Let us suppose that it is true for all trees containing at most n vertices. Suppose that T is a tree with $n+1$ vertices. From Lemma 11.2.2 we have that there is a vertex x of $V(T) - V(P)$ that belongs to exactly one edge of $E(T)$. Then $T \backslash \{x\}$ is a tree that contains n vertices. We list the vertices of this tree so that each vertex, after the first, shares an edge with exactly one earlier vertex of the list, and with the vertices of P heading the list. Then we add the vertex x to the end of the list. □

Lemma 13.1.2. *Let G be a connected graph and suppose that T is a maximal tree in G. Then each edge $\{u, v\}$ of $E(G)$ is in exactly one cycle C of G such that $E(C) \subseteq E(T) \cup \{\{u, v\}\}$.*

Proof. Suppose that $\{u, v\} \in E(G)$. As T is maximal, u and v are both in $V(T)$ and T contains a path P joining them. The addition of the edge $\{u, v\}$ to $E(P)$ gives a cycle of G. If C' were a second cycle of G containing $\{u, v\}$ and such that $E(C') \subseteq \{u, v\}$, then, from Lemma 11.5.1, we would have that T contains a cycle, and this not so. □

From Theorem 11.2.1 and Lemma 13.1.1 we have the existence of a maximal tree T of the graph $H(\mathbf{F}')$ that satisfies the requirement that each face image F'_j, other than the first, shares a common edge of $H(\mathbf{F}')$ with exactly one earlier face image F'_{i_j} in the list.

We choose an arbitrary viewpoint p not in the plane of the drawingand we choose P_1 to be the plane of the drawing. As P_2 we choose a plane that contains the line labeled 12 but not the point p. We then proceed inductively, supposing that, for the natural number $m - 1$, distinct planes $P_1, \ldots, P_{m-1}$ have been defined so that, for each $j < m$ the plane P_j contains the line $1j$, and so that the line $P_{i_j} \cap P_j$ is in the plane $p \vee ij$. Lemma 7.1.1, with the plane P_{i_m} in place of the plane P_2, the line labeled by $1m$ replacing M and the line labeled by $i_m m$ replacing N, enables us to define a plane P_m that contains both the lines $1m$ and $(p \vee i_m m) \cap P_{i_m}$. As in Theorem 7.1.1 it may be helpful to think of each plane P_m being "hinged" at the line labeled by $1m$ and swung about this hinge out of the plane P_1 until its intersection with P_{i_m} has an image, viewed from p, of the line labeled by $i_m m$. Thus we define a set $\{P_1, P_2, \ldots, P_f\}$ of planes.

For each j we select a polygonal region that is the image of the face image F'_j, in a perspective drawing drawn from the general viewpoint p in the plane P_j. These are the f polygonal regions that we hope will be the faces of a polyhedron that has $\mathbf{F}'$ as a perspective drawing.

For this to be the case, each pair of face cycle images F'_r and F'_s that share an edge E' must give rise to a pair, P_r and P_s, of planes so that the image of $P_i \cap P_j$ in a perspective drawing, drawn from p, is the line rs. We can be certain that this is true whenever F'_r and F'_s form an edge in the maximal tree T of $H(\mathbf{F}')$.

We call an edge $\{F'_r, F'_s\}$ of $H(\mathbf{F}')$ *flawed* if the line rs is not the image of

$P_r \cap P_s$ in a perspective drawing from p. Each cycle of $H(\mathbf{F}')$ that contains exactly one flawed edge is a *flawed cycle*. Let us suppose that a flawed edge $\{F'_r, F'_s\}$ exists. Then, from Lemma 13.1.2, we have that $\{F'_r, F'_s\}$ is an edge in at least one flawed cycle of $H(\mathbf{F}')$.

Let $\mathbf{F}''$ be a planar embedding of the graph of any polyhedron that has the drawing $\mathbf{F}'$. Then we may pair each face image in the drawing $\mathbf{F}'$ with the corresponding face of $\mathbf{F}''$. The union of the faces of $\mathbf{F}''$ that correspond to the faces of any cycle C_1 of $H(\mathbf{F}')$ has a well-defined interior. With each cycle C_1 we associate the number of faces in this interior in $\mathbf{F}''$. One of the vertices of C_1 may to correspond to the face that has the perimeter of the embedding $\mathbf{F}''$ as its boundary. In this case we consider the number of those faces that are on the right side during one traverse of the path in $\mathbf{F}''$ that is obtained by deleting this vertex from C_1. The number of these faces is then associated with C_1. In this way we associate a number with each flawed cycle of $H(\mathbf{F}')$, and we choose a flawed cycle in $H(\mathbf{F}')$ that is associated with a minimum number. There are two possibilities: either the number is zero or it is positive. We first prove that it cannot be zero.

Suppose that the flawed edge of the flawed cycle C_1 is $\{F'_r, F'_s\}$ and the vertices of the cycle in cyclic order are $F'_r = F'_{i_1}, F'_{i_2}, \ldots, F'_{i_m} = F'_s$. Then each three adjacent planes $P_{i_{t-1}}, P_{i_t}, P_{i_{t+1}}$ in the cycle share one point that is on the line containing p and the intersection of the lines $i_{t-i}i_t$ and $i_t i_{t+1}$. Therefore every plane of the list $P_r, P_{i_2}, \ldots, P_{i_{m-1}}, P_s$ contains the same point v, say, and in particular v is in both P_r and P_s. Also from their definition we have that the planes P_r and P_s intersect P_1 in the point of concurrence of the lines $1r$, $1s$, and rs. Therefore the image of the intersection $P_r \cap P_s$ in a perspective drawing from viewpoint p in P_1 is the line rs, contradicting the choice of $\{F'_r, F'_s\}$ as a flawed edge.

Now suppose that $\{F'_r, F'_s\}$ is the flawed edge, and $F'_r, F'_{i_2}, \ldots, F'_{i_{m-1}}, F'_s$ are the vertices in cyclic order, of a flawed cycle C_1 associated with a smallest, necessarily non-zero, number. Suppose further that v' is the vertex of $F'_r \cap F'_s$ that belongs to a region in the interior of the flawed cycle C_1. Then the cycle of $H(\mathbf{F}')$ whose vertices are those regions that contain the vertex v' is associated with the number zero. Therefore it is not flawed and so contains a second flawed edge that is interior to C_1. From Lemma 13.1.2, again applied to $H(\mathbf{D}')$, we have that this flawed edge $\{F'_{j_1}, F'_{j_n}\}$ belongs to a flawed cycle C_2 of $H(\mathbf{F}')$. We may write the vertices of C_2 in cyclic order, as $F'_{j_1}, F'_{j_2}, \ldots, F'_{j_{n-1}}, F'_{j_n}$. Consider the list $F'_r, F'_{i_2}, \ldots, F'_s, F'_{j_1}, F'_{j_2}, \ldots, F'_s, F'_{j_1}, F'_{j_2}, \ldots, F'_{j_n}$ of vertices of $H(\mathbf{F}')$. Suppose that F'_{j_a} is the first repeated member of the list, $F'_{j_a} = F'_{i_b}$ say. Suppose that $F'_{j_{a+k}}$

is the last repeated member of the list, $F'_{j_{a+k}} = F'_{i_c}$ say. Then, without loss of generality assuming that $1 < c$, the cycle C of $H(\mathbf{F}')$ with vertices, in cyclic order, F'_{j_1}, F'_{j_2}, ..., F'_{j_a}, $F'_{i_{b+1}}$, ..., F'_{j_a}, $F'_{i_{b+1}}$, ..., F'_{i_c}, $F'_{j_{a+k+1}}$, ..., F'_{j_n} is a flawed cycle contradicting the initial choice of C_1.

Therefore there is no flawed edge and, for each pair of face images F'_i and F'_j that share an edge, the image of $P_i \cap P_j$ in a perspective drawing, drawn from p, is the line ij. We note that for any point $x' \in F'_i \cap F'_j$ we have that $(p \vee x') \cap P_i = (p \vee x') \cap P_j$, ensuring that the intersection $F_i \cap F_j$ is a common edge that has image $F'_i \cap F'_j$ in a perspective drawing from the viewpoint p. Therefore the union of the polygonal regions F_1, F_2, ..., F_p is the required spherical polyhedron $\mathbf{F}$. □

13.2 Some applications of the theorem

In this section, we begin with simple applications of Theorem 13.1.1, eventually applying it to the question of the viability of an architectural proposal.

Example 13.2.1. Let us test our newly acquired technique on the drawing of the graph K_4 in Figure 13.1 and determine if it is a perspective drawing of a spherical polyhedron.

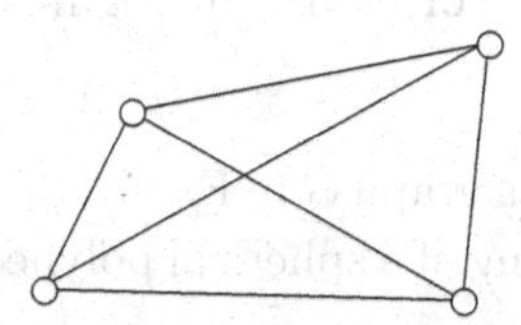

Fig. 13.1 A drawing of the graph K_4

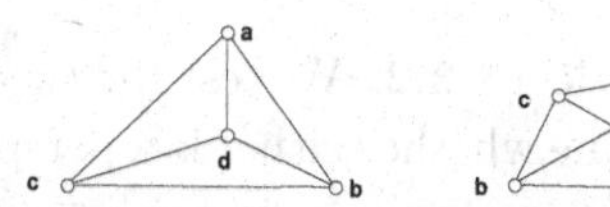

Fig. 13.2 Drawings with vertices labeled

We proceed by first deciding if the graph is planar and second deciding if it is three-connected. From any planar embedding that proves the graph planar we are able to list the faces, third we check whether each of these faces has a polygonal region as image in the original drawing. If the answer is affirmative in all three tests, then we apply Theorem 13.1.1. From the common labeling of the vertices of the two drawings in Figure 13.2 we deduce easily that each is a drawing of the same graph K_4, proving K_4 to be planar. Simple testing verifies that K_4 is three-connected. Therefore, from Steinitz' Theorem it is the graph of a spherical polyhedron. The faces of the polyhedron are the faces of the second drawing. The corresponding face images in the original drawing are triangular and therefore automatically polygonal regions, ensuring that the drawing is a drawing of the polyhedron. We now turn to Theorem 13.1.1, labeling the faces and edges of the embedding as in Figure 13.3. This gives the labeling of edges of the original drawing that is in Figure 13.3.

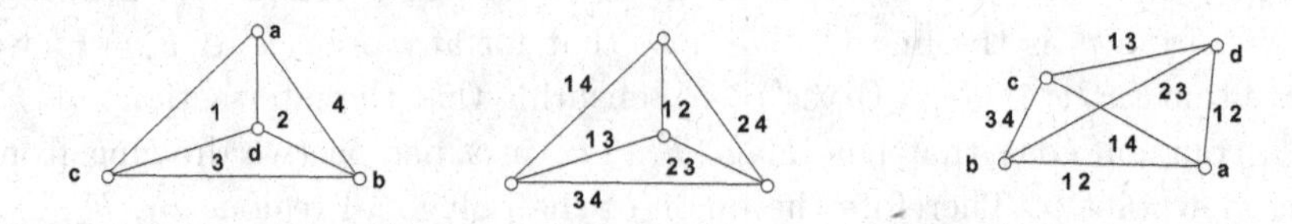

Fig. 13.3 Labeled edges of a planar embedding and of a drawing

There are no new lines to be drawn and labeled, as each line 12, 13, and 14 is determined by an edge. The labeling satisfies the requirements of Condition 2 of Theorem 13.1.1. As no special property of this drawing was needed in the above argument we have established the following lemma.

Lemma 13.2.1. *Any drawing of K_4 is a perspective drawing of a tetrahedron.*

Consequently, for example, the illustrations of figures in Figures 1.16 and 8.6 are perspective drawings.

Example 13.2.2. We test the two drawings of a graph G in Figure 13.4 and determine whether either is a perspective drawing of a spherical polyhedron.

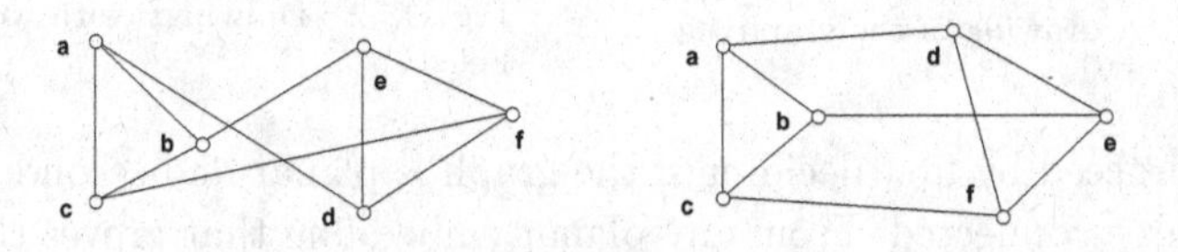

Fig. 13.4 Two drawings of a simple graph G

The graph G may be embedded as in Figure 13.5, the vertex labeling proving that the three drawings are of the same graph G, where $V(G) = \{a, b, c, d, e, f, g\}$ and $E(G) = \{\{a, b\}, \{b, c\}, \{c, a\}, \{d, e\}, \{e, f\}, \{f, d\}, \{a, d\}, \{b, e\}, \{c, f\}\}$. Verifying that G is three-connected is again a straightforward matter. Then from Steinitz' Theorem we have that G is the graph of a spherical polyhedron.

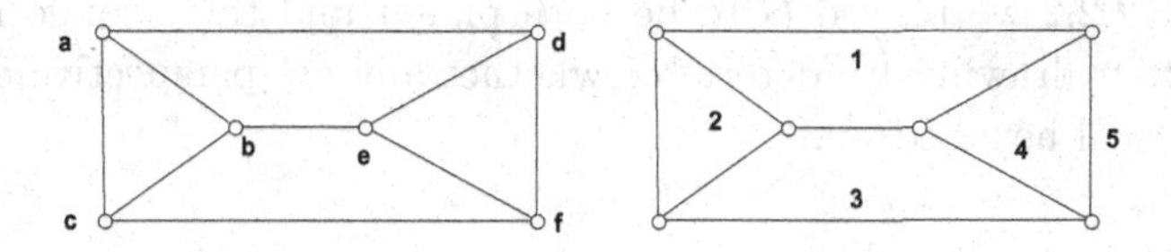

Fig. 13.5 An embedding of the same graph G

Labeling the faces of the embedding as shown in Figure 13.5 enables us to conclude that the first fails the test for a drawing of a polyhedron, as for example the face cycle image with vertex set $\{a, b, e, d\}$ is not a polygonal boundary. The second drawing satisfies the definition of a drawing of a polyhedron.

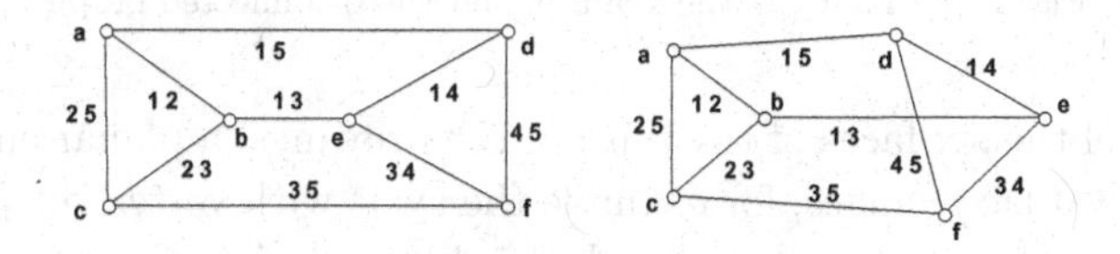

Fig. 13.6 Two drawings with edges labeled

The labeling of edges of the embedding induces a corresponding labeling of edges of the original drawing of a spherical polyhedron to which we apply Theorem 13.1.1. All required lines are already determined by edges. We need only check that $\{12, 23, 13\}^*$, $\{12, 25, 15\}^*$, $\{13, 34, 14\}^*$, $\{13, 35, 15\}$, and $\{14, 45, 15\}^*$, are concurrent triples. We notice the reduced number of checks required compared to those required for an application of Theorem 7.1.1 directly to planes. Those triples marked with an asterisk are clearly concurrent. Therefore the second drawing is a perspective drawing of a triangular column, but unsurprisingly the first drawing is not. From our argument we obtain the following lemma:

Lemma 13.2.2. *A drawing of a triangular column is a perspective drawing exactly when the lines containing the side edges of the column are concurrent.*

Exercise 13.2.1. Find necessary and sufficient conditions for a drawing of an n-spoked wheel W_n to be a perspective drawing of an n-sided pyramid.

Example 13.2.3. The common labeling of vertices shown below in Figure 13.7 indicates that each drawing in Figure 12.5 is of the same graph G. In Exercise 12.2.2 we proved G to be both planar and three-connected. We test the four drawings to determine whether any are perspective drawings of a spherical hexahedron.

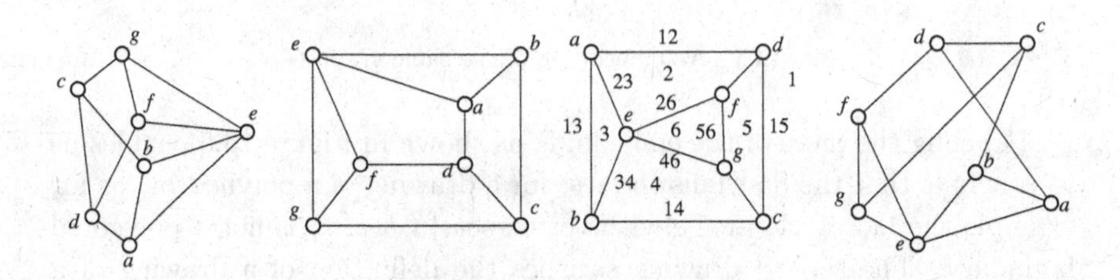

Fig. 13.7 Four drawings of a planar three-connected graph

We label the six faces of the third of the drawings, a planar embedding. The proposed face images, for example the cycle with vertex set $\{a, b, c, d\}$, in the fourth drawing are not all polygonal boundaries. Consequently, this is not a drawing of a spherical polyhedron. The common vertex labeling of the remaining three drawings enables us to label corresponding edges of each. To see if any is a perspective drawing of a spherical polyhedron we need only define a line 16 that is concurrent with 12 and 26, with 14 and 46; and with 15 and 56. The other concurrences occur at the vertices. In other words, we need only check that $12 \cap 26$, $14 \cap 46$ and $15 \cap 56$ are collinear in each case. This is so in the first and the third, but not in the second.

We may choose the order in the listing of embedded regions as we like. It is a good general rule to select a region with a large number of edges to be first, as this is likely to reduce the number of lines to be constructed in applying Theorem 13.1.1. All listings will give the same final answer, the equivalence of differing sets of conditions following from applications of Desargues' Theorem.

Figure 13.8 contains the line constructed in each of three applications of Theorem 13.1.1 to the first drawing in Figure 13.7, each with a different selection of a face as the first listed. Each dotted line represents the collinearity requirement of Theorem 13.1.1 following a choice of the region to be labeled by 1. Some applications of Desargues' theorem would verify the equivalence of all three conditions.

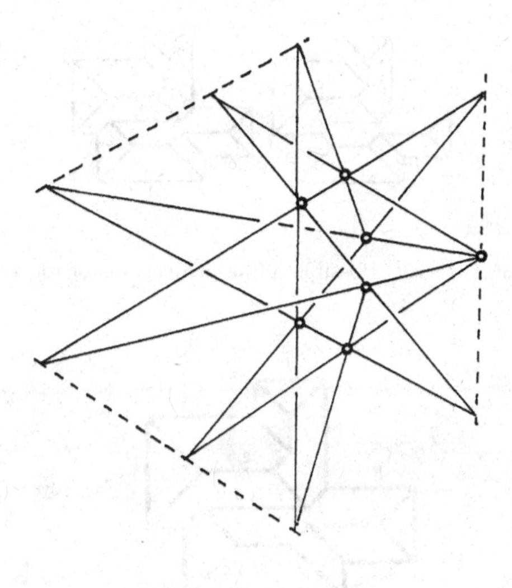

Fig. 13.8 Three conditions for the existence of a scene

Exercise 13.2.2. Decide which, if any, of the drawings of a graph in Figure 13.9 are perspective drawings of a spherical polyhedron.

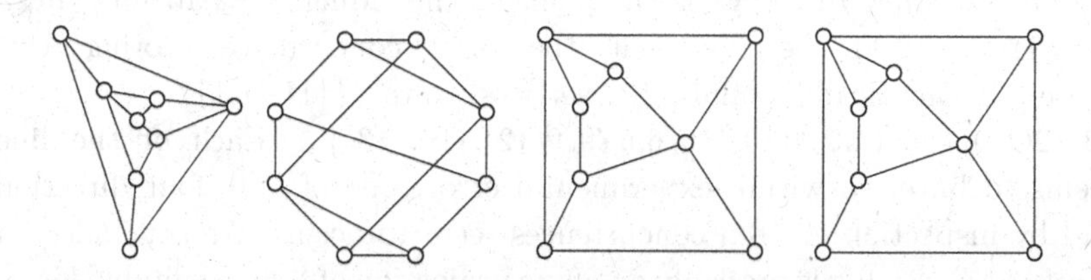

Fig. 13.9 Four drawings of a graph

Problem 13.2.1. *An architect has given me two sets of house-plans. I am having difficulty visualizing each proposal, particularly the plans of alternative roofs shown in Figure 13.10. Is this just my lack of skill, or is there something wrong with either of the two?*

Solution. Consider the spherical polyhedron modeled by the roof and the underlying plane ceiling in each case. The first drawing has a vertex that belongs to only two edges, preventing it from being a drawing of a three-connected graph. There is clearly an error in the plans.

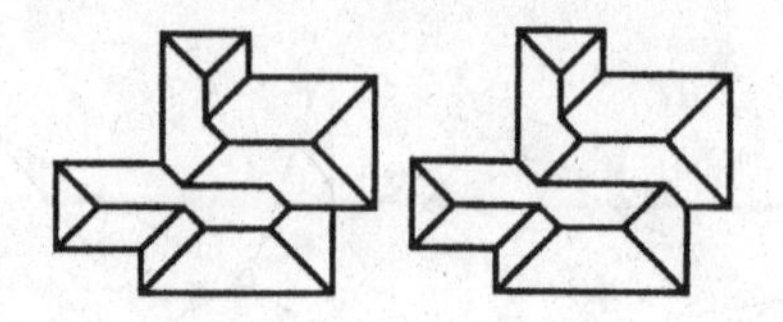

Fig. 13.10 Possible affine projections of roofs

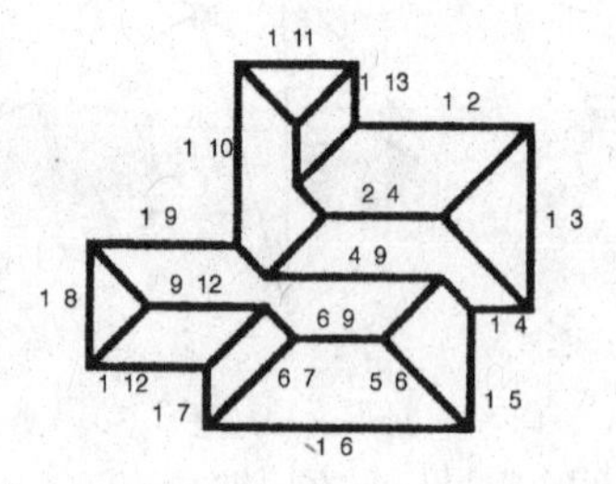

Fig. 13.11 A labeled drawing of a roof

The second drawing is a planar embedding. Labeling it as in Figure 13.11 we see that the only concurrences, other than those automatically valid at vertices, are $\{\{1i, ij, 1j\} : ij \in \{24, 2(10), 2(13), 49, 4(10), 59, 69, 79, 9(12), (10)(13)\}$. Each of the lines seems to have, to within experimental error, one of only four directions and by inspection the ten concurrences seem to occur. We conclude that the second drawing is probably an affine projection of a roof, suitable for use in an architectural plan. I should persevere with my attempts to visualize the roof.

Exercise 13.2.3. Do you think that each drawing of a spherical polyhedron in the list of Figure 12.15 is a perspective drawing? We could test each but it would be tedious. Choose one of the heptahedra (seven faces with six of the faces being interior faces) drawn in the list and prove that it is a perspective drawing.

Exercise 13.2.4. Each of the six drawings in Figure 13.12 purports to be an affine projection of a Platonic cube with a part sliced away by a plane cut. Which of the drawings are in fact affine projections?

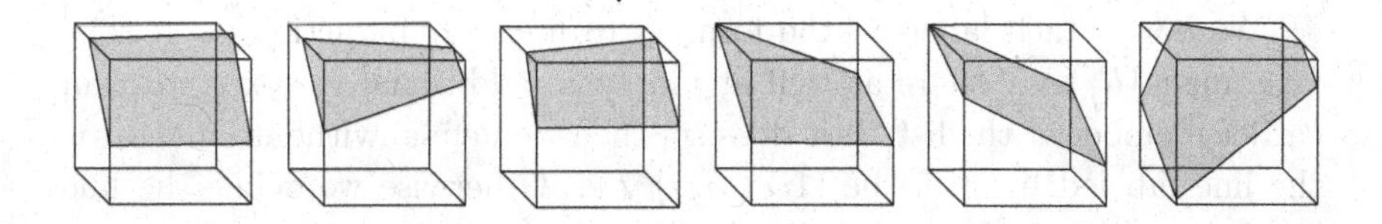

Fig. 13.12 Six figures that may be affine projections

13.3 Perspective drawings of a wider class of unions of polygonal regions

In some cases we may more easily know that an attempt is a perspective drawing. For example, it is straightforward to extend our definition of polyhedron to all figures modeled by hinged-panel models. The definition of the graph of such a figure is the natural generalization of the graph of a polyhedron. Then it is true that each drawing of a development is a perspective drawing of a development; that a drawing of a three-panel cycle is a perspective drawing if and only if the hinge-line images are concurrent; and that each drawing of a panel cycle that has at least four panels is a perspective drawing of a panel cycle.

As an example of the range of results possible we give the following theorem. It illustrates the special role that triangular faces play and, although we only state and prove it for polyhedra, it easily extends to a wider class of figures.

Theorem 13.3.1. *Let* $\mathbf{F}'$ *be a drawing of a polyhedron* $\mathbf{F}_1$ *that contains at most two non-triangular faces. Then* $\mathbf{F}'$ *is a perspective drawing of a polyhedron* $\mathbf{F}_2$ *that is combinatorially isomorphic to* $\mathbf{F}_1$*, drawn from a general viewpoint of the set of vertices of* $\mathbf{F}_2$*. No two panels of* $\mathbf{F}_2$ *are coplanar.*

Proof. Let the set $\{F_1', F_r'\}$ include all the face images of $\mathbf{F}'$ that are not triangular. We recall the graph $H(\mathbf{F}')$ introduced during the proof of Theorem 13.1.1. Choose any path P of least possible length r joining F_1' and F_r' in the graph $H(\mathbf{F}')$.

From Theorem 11.2.1 we have that the path P is contained in a maximal tree T of the graph $H(\mathbf{F}')$. Lemma 13.1.1 guarantees that we are able to complete a list of all the face images of $\mathbf{F}'$ in which F_1' is the first member and F_r' is the rth member. Each member F_j', after the first, of the list shares a common edge in T with exactly one earlier member F_{i_j}' of the list. We label the line that contains this common edge by $i_j j$ and by $j i_j$.

We now attach labels of the form $1j$ to lines, by induction on j. The face image F'_j may share as well at most one additional vertex v with an earlier member of the list. If it does so, then we define, without ambiguity, the line labeled by $1j$ to be $(1i_j \cap i_j j) \vee v$. Otherwise we define the line labeled by $1j$ to be any line that contains the point $1i_j \cap i_j j$.

We choose an arbitrary viewpoint p not in the plane of the drawing and choose P_1 to be the plane of the drawing. As P_2 we select a plane that contains the line labeled 12 but not the point p. We define further planes inductively, supposing that for the natural number $m - 1$ distinct planes $P_1, P_2, \ldots, P_{m-1}$ have been defined so that, for each $j < m$, the plane P_j contains the line $1j$ and so that the line $P_{i_j} \cap P_j$ is in the plane $p \vee i_j j$.

As the lines $1m$, $1i_m$, and $i_m m$ are concurrent we are able to apply Lemma 7.1.1, with the plane P_{i_m} in place of the plane P_2, the line labeled by $1m$ replacing M, and the line labeled by $i_m m$ replacing N, in order to define a plane P_m that contains both the lines $1m$ and $(p \vee i_m m) \cap P_{i_m}$. As in Theorem 7.1.1 it may be helpful to think of each plane P_m being "hinged" at the line labeled by $1m$ and swung about this hinge out of the plane P_1 until its intersection with P_{i_m} has an image, viewed from p, of the line labeled by $i_m m$. Thus we define a set $\{P_1, P_2, \ldots, P_f\}$ of planes.

For each j in $\{1, 2, \ldots, f\}$ the face F_j is chosen to be the image of F'_j in the plane P_j, in the perspective drawing that is drawn from viewpoint p. It follows that each pair F_{i_j} and F_j share an edge that has the required image when drawn from p; and that all edges of F_1 and F_r are included in this set; and further that if triangular face images F'_i and F'_j share an edge E', then F_i and F_j share an edge that has E' as image when drawn from p. □

Consequently we can be sure, without further ado, that Figure 10.14, Figure 10.22, and Figure 10.30 each contain scenes.

Summary of Chapter 13. We applied the scene analysis methods of Chapter 7 to drawings of spherical polyhedra, giving necessary and sufficient conditions for a drawing to be a perspective drawing. Finally, we gave an example showing how these methods generalize.

Chapter 14

MATROIDS

Combinatorial geometries and graphs have been central to the description of our world. By weakening the requirements of Definition 2.1.1, and restricting its application to finite sets, we define a structure that appears in every finite geometry and in every graph. It is also present in other, seemingly different, mathematical systems. Thus everything we learn about this structure contributes to our knowledge in these disparate areas. Hassler Whitney, in his seminal 1935 paper [62] called this structure a "matroid". His approach remains unsurpassed for clarity and insight, and we follow his treatment of independent sets, circuits, and the rank function. We then introduce minors, the natural substructures of matroids, and use them in a range of examples from graph theory, linear algebra, and geometry.

14.1 The definition of a matroid

We asked four requirements of a combinatorial geometry in Definition 2.1.1. If we relax the first condition and allow any non-empty set to be a circuit, and remove the fourth condition entirely, then we are left with the following three requirements of a finite set E and a collection $\mathcal{C}$ of its subsets:

MC1: The empty set is not a member of $\mathcal{C}$.

MC2: If C_1 and C_2 are members of $\mathcal{C}$ and $C_1 \subseteq C_2$, then $C_1 = C_2$.

MC3: If C_1 and C_2 are distinct members of $\mathcal{C}$ and $e \in C_1 \cap C_2$, then there is a member C_3 of $\mathcal{C}$ such that $e \notin C_3 \subseteq C_1 \cup C_2$.

Clearly each finite geometry satisfies these requirements. For each graph G containing only finitely many edges, Definition 11.2.2 and Lemma 11.5.1 imply that the pair $E = E(G)$ and $\mathcal{C} = \{C : C$ is the edge set of a cycle of $G\}$ satisfies Conditions **MC1**, **MC2**, and **MC3**.

It has been customary to initially focus attention on subsets of E that contain no member of $\mathcal{C}$ rather than on the members of $\mathcal{C}$ directly. The following lemma gives three properties of this collection of subsets of E.

Lemma 14.1.1. *Let $\mathcal{C}$ be a collection of subsets of a finite set E that satisfies Conditions* **MC1**, **MC2**, *and* **MC3**. *Suppose that $\mathbf{I} = \{I \subseteq E :$ no subset of I is in $\mathcal{C}\}$. Then $\mathbf{I}$ satisfies the following three conditions:*

MI1: *The empty set is a member of $\mathbf{I}$.*

MI2: *If I is a member of $\mathbf{I}$ and $J \subseteq I$, then J is a member of $\mathbf{I}$.*

MI3: *If I and J are members of $\mathbf{I}$ and $|I| > |J|$, then there is an element e of $I - J$ so that $J \cup \{e\}$ is a member of $\mathbf{I}$.*

Proof. From **MC1**, the empty set is not in $\mathcal{C}$ and so is in $\mathbf{I}$. Hence $\mathbf{I}$ satisfies **MI1**. If $I \in \mathbf{I}$ and $J \subseteq I$, then I contains no member of $\mathcal{C}$, ensuring that J contains no member of $\mathcal{C}$, giving **MI2**.

We now prove that Condition **MI3** holds for any two members, I and J, of $\mathbf{I}$ with $|J| < |I|$. Let I' be a subset of $I \cup J$ that belongs to $\mathbf{I}$, with $|I'| > |J|$, and of all such sets has $|J - I'|$ minimal. First suppose that $J - I'$ contains an element x. For each $y \in I' - J$, both $(I' \cup \{x\}) - \{y\} \subseteq J \cup I$ and $|J - ((I' \cup \{x\}) - \{y\})| < |J - I'|$. Thus $(I' \cup \{x\}) - \{y\}$ is not in $\mathbf{I}$ and contains a subset $C_y \in \mathcal{C}$. But $I' \in \mathbf{I}$, so that $x \in C_y$. Hence $x \in C_y \subseteq (I' \cup \{x\}) - \{y\}$. As $J \in \mathbf{I}$, C_y is not contained in $I' \cap J$. Hence there is some z in $C_y \cap (I' - J)$. Repeat the argument with z in place of y to obtain $C_z \in \mathcal{C}$ with $x \in C_z \subseteq (I' \cup \{x\}) - \{z\}$. From $z \in C_y - C_z$ and y not in C_y we have that $C_y \neq C_z$. Applying Condition **MC1** we have the existence of $C \in \mathcal{C}$ so that $C \subseteq (C_y \cup C_z) - \{x\} \subseteq I'$, a contradiction to our choice of I'. Thus $J - I' = \emptyset$ and J is a proper subset of I'. For any element e of $I' - J$ we have $J \cup \{e\} \in \mathbf{I}$ as required for **MI3** to hold. □

The requirements **MI1**, **MI2**, and **MI3** are properties of linearly independent sets of columns of any matrix. Whitney [62] took these properties as essential features of geometric structure. He also pointed out that many equivalent definitions give the same structure. From the motivating example of matrix columns, he coined the name "matroid" for such a structure.

Definition 14.1.1. *A* **matroid** *M is an ordered pair $(E, \mathbf{I})$ consisting of a finite set E and a collection $\mathbf{I}$ of subsets of E that satisfy the following three conditions.*

MI1: *The empty set is a member of $\mathbf{I}$.*

MI2: *If I is a member of $\mathbf{I}$ and $J \subseteq I$, then J is a member of $\mathbf{I}$.*

MI3: *(Augmentation Condition) If I and J are members of $\mathbf{I}$ and $|I| > |J|$, then there is an element e of $I - J$ so that $J \cup \{e\}$ is a member of $\mathbf{I}$.*

From this characterization it is immediate that the linear algebra of the columns of a matrix has an underlying matroid structure.

Theorem 14.1.1. *Let A be any matrix with entries from a field F. Then the ordered pair $(E, \mathbf{I})$, where E is the set of columns of A and each member of $\mathbf{I}$ is a linearly independent set of columns of A, is a matroid M.*

Proof. A set of columns of A is linearly independent exactly when none of its members is a linear combination of the others. Thus the linearly independent sets satisfy Conditions **MI1** and **MI2**. They also satisfy Condition **MI3** unless there are two linearly independent sets I and J, such that $|I| > |J|$ and $J \cup \{a_i\}$ is linearly dependent for every element $a_i \in I - J = \{a_1, a_2, \ldots, a_m\}$. Suppose that this is the case. Then J and $\{a_1\}$ are each linearly independent, and $J \cup \{a_1\}$ is not. Thus a_1 is a linear combination of some columns of J, with not all the coefficients zero. Thus some $x_1 \in J$, x_1 would be a linear combination of columns of $(J - \{x_1\}) \cup \{a_1\}$. Hence any linear combination of columns of J would also be a linear combination of $(J - \{x_1\}) \cup \{a_1\}$. In particular, a_2 would be a linear combination of $(J - \{x_1\}) \cup \{a_1\}$. Also $\{a_1, a_2\}$ would be linearly independent and so there would be a member x_2 of $J - \{x_1\}$ that was a linear combination of columns of $(J - \{x_1, x_2\}) \cup \{a_1, a_2\}$. Proceeding in this way, each a_r would be a linear combination of columns of $(J - \{x_1, x_2, \ldots x_{r-1}\}) \cup \{a_1, a_2, \ldots, a_{r-1}\}$, but for some $r \leq m$, $J - \{x_1, x_2, \ldots, x_{r-1}\} = \emptyset$, and a_r would be a linear combination of other members of I, which would contradict the choice of I as a linearly independent set. □

Let M be the matroid obtained from a matrix A with entries in a field F as in the statement of Theorem 14.1.1. We say that M is a vector matroid over F and write $M = M[A]$.

Example 14.1.1. Label the columns of the matrix A by the integers 1 through 7 from left to right. Each single column and each pair of columns of the real matrix

$$A = \begin{pmatrix} 1\,0\,0\,1\,0\,1\,1 \\ 1\,0\,1\,1\,1\,0\,0 \\ 0\,1\,0\,1\,1\,1\,0 \end{pmatrix}$$

is linearly independent. Each set of three columns except $\{1, 3, 7\}$, $\{1, 2, 4\}$, $\{2, 3, 5\}$, $\{3, 4, 6\}$, $\{4, 5, 7\}$, and $\{2, 6, 7\}$ is linearly independent. As A has only three rows, no set of more than three columns is linearly independent. Therefore the set $E = \{1, 2, 3, 4, 5, 6, 7\}$ and the collection

$\mathbf{I} = \{X \subseteq \{1,2,3,4,5,6,7\} : |X| \leq 3$ and X is not one of $\{1,3,7\}$, $\{1,2,4\}$, $\{2,3,5\}$, $\{3,4,6\}$, $\{4,5,7\}$, $\{2,6,7\}\}$ of subsets of E is the vector matroid $M[A]$.

Definition 14.1.2. *Let $M = (E, \mathbf{I})$ be a matroid. Each member of* $\mathbf{I}$ *is an* **independent set** *of M. Each subset of E that is not independent is called* **dependent**. *A* **circuit** *of M is a dependent set all of whose proper subsets are independent.*

If M is the matroid $(E, \mathbf{I})$, then we say that M is a *matroid on* E and we often write $E(M)$ for E, $\mathbf{I}(M)$ for $\mathbf{I}$, and $\mathbf{C}(M)$ or $\mathcal{C}$ for the collection of circuits of M.

14.2 Geometries, graphs, and matroids

In this section, we characterize each matroid by its set of circuits. It follows that each finite geometry is a matroid, and that each graph that has only finitely many edges defines a matroid.

The next theorem proves the equivalence of Definition 14.1.1 and an alternative definition given in terms of circuits.

Theorem 14.2.1. *Let $\mathcal{C}$ be a collection of subsets of a finite set E that satisfies the following three conditions:*

MC1: *The empty set is not a member of $\mathcal{C}$.*

MC2: *If C_1 and C_2 are members of $\mathcal{C}$ and $C_1 \subseteq C_2$, then $C_1 = C_2$.*

MC3: *(Elimination Condition) If C_1 and C_2 are distinct members of $\mathcal{C}$ and $e \in C_1 \cap C_2$, then there is a member C_3 of $\mathcal{C}$ such that $e \notin C_3 \subseteq C_1 \cup C_2$.*

Then the ordered pair E and $\mathbf{I} = \{I \subseteq E : I$ contains no member of $\mathcal{C}\}$ is a matroid M. The members of $\mathbf{C}$ *are the circuits of M.*

Conversely, the collection $\mathcal{C}$ of circuits of a matroid M on E satisfies Conditions **MC1**, **MC2**, *and* **MC3**.

Proof. Suppose that $\mathbf{I}$ is defined as above. It follows from Lemma 14.1.1 that Conditions **MI1**, **MI2**, and **MI3** hold. Therefore $(E, \mathbf{I})$ is a matroid M.

From Definition 14.1.2 we have that the set $\mathbf{C}(M)$ of circuits of M is $\{C \in E : C$ is not in $\mathbf{I}$ but every proper subset of C is in $\mathbf{I}\}$. Suppose that $C \in \mathcal{C}(M)$. Then C is not a member of $\mathbf{I}$ and so contains a subset $C' \in \mathcal{C}$. It follows from C' not being in $\mathbf{I}$ that $C = C' \in \mathcal{C}$ and $\mathcal{C}(M) \subseteq \mathcal{C}$. Suppose next that $C'' \in \mathcal{C}$. Then C'' is not a subset of any member of $\mathbf{I}$. In particular C'' is not in $\mathbf{I}$ and consequently contains a member C of $\mathcal{C}(M)$. Then $C = C''$ and $\mathcal{C} = \mathcal{C}(M)$.

Conversely, the set $\mathcal{C}$ of circuits of a matroid M on E does not contain the empty set, and so satisfies **MC1**. If C_1 and C_2 are circuits and $C_1 \subseteq C_2$, then the fact that C_2 is a minimal dependent set implies that $C_1 = C_2$ and $\mathcal{C}$ satisfies **MC2**.

Suppose that C_1 and C_2 are distinct circuits, both containing the element e. We need only eliminate the possibility that $(C_1 \cup C_2) - \{e\}$ is in $\mathbf{I}$ to prove the validity of **MC3**. Supposing that f is any element of $C_2 - C_1$, we know that $C_2 - \{f\} \in \mathbf{I}$. Let J be any subset of $C_1 \cup C_2$ that is maximal with respect to both containing $C_2 - \{f\}$ and belonging to $\mathbf{I}$. Clearly $f \notin J$ and there is an element g of C_1 that is not in J. Thus $|J| \leq |(C_1 \cup C_2) - \{f, g\}| = |C_1 \cup C_2| - 2 < |(C_1 \cup C_2) - \{e\}|$. If $(C_1 \cup C_2) - \{e\}$ were in $\mathbf{I}$, then applying the Augmentation Condition **MI3** to J and $I = (C_1 \cup C_2) - \{e\}$ would give a contradiction to the maximality of J. Thus $(C_1 \cup C_2) - \{e\} \notin \mathbf{I}$ and Condition **MC3** holds for the collection of circuits of M. □

Consequently, a matroid M on a set E may be thought of as a collection of independent sets that satisfies **MI1**, **MI2**, and **MI3**, or equivalently as a collection of circuits that satisfies **MC1**, **MC2**, and **MC3**. The matroid is completely specified either by $\mathbf{I}(M)$ or by $\mathcal{C}(M)$.

Corollary 14.2.1. *Let M be a matroid. Then*
$\mathbf{I}(M) = \{I \subseteq E : I$ *contains no member of* $\mathcal{C}(M)\}$, *and*
$\mathcal{C}(M) = \{C \subseteq E : C$ *is not in* $\mathbf{I}(M)$ *but every proper subset of* C *is in* $\mathbf{I}(M)\}$.

There are four other common characterizations of a matroid. Two of these other common characterizations are given in this chapter. These are the characterizations of a matroid by its bases and by its rank function. The remaining two other common characterizations of a matroid are by its hyperplanes and by its closure function. Other useful equivalent characterizations of matroids are given in [8], [9], [16], and [49]. Thus a matroid consists of an underlying set with several associated structures - any one of which suffices to determine the others. We take advantage of this freedom to specify matroids in ways appropriate to the occasion. Until now we have leaned towards circuits as our descriptive medium. In Chapters 1 and 2 we showed that the simplest point-sets that non-trivially exhibit properties of collinearity and coplanarity are circuits of matroids. These were the building blocks of our treatment of geometry. For obvious reasons, those who approach matroids from a graph-theory background also find circuits a particularly appropriate way to describe a matroid. For example, the edge set of a three-edge cycle of a graph is called a "triangle" of the graph. Likewise, a three-element circuit of a matroid is called a "triangle" of the matroid. This use of the word "triangle" for matroids doesn't agree with its use for combinatorial geometries where it represents a figure of three non-collinear points. So the concept of a circuit is quite natural when considering matroids in the context of graphs and combinatorial geometries even though we have to be careful when borrowing our terminology from these areas. If we think of a matroid in the context of linear algebra, then independent sets may be more natural to consider than circuits.

The many faces, and underlying simplicity, of matroid theory are an attractive and unusual combination. We now verify that Definition 2.1.2 and Definition 14.1.2 agree in the context of finite geometries.

Theorem 14.2.2. *If $(E, \mathcal{C})$ is a finite geometry, then $\mathcal{C}$ is the set of circuits of a matroid on the set E.*

Proof. From Definition 2.1.1 we have that $\mathcal{C}$ satisfies Conditions **MC1**, **MC2**, and **MC3**. From Theorem 14.2.1 we have that $\mathcal{C}$ is therefore the set of circuits of a matroid on E. □

Example 14.2.1. Let $M[A]$ be the real vector matroid of Example 14.1.1. This matroid is called the non-Fano matroid (see Definition 2.3.2). Then $\mathcal{C}(M[A])$ consists of the members of $\{\{1,3,7\}, \{1,2,4\}, \{2,3,5\}, \{3,4,6\}, \{4,5,7\}, \{2,6,7\}\}$ as well as each four-element subset of E that does not contain one of the above three-element subsets.

Since the non-Fano matroid is a combinatorial geometry, it can be sketched (use arcs that are straight or curved) similar to the sketches given in Chapter 2. Such a sketch for a matroid is called a geometric representation (see [46] or [49]). A geometric representation for the non-Fano matroid is given in Figure 14.1.

Exercise 14.2.1. Label the points of Figure 14.1 so that each set of three collinear points of the figure is one of the six three-element subsets mentioned in Example 14.2.1.

Lemma 14.2.1. *The independent sets of a finite planar geometry are: the empty set, each singleton, each two-point set, and each non-collinear three-*

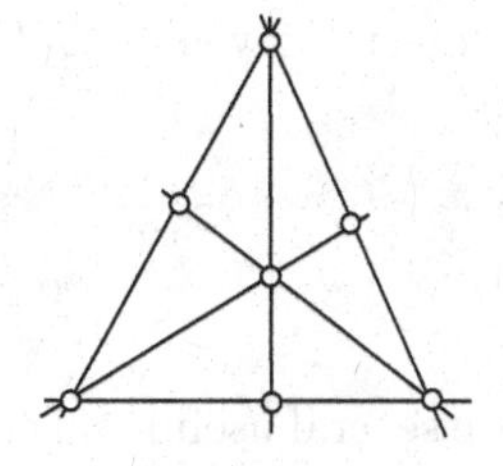

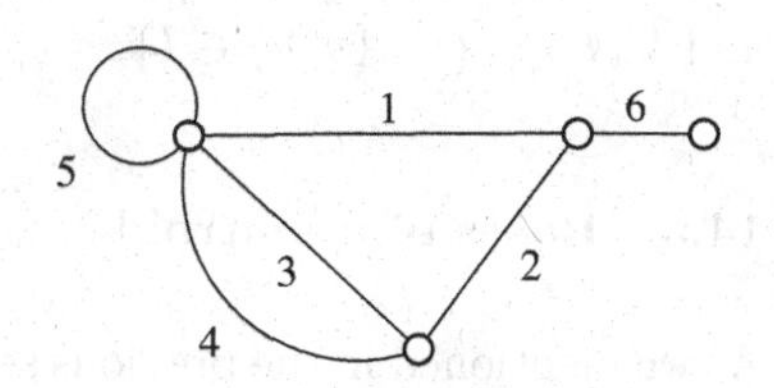

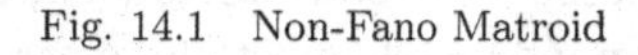

Fig. 14.1 Non-Fano Matroid

Fig. 14.2 A non-simple graph

point set. The independent sets of a finite non-planar geometry include also the non-coplanar four-point sets.

Proof. The sets listed in the statement are exactly those sets that do not contain a circuit. □

Example 14.2.2. The independent sets of the Vámos cube of Definition 2.3.6 are all the subsets of $\{0,1,2,3,4,5,6,7\}$ that contain at most four points, except the sets $\{0,1,3,5\}$, $\{0,2,3,6\}$, $\{1,4,5,7\}$, $\{2,4,6,7\}$, and $\{0,3,4,7\}$. (see Figure 2.15 for an attempted drawing of the Vámos geometry - in Chapter 4 we showed that the Vámos cube is not a figure and so does not have a drawing)

Theorem 14.2.3. *Let G be a graph that has only finitely many edges. Then $\mathcal{C} = \{C : C \text{ is the edge set of a cycle of } G\}$ is the set of circuits of a matroid M on the set $E(G)$. We call M the cycle matroid of G and we denote it by $M(G)$.*

Proof. From Lemma 11.5.1 and Theorem 14.2.1 we have that the collection of edge-sets of cycles of G satisfies Conditions **MC1**, **MC2**, and **MC3**. □

Example 14.2.3. Let $M(G)$ be the cycle matroid defined by the graph G of Figure 14.2. Then $\mathbf{I}(M(G)) = \{\emptyset$, each single edge other than 5, each set of two edges not containing 5 and not equal to $\{3,4\}$, each set of three edges that includes 6 but not 5 and is not equal to $\{3,4,6\}\}$, and $\mathcal{C}(M(G)) = \{\{5\},\{3,4\},\{1,2,3\},\{1,2,4\}\}$.

Lemma 14.2.2. *The independent sets of the cycle matroid of the graph G are the edge-sets of forests of G.*

Proof. The edge-set of a forest does not contain the edge-set of any cycle and by Corollary 14.2.1 is independent in $M(G)$. Conversely, an independent set I of $M(G)$ is the edge-set of the forest F given by $E(F) = I$ and $V(F) = \{u : \{u, v\} \in I\}$. □

14.3 Bases of a matroid

As we mentioned in the previous section, there are several useful and equivalent characterizations of a matroid. We have used two - one via independent sets, and one via circuits. As each subset of an independent set of a matroid M is also independent, the maximal members of $\mathbf{I}(M)$ suffice to uniquely determine $\mathbf{I}(M)$. In this section we examine these maximal independent sets and characterize them.

Lemma 14.3.1. *Let M be a matroid. Then any two maximal independent subsets of any $A \subseteq E(M)$ are equicardinal. In particular, any two maximal independent subsets of E are equicardinal.*

Proof. Suppose that I and J are maximal independent subsets of A and $|J| < |I|$. Then Condition **MI3** guarantees the existence of an independent set $I' \subseteq I \cup J \subseteq A$, so that $|J| < |I'|$, contradicting the choice of J. Therefore $|J| \geq |I|$, and repeating the argument above with the roles of I and J interchanged gives $|I| = |J|$ as the only possibility. □

Definition 14.3.1. *We call each maximal independent subset of a matroid M a* **basis** *of M.*

Lemma 14.3.1 gives enough information to nicely characterize the set $\mathbf{B}(M)$ of bases of a matroid M.

Lemma 14.3.2. *The collection of bases of any matroid M satisfies the two conditions:*

MB1: $\mathbf{B}(M) \neq \emptyset$.

MB2: *If B_1 and B_2 are two bases of M and $e \in B_1 - B_2$, then there is an element f of $B_2 - B_1$ such that $(B_1 - \{e\}) \cup \{f\}$ is a basis of M.*

Proof. The first requirement follows from **MI1** directly. In order to prove the second condition we observe that $B_1 - \{e\}$ and B_2 are independent, and from Lemma 14.3.1 $|B_1 - \{e\}| < |B_2|$. Therefore, from Condition **MI3** there is an element f of $B_2 - (B_1 - \{e\})$ so that $(B_1 - \{e\}) \cup \{f\}$ is independent. From $|(B_1 - \{e\}) \cup \{f\}| = |B_1|$ we have that $(B_1 - \{e\}) \cup \{f\}$ is a maximal independent set, as required. □

Lemma 14.3.3. *Let* $\mathbf{B}$ *be a collection of subsets of a finite set* E *that satisfies the two conditions:*

MB1: *Th matroid* M *has at least one basis.*

MB2: *If* B_1 *and* B_2 *are members of* $\mathbf{B}$ *and* $e \in B_1 - B_2$*, then there is an element* f *of* $B_2 - B_1$ *such that* $(B_1 - \{e\}) \cup \{f\}$ *is in* $\mathbf{B}$.

Then the members of $\mathbf{B}$ *are equicardinal.*

Proof. Suppose that B_1 and B_2 are distinct members of $\mathbf{B}$ for which $|B_1| > |B_2|$, such that, among all such pairs, $|B_1 - B_2|$ is minimal. Clearly, $B_1 - B_2$ is not empty. Choosing $e \in B_1 - B_2$, we can find an element f of $B_2 - B_1$ so that $(B_1 - \{e\}) \cup \{f\} \in \mathbf{B}$. But $|(B_1 - \{e\}) \cup \{f\}| = |B_1| > |B_2|$ and $|((B_1 - \{e\}) \cup \{f\}) - B_2| < |B_1 - B_2|$. This contradicts the choice of B_1 and B_2. □

Theorem 14.3.1. *Let* $\mathbf{B}$ *be a collection of subsets of a finite set* E *that satisfies the following two conditions:*

MB1: $\mathbf{B} \neq \emptyset$.

MB2: *If* B_1 *and* B_2 *are two members of* $\mathbf{B}$ *and* $e \in B_1 - B_2$*, then there is an element* f *of* $B_2 - B_1$ *such that* $(B_1 - \{e\}) \cup \{f\}$ *is a member of* $\mathbf{B}$.

Then the ordered pair E *and* $\mathbf{I} = \{I \subseteq E : I$ *is a subset of a member of* $\mathbf{B}\}$ *is a matroid* M. *The members of* $\mathbf{B}$ *are the bases of* M.

Conversely, the collection $\mathbf{B}$ *of bases of a matroid* M *satisfies Conditions* **MB1** *and* **MB2**.

Proof. Clearly the set $\mathbf{I}$ satisfies Conditions **MI1** and **MI2**

Suppose that I and J belong to $\mathbf{I}$ and that $|J| < |I|$. If **MI3** fails for these two sets, then, for all e in $I - J$, $J \cup \{e\}$ is not in $\mathbf{I}$. Therefore $I - B_1 = I - J$.

From the definition of $\mathbf{I}$, it follows that the set $\mathbf{B}$ has two members B_1 and B_2 such that $J \subseteq B_1$ and $I \subseteq B_2$. We suppose that B_2 is chosen such that $|B_2 - (I \cup B_1)|$ is minimal. If there is an $x \in B_2 - (I \cup B_1)$, then Condition **MB2** implies that there is some $y \in B_1 - B_2$ such that $(B_1 - \{x\}) \cup \{y\}$ is in $\mathbf{B}$. This contradicts the minimality of $|B_2 - (I \cup B_1)|$. Therefore $B_2 - (I \cup B_1) = \emptyset$ and $B_2 - B_1 = I - B_1$. Furthermore, $B_2 - B_1 = I - J$.

Next suppose that $x \in B_1 - (I - B_2)$. From **MB2** there is an element y of $B_2 - B_1$ such that $(B_1 - \{x\}) \cup \{y\} \in \mathbf{B}$. Then $J \cup \{y\}$ is in $\mathbf{I}$, and as $y \in B_2 - B_1 = I - J$, this contradicts our choice of J and I. Thus $B_1 - (I - B_2) = \emptyset$ and so $B_1 - B_2 = I - B_2 \subseteq I - J$. From Lemma 14.3.3 $|B_1| = |B_2|$. Thus $|B_1 - B_2| = |B_2 - B_1|$

Therefore $|I - J| \geq |J - I|$ and consequently $|J| \geq |I|$. This contradicts the possibility of choosing a pair I and J for which **MI3** fails. Therefore the ordered pair $(E, \mathbf{I})$ is a matroid M.

Each member of $\mathbf{B}$ is independent and therefore a subset of a basis. Each basis is independent and therefore a subset of a member of $\mathbf{B}$. It follows from the members of $\mathbf{B}$ being equicardinal and the bases being equicardinal that $\mathbf{B} = \mathbf{B}(M)$.

The converse is Lemma 14.3.2. □

Corollary 14.3.1. *Let M be a matroid. Then*
$\mathbf{I}(M) = \{I \subseteq E : I$ *is a subset of a member of* $\mathbf{B}(M)\}$, *and*
$\mathbf{B}(M) = \{B \subseteq E : B$ *is in* $\mathbf{I}(M)$ *but no set containing* B *is in* $\mathbf{I}(M)\}$.

Example 14.3.1. Let A be the matrix of Example 14.1.1. Then $\mathbf{B}(M[A]) = \{X \subseteq \{1,2,3,4,5,6,7\} : |X| = 3$ and X is not one of $\{1,3,7\}$, $\{1,2,4\}, \{2,3,5\}, \{3,4,6\}, \{4,5,7\}, \{2,6,7\}\}$.

Lemma 14.3.4. *Let $(E, \mathcal{C})$ be a finite geometry. If E is a point, then the point is a basis of the geometry. If E is a line, then each two-point subset of E is a basis. If E is a plane, then each set of three non-collinear points is a basis. If $(E, \mathcal{C})$ is non-planar, then each set of four non-coplanar points is a basis.*

Proof. In each case the set above consists of exactly the maximal members of the corresponding set given in Lemma 14.2.1. □

From Chapter 11 we have that each basis of the cycle matroid $M(G)$ of a graph G is the edge set of a maximal forest of G. Therefore we have the following corollary of Lemma 14.3.1.

Exercise 14.3.1. Let G be a connected graph containing only finitely many edges. Prove that the bases of $M(G)$ are the edge-sets of maximal forests of G, and each basis has $|V(G)| - 1$ members. Draw each maximal forest of the graph drawn in Figure 14.2.

14.4 The rank function of a matroid

In this section, we continue our discussion of equivalent definitions of matroids. Lemma 14.3.1 enables us to associate a well-defined number with each set of points of a matroid. In this way we introduce the notion of a dimension, or rank, function of a matroid.

Definition 14.4.1. *Let M be a matroid on E. With each subset A of E we associate the number of elements in any maximal independent subset of A. We call this number, $rk_M(A)$, the* **rank** *of A. The function, rk_M, defined in this way for all subsets of E is the rank function of M. The value the rank function takes on E, $rk_M(E)$, is called the rank of M.*

Lemma 14.4.1. *Let M be a matroid and suppose that A and B are subsets of $E(M)$ and that a is an element of $E(M)$. Then the rank function, rk_M, of M satisfies the three conditions:*

MR1: $rk_M(\emptyset) = 0$.

MR2: $rk_M(A) \leq rk_M(A \cup a) \leq rk_M(A) + 1$.

MR3: $rk_M(A) + rk_M(B) \geq rk_M(A \cup B) + rk_M(A \cap B)$.

Proof. The first condition follows from $|\emptyset| = 0$. The second condition follows from the observation that if I is a maximal independent subset of A, then either I or $I \cup \{a\}$ is maximally independent in $A \cup a$.

For any subsets A and B of E, we choose a maximal independent subset J of $A \cap B$, and a maximal independent subset I of $A \cup B$ that contains J. Then $|I \cap A| + |I \cap B| = |(I \cap A) \cap (I \cap B)| + |(I \cap A) \cup (I \cap B)| = |I \cap (A \cap B)| + |I \cap (A \cup B)| = |J| + |I|$. From $rk_M(A) \geq |I \cap A|$ and $rk_M(B) \geq |I \cap B|$, we have $rk_M(A) + rk_M(B) \geq rk_M(A \cap B) + rk_M(A \cup B)$. □

Lemma 14.4.2. *Let r be a function defined on the power set of a finite set E, taking integer values, and satisfying the following three conditions:*

MR1: $r(\emptyset) = 0$.

MR2: *If $A \subseteq E$ and $a \in E$, then $r(A) \leq r(A \cup a) \leq r(A) + 1$.*

MR3: *(Submodularity Condition) If A and B are subsets of E, then $r(A) + r(B) \geq r(A \cup B) + r(A \cap B)$.*

If $A \subseteq E$, then $r(A) \leq |A|$. If $B \subseteq A \subseteq E$ and $r(B \cup \{a\}) = r(B)$ for each element a of $A - B$, then $r(A) = r(B)$.

Proof. A simple induction argument, starting with $r(\emptyset) = 0$ and using $r(A' \cup \{a\}) \leq r(A') + 1$, proves the first result.

Now suppose that $B \subseteq A \subseteq E$ and $r(B \cup \{a\}) = r(B)$. We write $A - B = \{a_1, a_2, \ldots, a_m\}$. It follows from $r(B \cup \{a_i\}) = r(B)$, for all i, that by applying the inequality **MR3**, we have $r(B \cup \{a_1, a_2, \ldots, a_{i-1}) + r(B \cup \{a_i\}) \geq r(B \cup \{a_1, a_2, \ldots, a_i\}) + r(B)$. From this we have that $r(B \cup \{a_1, a_2, \ldots, a_{i-1}) \geq r(B \cup \{a_1, a_2, \ldots, a_i\})$, and from the first result, we have that $r(B \cup \{a_1, a_2, \ldots, a_{i-1}) = r(B \cup \{a_1, a_2, \ldots, a_i\})$. Applying this result $m - 1$ times we have that $r(B) = r(A)$. □

Theorem 14.4.1. *Let r be a function defined on the power set of a finite set E, taking integer values, and satisfying the following three conditions:*

MR1: $r(\emptyset) = 0$.

MR2: *If $A \subseteq E$ and $a \in E$, then $r(A) \leq r(A \cup a) \leq r(A) + 1$.*

MR3: *(Submodularity Condition) If A and B are subsets of E, then $r(A) + r(B) \geq r(A \cup B) + r(A \cap B)$.*

Then the ordered pair E and $\mathbf{I} = \{I \subset E : r(I) = |I|\}$ is a matroid M. The function r is the rank function of M.

Conversely, the rank function r of a matroid M satisfies Conditions **MR1**, **MR2**, *and* **MR3**.

Proof. That the set $\mathbf{I}$ satisfies Condition **MI1** is a direct consequence of **MR1** and the definition of $\mathbf{I}$.

Suppose that $I \in \mathbf{I}$ and $J \subseteq I$. From **MR3** we have that $r(J) + r(I - J) \geq r(I) + r(\emptyset)$. This together with Lemma 14.4.2 gives $r(J) + |I - J| \geq |I|$, and therefore $r(J) \geq |J|$. But from Lemma 14.4.2 we have that $r(J) = |J|$, giving J in $\mathbf{I}$ and Condition **MI2** holds for $\mathbf{I}$.

Now suppose that I and J are in $\mathbf{I}$ with $|J| < |I|$. We write $I - J = \{a_1, a_2, \ldots, a_m\}$. If $r(J \cup \{a_i\}) = r(J) + 1$ for some a_i, then $J \cup \{a_i\} \in \mathbf{I}$ and **MI3** holds for I and J. On the other hand if $r(J \cup \{a_i\}) = r(J)$, for all i, then from **MR3** we would have $r(J \cup \{a_1, a_2, \ldots, a_{i-1}) + r(J \cup \{a_i\}) \geq r(J \cup \{a_1, a_2, \ldots, a_i\}) + r(J)$. From this we would have that $r(J \cup \{a_1, a_2, \ldots, a_{i-1}) > r(J \cup \{a_1, a_2, \ldots, a_i\})$, and thus $r(J \cup \{a_1, a_2, \ldots, a_{i-1}) = r(J \cup \{a_1, a_2, \ldots, a_i\})$. Applying this result $m - i$ times we would have that $r(J) = r(I) = |I| > |J|$, which is false. Therefore **MI3** holds for $\mathbf{I}$.

Consequently, $\mathbf{I}$ is the collection of independent sets of a matroid M on E. Suppose that $A \subseteq E$. Suppose further that I is a maximal independent subset of A. From the definition of the rank function of M, the definition of $\mathbf{I}$, and by applying Lemma 14.4.2 we have $rk_M(A) = |I| = r(I) = r(A)$. Thus $r = rk_M$.

The converse is Lemma 14.4.1. □

Corollary 14.4.1. *Let M be a matroid. Then*
$\mathbf{I}(M) = \{I \subset E : rk_M(I) = |I|\}$, *and for each $A \subseteq E(M)$,*
$rk_M(A) = max\{|I| : I \subseteq A, \text{ and } I \in \mathbf{I}(M)\}$.

Lemma 14.4.3. *Let A be a matrix. Then the rank of $M[A]$ is equal to the row rank and column rank of A.*

Lemma 14.4.4. *Let $(E, \mathcal{C})$ be a finite geometry. The rank of a point of E is 1, the rank of a line is 2, and the rank of a plane is 3. If the geometry is planar, then its rank is at most 3. If it is non-planar, then it has rank 4.*

Proof. This a consequence of the characterization of bases given in Lemma 14.3.4. □

Example 14.4.1. Let G be the graph drawn in Figure 14.2. Then $rk_{M(G)}\{5\} = 0$, $rk_{M(G)}\{3,4\} = 1$, and $rk_{M(G)}\{1,2,6\} = rk_{M(G)}\{1,2,3,4,5,6\} = 3$.

Exercise 14.4.1. Prove that, for each finite set E of n points and non-negative integer $k \leq n$, $r(A) = \min\{|A|, k\}$, $\forall A \subseteq E$, is the rank function of a matroid $U_{k,n}$. Calculate $\mathcal{C}(U_{k,n})$, $\mathbf{I}(U_{k,n})$ and $\mathbf{B}(U_{k,n})$.

Definition 14.4.2. *The matroid $U_{k,n}$ is a* **uniform matroid**. *If $k = n$, then $\mathcal{C}(U_{k,n}) = \emptyset$ and $U_{n,n}$ is the* **free matroid** *on E.*

14.5 Isomorphism and representable matroids

Just as we did in Definition 2.7.2 for geometries, we explore the meaning of the phrase "the same" in relation to matroids. This enables us to pursue a goal that we have previously alluded to in Chapters 3 and 4, namely replacing geometric questions by routine arithmetic questions.

Definition 14.5.1. *Two matroids are* **isomorphic** *if the points of the first can be paired with the points of the second, and paired points given the same label, so that a list of independent sets of the first matroid is identical to a list of independent sets of the second.*

As we would hope, each of the four ways that characterize matroids in this chapter lead to a characterization of isomorphic matroids (we used independent sets in the definition).

Lemma 14.5.1. *Two matroids are isomorphic if the points of the first can be paired with the points of the second, and paired points given the same label, so that*

(i) a list of circuits of the first matroid is identical to a list of circuits of the second, or

(ii) a list of bases of the first matroid is identical to a list of bases of the second, or

(iii) the two rank functions act identically on the set of labels.

Proof. Corollary 14.2.1 guarantees that identical lists of circuits lead to identical lists of independent sets. Therefore Condition (i) ensures the existence of the required isomorphism. Similarly Corollaries 14.3.1 and 14.4.1 enable us to complete the proof. □

Definition 14.5.2. *A matroid is* **representable** *over the field F if it is isomorphic to a vector matroid over F. A matroid is graphic if it is isomorphic to the cycle matroid of a graph G.*

Example 14.5.1. By relabelling the point 0 of the non-Fano geometry by 7 and then verifying that a list of its circuits is identical to the list of circuits in Example 14.2.1, we prove that the non-Fano geometry is representable over the real number field.

The field F of Definition 14.5.2 plays a crucial role in the notion of a representable matroid. We see this by allowing the entries of the matrix A of Example 14.1.1 to come from a field other than the field of real numbers.

Theorem 14.5.1. *The Fano geometry is representable over a field F if and only if $1+1=0$ in F. The non-Fano geometry is representable over a field F if and only if $1+1 \neq 0$ in F.*

Proof. Let $A = \begin{pmatrix} 1\,0\,0\,1\,0\,1\,1 \\ 1\,0\,1\,1\,1\,0\,0 \\ 0\,1\,0\,1\,1\,1\,0 \end{pmatrix}$. Label the columns of the matrix by 1 through 7 from left to right. If the entries of A are from a field F in which $1+1=0$, then the set $\{1,5,6\}$ is a circuit and we obtain the Fano geometry sketched in Figure 2.10. If on the other hand $1+1 \neq 0$, then $\{1,5,6\}$ is not a dependent set and the matroid $M[A]$ is isomorphic to the non-Fano geometry.

Conversely, let us suppose that the Fano geometry is isomorphic to a vector matroid $M[A]$. We write 0 to stand for a zero column, and $i \neq 0$ to mean the ith column. Any column in A may be replaced by a non-zero scalar multiple of the column, so without loss of generality the following dependencies occur in the columns of A: $4+5+7=0, 1=3+7, 6=3+4$. Therefore $1+5+6=(3+7)+5+(3+4)=(3+3)+(4+5+7)=3+3$. Therefore the first, fifth, and sixth columns are linearly dependent if and only if $1+1=0$ in F. □

We saw in Chapter 11 that a geometry and a graphic matroid may be isomorphic. The following Example proves that each may also be representable and illustrates the desirability of further investigation into the relations between geometries, representable matroids, and graphic matroids.

Example 14.5.2. By relabelling the edge 0 of the drawing of K_5 in Figure 11.16 by 10 and then verifying that a list of the circuits of the cycle matroid $M(K_5)$ is identical to a list of the minimally linearly dependent sets of columns of the matrix A below we prove that $M(K_5)$ is representable over the real number field. From the solution to Problem 11.5.1 we have that the the non-planar Desargues geometry is also isomorphic to $M[A]$, for the real matrix

$$A = \begin{pmatrix} 1\,0\,0\,0\,1\,1 & 0 & 1 & 1 & 1 \\ 0\,1\,0\,1\,0\,1 & 1 & 0 & -1 & 1 \\ 0\,0\,1\,1\,1\,0 & -1 & -1 & 0 & 1 \\ 0\,0\,0\,1\,1\,1 & 0 & 0 & 0 & 1. \end{pmatrix}.$$

We may go further:

Theorem 14.5.2. *Each graphic matroid is representable over every field.*

Proof. Let G be a graph. We construct a $|V(G)| \times |E(G)|$ matrix A with entries from a field F as follows. Each row of A is indexed by a vertex of G and each column of A by an edge of G.

We order the vertices within each edge of G, calling the first vertex the tail, and the second the head. Then the entry $a_{v,e}$ in the vth row and eth column of A is -1 if v is the tail of e and e is not a loop, it is 1 if v is the head of e and e is not a loop, and it is 0 otherwise.

A single column of A is a circuit of $M[A]$ if and only if it is a zero column. This is so if and only if the edge indexing the column is a loop in G. Suppose that $m > 1$ and $\{\{v_1, v_2\}, \{v_2, v_3\}, \ldots, \{v_{m-1}, v_m\}, \{v_m, v_1\}\} = C$ is a member of $\mathcal{C}(M(G))$.

For each $i \in \{1, 2, \ldots, m-1\}$ we multiply the column that has index $\{v_i, v_{i+1}\}$ by 1 if v_i is the tail of $\{v_i, v_{i+1}\}$, and by -1 if v_i is the head of $\{v_i, v_{i+1}\}$. Adding the resulting columns gives a zero column. Thus C is linearly dependent and C contains a member of $\mathbf{C}(M[A])$.

Conversely, we suppose that C is a circuit of $M[A]$. Then there is a linear combination of the members of C, in which each column has a non-zero coefficient, that sums to zero. Consequently, each row of A contains non-zero entries in at least two columns of C. Two such entries represent a vertex belonging to two edges of G. Proceeding from one vertex to another in this way gives the vertices of a circuit of G. Therefore C contains a circuit of $M(G)$.

Therefore each matroid has an identical set of circuits and the two are isomorphic. □

Exercise 14.5.1. Let G be the graph drawn in Figure 14.2. Find a real matrix A so that $M[A]$ is isomorphic to $M(G)$.

Exercise 14.5.2. Prove that a figure of four collinear points is not graphic. Give an example of a graphic matroid that is not a geometry.

Exercise 14.5.3. Specify the values of k and n for which the uniform matroid $U_{k,n}$ is a geometry.

Exercise 14.5.4. Prove that the geometry sketched in Figure 14.3 is representable over the complex number field but not over the real number field.

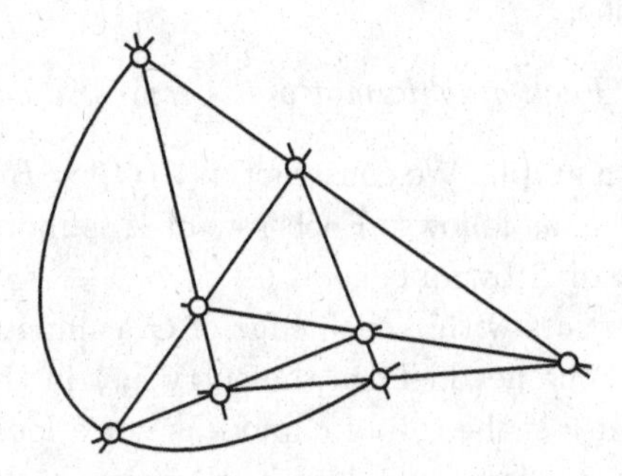

Fig. 14.3 An eight-point planar geometry that is not a figure

14.6 Projective and affine geometry

In this section, we generalize our notion of a figure, in order to allow figures that have a rank greater than 4 and also to allow coordinates from fields other than the real number field. The notion of a figure containing only Euclidean points is carried over to this more general setting for affine and projective geometry.

Let A be a $3 \times n$ matrix in which the i-th column is the transpose of (a_{1i}, a_{2i}, a_{3i}). If no column of A is a scalar multiple of another, then we may consider the set E containing each point of an extended Euclidean plane that has the transpose of a column as a set of homogeneous coordinates, distinct columns giving distinct points. Then, from Proposition 3.6.2 and Theorem 14.1.1, we have that the figure E is isomorphic to the vector matroid $M[A]$. This isomorphism motivates the following definition of a projective geometry.

Definition 14.6.1. *Let F be a field and r a natural number. Suppose that A is an $(r+1) \times n$ matrix with entries from F. Suppose further that no column of A is a scalar multiple of another column. Then $M[A]$ is a* **projective geometry** *over the field F. The elements of the projective geometry are commonly called points of projective r-space over F.*

Example 14.6.1. The matrix $A = \begin{pmatrix} 1\,0\,0\,1\,0\,1\,1 \\ 0\,1\,0\,1\,1\,1\,0 \\ 0\,0\,1\,0\,1\,1\,1 \end{pmatrix}$ defines a vector matroid $M[A]$ that is a projective geometry. As we saw in Theorem 14.5.1, if the entries in A are from a field for which $1+1 \neq 0$, then $M[A]$ is the non-Fano geometry. If they are from a field for which $1+1=0$, then $M[A]$ is the Fano geometry.

If we had not allowed ourselves the pleasure of removing the special nature of parallel lines in Chapters 3 and 4, then figures that we discuss would contain only Euclidean points and be the subjects of the Cartesian coordinate geometry that is familiar to us from college. We recall from plane Cartesian coordinate geometry that a point (x_1, y_1) belongs to the Euclidean line containing (x_2, y_2) and (x_3, y_3) if α and β exist such that $x_1 = \alpha x_2 + \beta x_3$ and $y_1 = \alpha y_2 + \beta y_3$; equivalently if $(1-\alpha-\beta)(x_1, y_1) + \alpha(x_2, y_2) + \beta(x_3, y_3) = (0, 0)$.

The triple $(1, x, y)$ is a set of homogeneous coordinates of each Euclidean point (x, y) whereas the first member of each triple that labels any ideal point is 0. Thus insisting that a finite planar figure E contain only Euclidean points is equivalent to asking thatthe figure is isomorphic to a vector matroid $M[A]$, for which A is a real matrix in which each column has a transpose of the form $(1, x, y)$. We make use of the fact that each entry of the first row of this matrix is 1 in order to extend our notions of affine and projective geometry.

Definition 14.6.2. *Let F be a field and r a natural number. Suppose that A is an $(r+1) \times n$ matrix with entries from F and in which, for each $i \in \{1, 2, \ldots, n\}$, the transpose of $(1, a_{1i}, a_{2i}, \ldots, a_{ri})$ is the i-th column.*

Suppose further that no two columns of A are the same. Then the matroid $M[A]$ is an **affine geometry** *over the field F. The elements of the affine geometry are commonly called points of affine r-space over F.*

A routine verification shows that the calculations in an affine geometry generalize those in the Cartesian coordinate calculations of the usual college geometry course gives the following lemma:

Lemma 14.6.1. *It is usual to write $(a_{1i}, a_{2i}, \ldots, a_{ri})$ rather than $(1, a_{1i}, a_{2i}, \ldots, a_{ri})^T$ as notation for the i-th point of an affine geometry. Let M be an affine geometry over the field F. Then a subset $A = \{(a_{1i}, a_{2i}, \ldots, a_{ri}) : i \in \{j(1), j(2), \ldots, j(m)\}$ of $E(M)$ is dependent in the geometry if there exist m members α_i of F, not all zero, so that both $\sum_{i=1}^{m} \alpha_{j(i)}(a_{1j(i)}, a_{2j(i)}, \ldots, a_{rj(i)}) = (0, 0, \ldots, 0)$ and $\sum_{i=1}^{m} \alpha_j(i) = 0$.*

Proof. We have that $\sum_{i=1}^{m} \alpha_{j(i)}(1, a_{1j(i)}, a_{2j(i)}, \ldots, a_{rj(i)})^T = (0, 0, 0, \ldots, 0)^T$ if and only if both conditions $\sum_{i=1}^{m} \alpha_{j(i)}(a_{1j(i)}, a_{2j(i)}, \ldots, a_{rj(i)}) = (0, 0, \ldots, 0)$ and $\sum_{i=1}^{m} \alpha_j(i) = 0$ are met. □

Lemma 14.6.2. *Each affine geometry over a field F is representable over F.*

Proof. Each affine geometry over F is by definition a vector matroid. □

Lemma 14.6.3. *No affine matroid over the field $GF(2)$ contains a 3-point circuit.*

Proof. Suppose that $\{(a_{1i(j)}, a_{2i(j)}, \ldots, a_{ri(j)}) : j = 1, 2, 3\}$ was a circuit of an affine matroid over $GF(2)$. Then there would exist three members α_1, α_2 and α_3 of F, not all zero, so that both $\sum_{j=1}^{3} \alpha_j(a_{1i(j)}, a_{2i(j)}, \ldots, a_{ri(j)}) = (0, 0, \ldots, 0)$ and $\alpha_1 + \alpha_2 + \alpha_3 = 0$. No α_j is zero, as that would imply a subset of the circuit was dependent. Therefore $\alpha_1 = \alpha_2 = \alpha_3 = 1$. But $1 + 1 + 1 \neq 0$ in $GF(2)$. □

We referred earlier, in Chapters 3 and 4, to the fact that each finite real representable geometry is also an affine figure. The above lemma shows that this is not true in general, for example the three-point line is representable over $GF(2)$, but it is not an affine matroid over $GF(2)$.

14.7 Dual matroids

In this section, we introduce a notion of duality for matroids. We could, with more difficulty, have developed a theory of matroids on not-necessarily-finite sets. This is done in [16]. Possibly the extra structures included in such a treatment, for example EE^2 and EE^3, justify the increased complexity of the proofs. But as matroid duality provides such a powerful tool, and is easily available only for matroids on finite sets, we restrict our investigations to matroids on finite sets.

Lemma 14.7.1. *Let B be a basis of a matroid M. Suppose that e is an element of $E(M)-B$. Then e belongs to exactly one circuit that is contained in $B \cup \{e\}$.*

Proof. As B is a maximal independent set it follows that $B \cup \{e\}$ contains a circuit that is not contained in B. Suppose that C_1 and C_2 are two such circuits. Then Condition **MC3** would ensure the existence of a circuit C such that $C \subseteq (C_1 \cup C_2) - \{e\} \subseteq B$, contradicting the independence of B. □

Lemma 14.7.2. *The set $\mathbf{B}(M)$ of bases of a matroid M satisfies the following condition:*

MB*2: *If B_1 and B_2 are two members of $\mathbf{B}(M)$ and e is an element of $B_2 - B_1$, then there is an element f of $B_1 - B_2$ such that $(B_1 - \{f\}) \cup \{e\}$ is a member of $\mathbf{B}(M)$.*

Proof. From Lemma 14.7.1, $B_1 \cup \{e\}$ contains exactly one circuit C. There is an element f of $C - B_2$. As $e \in B_2$, then $f \neq e$ and f is in B_1. The set $(B_1 - \{f\}) \cup \{e\}$ does not contain C and is therefore independent. Since B_1 and $(B_1 - \{f\}) \cup \{e\}$ each have the same number of elements, the latter is a basis. □

We are now able to associate with each matroid M a matroid, called the dual of M, on the same set of elements as M.

Theorem 14.7.1. *Let M be a matroid and let $\mathbf{B}^* = \{E(M) - B : B \in \mathbf{B}(M)\}$. Then $\mathbf{B}^*$ is the set of bases of a matroid.*

Proof. As $\mathbf{B}(M)$ is non-empty, $\mathbf{B}^*$ is non-empty and satisfies Condition **MB1**.

Suppose that both $B_1^* = E(M) - B_1$ and $B_2^* = E(M) - B_2$ are in $\mathbf{B}^*$ and that $e \in B_1^* - B_2^* = B_2 - B_1$. Then both B_1 and B_2 are in $\mathbf{B}$(M) and by **MB*2** there is an element f of $B_1 - B_2$ such that $(B_1 - \{f\}) \cup \{e\}$ is a member of $\mathbf{B}(M)$. Consequently, $f \in B_2^* - B_1^*$ and $E(M) - ((B_1 - \{f\}) \cup \{e\}) \in \mathbf{B}^*$. But $E(M) - ((B_1 - \{f\}) \cup \{e\}) = ((E(M) - B_1) - \{e\}) \cup \{f\} = (B_1^* - \{e\}) \cup \{f\}$. Therefore $\mathbf{B}^*$ satisfies Condition **MB2** and is the set of bases of a matroid on $E(M)$. □

Definition 14.7.1. *The matroid M^* defined by $\mathbf{B}(M^*) = \{E(M) - B : B \in \mathbf{B}(M)\}$ is the* **dual matroid** *of M.*

Lemma 14.7.3. *The dual matroid of the dual matroid of a matroid M is M, that is, $(M^*)^* = M$.*

Proof. We have $\mathbf{B}(M^*) = \{E - B : B \in \mathbf{B}(M)\}$, and again, $\mathbf{B}((M^*)^*) = \{E - (E - B) = B : E - B \in \mathbf{B}(M^*)$. □

Lemma 14.7.4. *Let M be a matroid on E. Then for each subset A of E, $rk_{M^*}(A) = |A| + rk_M(E - A) - rk_M(E)$.*

Proof. Let A be a subset of E. Suppose that B is a basis of M that has a smallest possible intersection with A. Then $E - B$ is a cobasis of M that has a largest possible intersection with A, and $rk_{M^*}(A) = |A \cap (E - B)|$. Similarly $rk_M(E - A) = |(E - A) \cap B|$. Thinking of B as the union of two disjoint subsets we obtain $(|A| - rk_{M^*}(A)) + rk_M(E - A) = |B|$. □

Exercise 14.7.1. Let M be a combinatorial cube. Prove that the matroid dual of M is also a combinatorial cube.

It would be interesting to know exactly when the matroid dual of a geometry (or figure) is a geometry (or figure).

Exercise 14.7.2. Prove that the dual of any uniform matroid is also a uniform matroid.

The following exercise provides another example of the interrelationships between the various aspects of the geometry that we have pursued.

Exercise 14.7.3. Prove that the cycle matroid of the hinge-graph of the octahedral models of Chapter 10 is dual to the cycle matroid of the graph of the octahedra. Prove also that the cycle matroid of the hinge-graph of the model of a sealed cube is dual to cycle matroid of the graph of the cube.

14.8 Restrictions and contractions of a matroid

In Theorem 2.6.1 we were delighted to find that every subset of the point-set of a geometry has an induced geometric structure on it. This carries over immediately to matroids. In this section we use the results of the previous section to prove that there are two structures on each subset of elements of a matroid. One is the obvious generalization of a subgeometry and the other has its origins in perspective drawings.

Definition 14.8.1. *Let M be a matroid and T be a subset of $E(M)$. The ordered pair T and the collection $\mathbf{I} \cap 2^E$ of subsets of T is called the* **restriction** *of M to T, or the* **deletion** *of $E - T$ from M, and is written either as $M|T$ or as $M\backslash(E-T)$.*

The details of the proof that the restriction of M to T is indeed a matroid are exactly as in the proof of Theorem 2.6.1.

Lemma 14.8.1. *Let M be a matroid on E and T be a subset of E. Then the restriction $M|T$ is a matroid.*

Lemma 14.8.2. *Let $M|T$ be a restriction of the matroid M. Then;*
(i) $\mathbf{I}(M|T) = \{I \subseteq T : I \in \mathbf{I}(M)\}$,
(ii) $\mathcal{C}(M|T) = \{C \subseteq T : C \in \mathcal{C}(M)\}$,
(iii) for each $A \subseteq T$, $rk_{M|T}(A) = rk_M(A)$.

Proof. The first statement is the definition of $\mathbf{I}(M|T)$. The specification of $\mathcal{C}(M|T)$ in (ii) follows from Corollary 14.2.1, $\mathcal{C}(M|T) = \{C \subseteq T : C$ is not in $\mathbf{I}(M|T)$ but every proper subset of C is in $\mathbf{I}(M|T)\} = \{C \subseteq T : C$ is not in $\mathbf{I}(M)$ but every proper subset of C is in $\mathbf{I}(M)\} = \{C \subseteq T : C \in \mathcal{C}(M)\}$. The specification of the rank function of $M|T$ follows from the observation that any $I \subseteq A$ is in $\mathbf{I}(M|T)$ if and only if it is in $\mathbf{I}(M)$. □

In the case that a matroid M is also a finite geometry it is clear that any restriction $M|T$ is a subgeometry of M. It is also straightforward to see that any restriction $M|T$ of a matroid M that is representable over a field F is also representable over F. If $M = M[A]$, then $M|T = M[A']$, where A' is obtained from A by deleting the columns not in T. In particular, if A' is the matrix obtained by deleting the i-th column of A, then the vector matroid $M[A']$ is $M[A]\backslash\{i\}$.

Exercise 14.8.1. Prove that the deletion of a point from the Fano geometry gives a matroid that is isomorphic to $M(K_4)$.

Exercise 14.8.2. Prove that matroid obtained by the deletion of a point from a uniform matroid is also a uniform matroid.

Example 14.8.1. The fourteen-point planar geometry drawn in Figure 14.4 cannot be representable over any field as the subgeometry on $\{1,2,3,4,5,6,7\}$ is representable over a field only if $1+1=0$, whereas the subgeometry on $\{8,9,10,11,12,13,14\}$ is representable over a field only if $1+1\neq 0$.

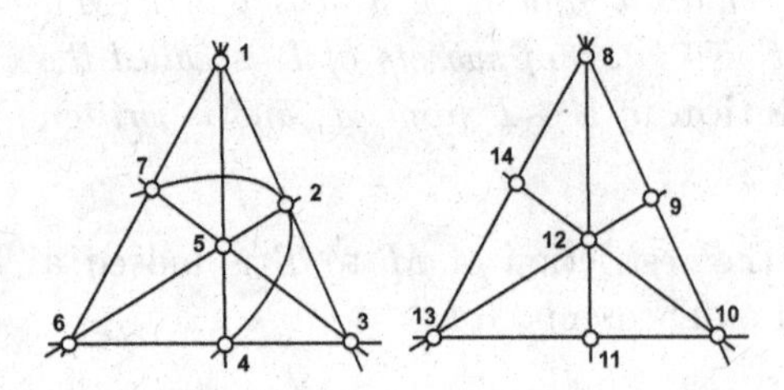

Fig. 14.4 A fourteen-point planar geometry that is not representable

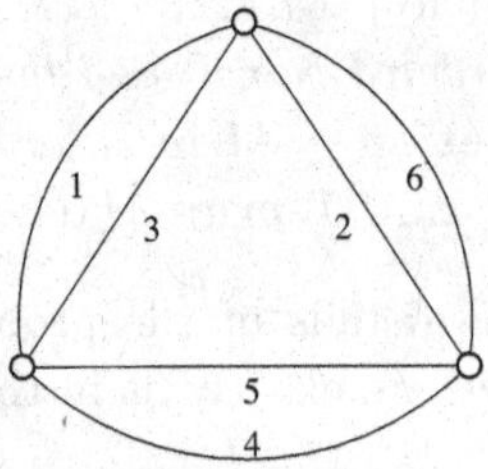

Fig. 14.5 A graphic minor of the planar Fano geometry

We are now begin to see the relationships between some important classes of matroids. For example; representable matroids encompass exactly those matroids whose structure can be analyzed with the convenience of linear algebra, graphic matroids have the advantage of a clear pictorial representation, and geometries are by now familiar to us. From Example 14.5.2 we have that all three classes have common members. From Theorem 14.5.2 we have that each graphic matroid is representable over every field. The results of Exercise 14.5.2 ensure that neither the class of geometries nor the class of graphic matroids is a subclass of the other. The fourteen-point planar geometry of Figure 14.4 demonstrates that not every geometry is representable. A matrix consisting of one column of zeros provides an example of a representable matroid that is not a geometry.

Definition 14.8.2. *Let M be a matroid and T be a subset of $E(M)$. The matroid $(M^*|T)^*$ is the* **contraction** *of M to T or the contraction of $E-T$ from M, and is written either as $M.T$ or as $M/(E-T)$.*

Lemma 14.8.3. *Let $M.T$ be a contraction of the matroid M. Then:*

(i) $\mathbf{I}(M.T)=\{I\subseteq T : M|T \text{ has a basis } B' \text{ such that } B'\cup I\in \mathbf{I}(M)\}$,

(ii) $\mathcal{C}(M.T)$ *consists of the minimal non-empty members of* $\{C\cap T : C\in\mathcal{C}(M)\}$,

(iii) for each $A \subseteq T$, $rk_{M.T}(A) = rk_M(A \cup (E-T)) - rk_M(E-T)$.

Proof. We suppose that $I \in \mathbf{I}(M.T)$ and $X \subseteq E-T$ is in $\mathbf{I}(M)$. Then if $I \cup X$ were not in $\mathbf{I}(M)$, for some circuit C of M, we would have $C \subseteq I \cup X$. But if $C \cap I \neq \emptyset$, then $C \cap T$ would contain a circuit of $M.T$, contradicting the choice of I. So $C \subseteq X$, contradicting the choice of X. Hence $I \cup X \in \mathbf{I}(M)$.

Conversely, we suppose that $I \subseteq T$, X is a maximal subset of $E-T$ that is in $\mathbf{I}(M)$ and $I \cup X \in \mathbf{I}(\mathrm{M})$. If there were a circuit of M that is contained in $I \cup (E-T)$, then we choose such a circuit C_1 so that $|C_1 - X|$ is minimal.

Either each element $a \in C_1 - X$ is in I, contradicting the independence of $I \cup X$, or there is an element $a \in (C_1 - X) \cap (E-T)$. In this case $X \cup \{a\}$ is dependent, as X is maximally independent in $E-T$, and there is a member C_2 of $\mathcal{C}(M)$ so that $a \in C_2 \subseteq X \cup \{a\}$.

If $C_1 \cap I \neq \emptyset$, then $C_1 \neq C_2$ and the Elimination Condition MC3 applied to C_1 and C_2 would give the existence of a circuit C of M so that $a \notin C \subseteq C_1 \cup C_2$. Then $|C-X| < |(C_1 \cup C_2) - X| = |C_1 - X|$, contradicting the choice of C_1. Therefore $C_1 \cap I = \emptyset$ and so $I \in \mathbf{I}(M|T)$. Thus condition (ii) is correct.

Suppose that $A \subseteq T$ and X is a maximal independent subset of $E-T$, and $I \subseteq A$ so that $I \cup X$ is a maximal independent subset of $A \cup (E-T)$. Then $rk_{M.T}(A) = |I| = |I \cup X| - |X| = rk_M(A \cup (E-T)) - rk_M(E-T)$. □

Example 14.8.2. Let M be the Fano geometry sketched in Figure 2.10. Then $\mathcal{C}(M/\{0\}) = \{\{1,3\}, \{2,6\}, \{4,5\}, \{1,2,4\}, \{2,3,5\}, \{3,4,6\}, \{1,5,6\}, \{1,2,5\}, \{1,4,6\}, \{2,3,4\}, \{3,5,6\}\}$. Each of the first three circuits is the remainder of a 3-point Fano circuit containing 0, the next four are 3-point Fano circuits not containing 0, and each of the last four is the remainder of a 4-point Fano circuit that contains 0. Each 4-point Fano circuit, that does not contain 0, properly contains one of the sets in $\mathcal{C}(M/\{0\})$. This list of circuits proves that $M/\{0\}$ is isomorphic to $M(G)$, where G is the graph drawn in Figure 14.5.

One of the motivations for matroid contraction came from the process of contracting an edge from a graph.

Definition 14.8.3. *Let G be a graph and suppose that $e = \{u_1, u_2\}$ is an edge of G. The graph $G/\{e\}$ is defined by $V(G/\{e\}) = (V(G) - \{u_1, u_2\}) \cup \{u\}$ and the list $E(G/\{e\})$ is obtained from $E(G)$ by deleting e and replacing each appearance of u_1 and u_2 in the remaining list by u. We call $G/\{e\}$ the* **contraction** *of e from G.*

We easily visualize contraction of e from G in any drawing of G by identifying u_1 with u_2 and erasing the arc representing e to give a drawing of $G/\{e\}$. Each of matroid restriction and matroid contraction is motivated by the corresponding construction for graphs, giving the following lemma.

Lemma 14.8.4. *Let G be a graph and e an edge of G. Then the cycle matroid $M(G/\{e\})$ of the graph $G/\{e\}$ is the restriction $M(G)/\{e\}$ of the cycle matroid $M(G)$. Also, the cycle matroid $M(G\backslash\{e\})$ of the graph $G\backslash\{e\}$ is the contraction $M(G)\backslash\{e\}$ of the cycle matroid $M(G)$.*

Exercise 14.8.3. Prove that the matroid obtained by the contraction of an element from a uniform matroid is also a uniform matroid.

Our experience of the world relies not only on examining figures, but also picturing them via the perspective rendition discussed in Chapter 5. There we hinted at a role for "perspective drawings" in any geometry in Theorem 5.2.1 and we can now formalize the situation.

Theorem 14.8.1. *Let M be a finite figure. Suppose that p is any point of the figure that does not belong to a 3-point circuit of M. Then the contraction, $M/\{p\}$ is isomorphic to the perspective drawing of the restriction $M\backslash\{p\}$ drawn from the general viewpoint p.*

Proof. From the condition on p it follows that perspective rendition from the viewpoint p pairs the points of $M\backslash\{p\}$ with their images in the perspective drawing. If we label each point of $E(M) - \{p\}$ by its image, then the result follows from a comparison of the list of circuits of $M/\{p\}$ given by Lemma 14.8.3 and a list of circuits of the perspective drawing given by Proposition 5.1.2 and Lemma 5.1.1. □

We notice that unfortunately the concept of matroid duality is not the same as that specified in Definition 3.2.2 for a matroid that is a projective plane. Similarly the concept differs from that of Definition 4.1.3 for those matroids that are projective spaces.

14.9 Minors of a matroid

As any restriction or contraction of a matroid is itself a matroid we may construct a restriction or contraction of this matroid. In this section we prove that matroids resulting from any sequence of restrictions and contractions are the natural "subobjects" of a matroid.

Example 14.9.1. Suppose that M is the figure drawn in Figure 14.6. Then first deleting the point 1 and then contracting 4 leaves a three-point line. On the other hand contracting 4 and then deleting 1 from the figure gives the same three-point line.

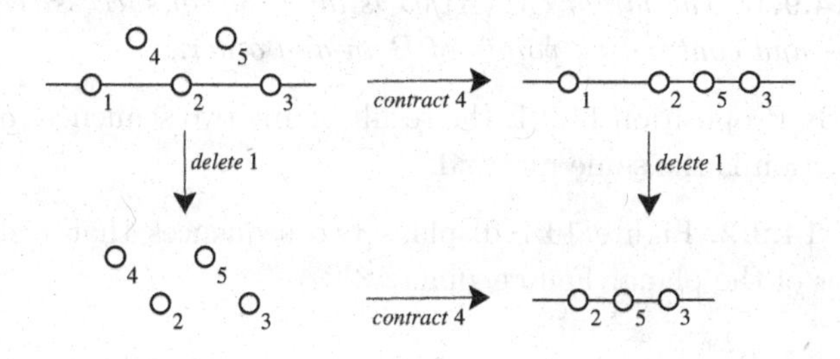

Fig. 14.6 A figure and three minors

The following Proposition shows that Example 14.9.1 is not an isolated case and enables us to give well-defined meaning to the term "subobject of a matroid".

Proposition 14.9.1. *Let M be a matroid on E. Then, for any distinct elements e and f of E, $(M\backslash\{e\})\backslash\{f\}) = M\backslash\{e,f\})$, $(M/\{e\})/\{f\}) = M/\{e,f\}$, and*
$(M\backslash\{e\})/\{f\}) = (M/\{f\})\backslash\{e\}$.

Proof. First, for any $A \subseteq E - \{e,f\}$ we have $rk_{(M\backslash\{e\})\backslash\{f\}}(A) = rk_{M\backslash\{e\}}(A) = rk_M(A) = rk_{M\backslash\{f\}}(A) = rk_{M\backslash\{e,f\}}(A)$. Second, $(M/\{e\})/\{f\} = ((M/\{e\})^*\backslash\{f\})^* = (((M^*\backslash\{e\})^*)^*\backslash\{f\})^* = ((M^*\backslash\{e\})\backslash\{f\})^* = (M^*\backslash\{e,f\})^* = M/\{e,f\}$.

Third, $rk_{(M\backslash\{e\})/f}(A) = rk_{M\backslash\{e\}}(A \cup \{f\}) = rk_M(A \cup \{f\})$ - $\mathrm{rk}_M(\{f\}) = rk_{M/\{f\}}(A) = rk_{(M/\{f\})\backslash\{e\}}(A)$. □

Let M be a matroid on E. By Proposition 14.9.1 any matroid obtained from M by a sequence of deletions and contractions may also be constructed by, for example, a single deletion followed by a single contraction - or equally well - by a single contraction followed by a single deletion. Or we could construct it by a sequence of 1-point deletions and 1-point contractions made in any order. The resultant matroid depends only on the subset of deleted elements and the subset of contracted elements. Finally, we have arrived at our "subobjects".

Definition 14.9.1. *Let M be a matroid. Suppose that A and B are two disjoint subsets of $E(M)$. Then $(M\backslash A)/B$ is a* **minor** *of M.*

Lemma 14.9.1. *The minor $(M\backslash A)/B$ is the result of successively deleting points of A and contracting points of B in any order.*

Proof. By Proposition 14.9.1, the result of any two sequences of deletion and contraction is the same matroid. □

Example 14.9.2. Figure 14.7 displays two sequences that result in the same minor of the planar Fano geometry.

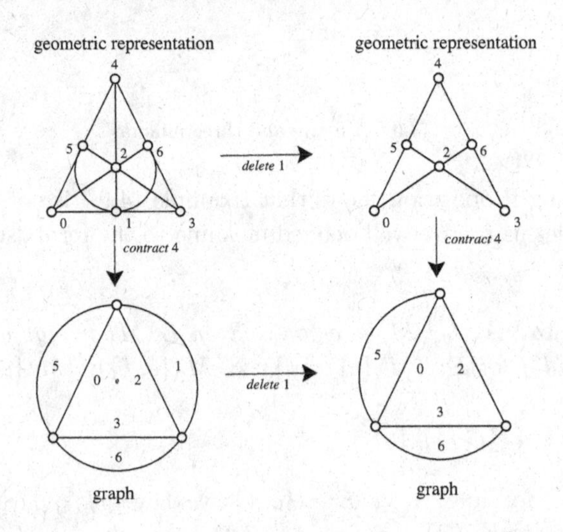

Fig. 14.7 Minors of the planar Fano geometry

Minors have two important roles. Many calculations within a matroid can be carried out recursively using its minors [10]. As well we describe any class of matroids, that with each member M also contains each minor of M, in terms of minimal non-members of the class.

Definition 14.9.2. *Any class of matroids, that with each member M also contains each matroid isomorphic to M and each minor of M, is* **hereditary**.

We have seen that graphic matroids have deletions and contractions which are graphic. Consequently all their minors are graphic, and graphic matroids form an hereditary class. It follows from Exercises 14.8.2, 14.8.3 and Proposition 14.9.1 that uniform matroids form an hereditary class.

Exercise 14.9.1. Prove that free matroids form an hereditary class.

Definition 14.9.3. *Let* **H** *be an hereditary class of matroids. Suppose that a matroid M is not in* **H**, *but all of the minors of M are in* **H**. *Then M is a* **forbidden,** *or* **excluded,** *minor of* **H**.

Theorem 14.9.1. *The members of any hereditary class* **H** *are characterized by having no minor isomorphic to a member of a maximal set S of pairwise non-isomorphic excluded minors.*

Proof. If $M \in \mathbf{H}$, then so are all its minors, and no excluded minor is a minor of M. Conversely, if M is not in **H**, then one of its minors is minimally not in the class and so is isomorphic to an excluded minor. □

It follows from the fact that each two-dimensional vector space over $GF(2)$ contains only three distinct non-zero elements that a four-point line $U_{2,4}$ is not representable over the field $GF(2)$. But the more difficult converse was proved in 1958 by the great graph- and matroid-theorist W.T. Tutte.

Theorem 14.9.2. (Tutte's Theorem) *The class of matroids representable over $GF(2)$ is hereditary and has the single excluded minor $U_{2,4}$.*

Exercise 14.9.2. Inspection verifies that $U_{2,4}$ is not among the minors of the Fano geometry, providing another proof of the representability of this geometry over GF(2). Construct a $U_{2,4}$ minor of the non-Fano geometry, thus providing an alternative proof of the non-representability of the non-Fano geometry over $GF(2)$.

Let us investigate the hereditary class of free matroids. If a matroid M is not free, then it has a circuit C. The minor $M|C$ also has C as a circuit. If $|C| > 1$, then, for any element e of C, we have that $(M|C).\{e\}$ has the set $\{e\}$ as a circuit. Conversely, if M is free, then $\mathcal{C}(M) = \emptyset$. Thus (to within isomorphism) the one-element matroid $U_{0,1}$ is the unique minimal excluded minor of this class and we have proved the following:

Theorem 14.9.3. *The class of free matroids has the single excluded minor $U_{0,1}$.*

Exercise 14.9.3. Verify that a matroid M is not uniform exactly if $E(M)$ has two equicardinal subsets I and C, with $I \in \mathbf{I}(M)$ and $C \in \mathcal{C}(M)$. Hence deduce that the class of uniform matroids is $Ex\{M\}$, where $E(M) = \{1,2\}$ and $\mathbf{I}(M) = \{\emptyset, \{1\}\}$.

Ralph Reid (unpublished result), R.E. Bixby [3], and P.D. Seymour [56] obtained the following result in the 1970s. We denote by F_7 and F_7^*, respectively, the matroid of the Fano geometry and its dual.

Theorem 14.9.4. *The class of matroids that are representable over $G(3)$, the field of three elements, is hereditary and has $U_{2,5}$, $U_{3,5}$, F_7, and F_7^* as a maximal set of pairwise non-isomorphic excluded minors.*

The characterization of the quaternary matroids (those matroids that are representable over $GF(4)$) is due to Geelen, Gerards, and Kapoor [26] from 2000. Geometric representations of the matroids listed in this theorem are given in Figure 14.8. A matrix representation for the matroid P_8 over a field of characteristic different from two is shown in Figure 14.9. The matroid P_8'' is obtained from P_8 by relaxing the two complementary circuit-hyperplanes.

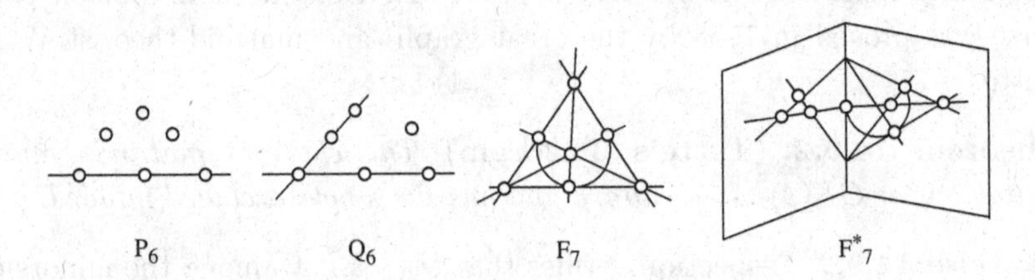

Fig. 14.8 Some excluded minors for $GF(4)$−representability

$$\begin{pmatrix} 1\,0\,0\,0\,0\,1\,1\,2 \\ 0\,1\,0\,0\,1\,0\,1\,1 \\ 0\,0\,1\,0\,1\,1\,0\,1 \\ 0\,0\,0\,1\,2\,1\,1\,0 \end{pmatrix}$$

Fig. 14.9 The matroid P_8

Theorem 14.9.5. *The class of matroids that are representable over $G(4)$, the field of four elements, is hereditary and has $U_{2,6}$, $U_{4,6}$, P_6, F_7^-, $(F_7^-)^*$,*

P_8, and P_8'' as a maximal set of pairwise non-isomorphic excluded minors.

In general it is not known whether the hereditary class of matroids representable over a finite field is characterized by a finite maximal set of pairwise non-isomorphic excluded minors. In 1959 W. Tutte also demonstrated exactly what stops a matroid from being graphic. Let us write $K_{3,3}$ for the graph with vertex set $V = \{1,2,3,4,5,6\}$ and edge set $E = \{\{i,j\} : i \in \{1,2,3\} \text{ and } j \in \{4,5,6\}\}$.

Theorem 14.9.6. *The class of graphic matroids is editary and has $U_{2,4}$, F_7, F_7^*, $M^*(K_5)$, and $M^*(K_{3,3})$ as a maximal set of pairwise non-isomorphic excluded minors.*

14.10 Paving matroids

In this section, we test our understanding of the aspects of matroid theory that we have canvassed by applying them to matroids that arise in the following way.

Definition 14.10.1. *Suppose that $\mathbf{T} = \{T_1, T_2, \ldots, T_k\}$ is a set of subsets of a finite set E so that each member of $\mathbf{T}$ contains at least m members and each m-element subset of E is in exactly one member of $\mathbf{T}$. We call $\mathbf{T}$ an* **m-partition** *of E and each member of $\mathbf{T}$ a block of the partition.*

Exercise 14.10.1. Let $\mathbf{T}$ be an m-partition of E. Writing $\mathbf{B}_T = \{B \subseteq E : |B| = m+1, B \text{ is not a subset of any block }\}$, prove that $\mathbf{B}_T$ is the set of bases of a matroid M on E.

Definition 14.10.2. *Let $\mathbf{T}$ be an m-partition of E. We call the matroid M defined by $\mathbf{B}(M) = \mathbf{B}_T$ a* **paving matroid.**

We have already met examples of paving matroids. The lines of any planar geometry are the blocks of a 2-partition on the set of points of the geometry, and the geometry itself is the paving matroid associated with this partition. The planes of a combinatorial cube are the blocks of a 3-partition on the set of eight points of the cube, and the cube is the associated paving matroid.

Exercise 14.10.2. Prove that each uniform matroid is a paving matroid.

Exercise 14.10.3. Suppose that M is a paving matroid. Specify $\mathbf{I}(M)$, $\mathcal{C}(M)$, and the value of rk_M on each subset of $E(M)$, in terms of the blocks of the m-partition that gives rise to M.

We conclude by examining the minors of paving matroids.

Exercise 14.10.4. Let M be a paving matroid defined by the m-partition $\mathbf{T}$ on the set E. Suppose that e is any element of E. By examining the collection $\{T - \{e\} : T \in \mathbf{T}\}$ prove that $M\backslash\{e\}$ is a paving matroid. By examining the collection $\{T \in \mathbf{T} : e \text{ is not in } T\}$ prove that $M/\{e\}$ is a paving matroid.

From Definition 14.9.2, Lemma 14.9.1 and the above Exercise 14.10.4 we deduce that paving matroids form an hereditary class.

Exercise 14.10.5. As a final test of our grasp of matroid techniques, find a maximal set of pairwise non-isomorphic excluded minors of the hereditary class of paving matroids.

The reader is referred to Oxley's "Matroid Theory" book [49] for a hint to the solution of this exercise. That book is also an excellent source for the reader whose interest in matroids was stimulated by this book.

Summary of Chapter 14. We proved several definitions of matroids equivalent, and used these alternative definitions to find many examples of matroids. In particular, we investigated graphic matroids and used vector matroids to convert geometric problems to arithmetic problems via homogeneous coordinates. Finally we defined minors, the natural subobjects of matroids, and obtained some of the properties that give them their important place in the theory of matroids.

In this book we have taken a journey through the lands of combinatorial geometries, projective planes, projective spaces, extended Euclidean space, perspective drawings, art, mechanisms, graphs, and polyhedra and managed to come out of the other side of the "looking glass" in the land of matroids. We hope that the readers imagination and geometric intuition have been stimulated, and that their appreciation of the beautiful geometric world that we live in has been enhanced.

Bibliography

[1] Auckly, D. and Cleveland, J. (1995). Totally real origami and impossible paper folding, *Amer. Math. Monthly* **102**, 3, pp. 215–226.

[2] Beyer, R. (1931). *Technische Kinematik* (Johann Ambrosius Barth, Leipzig).

[3] Bixby, R. E. (1979). On Reid's characterization of the ternary matroids, *J. Combin. Theory Ser. B* **26**, 2, pp. 174–204.

[4] Blackburn, J. E., Crapo, H. H. and Higgs, D. A. (1973). A catalogue of combinatorial geometries, *Math. Comp.* **27**, pp. 155–166; addendum, ibid. 27 (1973), no. 121, loose microfiche suppl. A12–G12.

[5] Bogart, K. P. (2000). *Introductory combinatorics*, 3rd edn. (Harcourt/Academic Press, San Diego, CA), ISBN 0-12-110830-9.

[6] Bondy, J. A. and Murty, U. S. R. (1976). *Graph theory with applications* (American Elsevier Publishing Co., Inc., New York).

[7] Bricard, R. (1897). Mémoire sur la théorie de l'octaèdre articulé, *J. Math. pures é appl.* **3**, pp. 113–148.

[8] Brylawski, T. (1986a). Appendix of matroid cryptomorphisms, in *Theory of matroids, Encyclopedia Math. Appl.*, Vol. 26 (Cambridge Univ. Press, Cambridge), pp. 298–316.

[9] Brylawski, T. (1986b). Constructions, in *Theory of matroids, Encyclopedia Math. Appl.*, Vol. 26 (Cambridge Univ. Press, Cambridge), pp. 127–223.

[10] Brylawski, T. and Oxley, J. (1992). The Tutte polynomial and its applications, in *Matroid applications, Encyclopedia Math. Appl.*, Vol. 40 (Cambridge Univ. Press, Cambridge), pp. 123–225.

[11] Chakerian, G. D. (1970). Sylvester's problem on collinear points and a relative, *Amer. Math. Monthly* **77**, pp. 164–167.

[12] Connelly, R. (1979). The rigidity of polyhedral surfaces, *Math. Mag.* **52**, 5, pp. 275–283.

[13] Coxeter, H. S. M. (1961). *Introduction to geometry* (John Wiley & Sons Inc., New York).

[14] Crapo, H. (1979). Structural rigidity, *Structural Topology* , 1, pp. 26–45, 73.

[15] Crapo, H. and Whiteley, W. (1982). Statics of frameworks and motions of panel structures, a projective geometric introduction, *Structural Topology* , 6, pp. 43–82, with a French translation.

[16] Crapo, H. H. and Rota, G.-C. (1970). *On the foundations of combinatorial theory: Combinatorial geometries*, preliminary edn. (The M.I.T. Press, Cambridge, Mass.-London).

[17] Cronin, D. (1995). Single-image stereograms, *Dr. Dobb's Journal* , pp. 18–27.

[18] Cundy, H. M. and Rollett, A. P. (1961). *Mathematical models* (2nd ed. Clarendon Press, Oxford).

[19] da Vinci, L. (1519). *Codex Atlanticus* (Art Resources, Inc.).

[20] Dorwart, H. L. and Grünbaum, B. (1992). Are these figures oxymora? *Math. Mag.* **65**, 3, pp. 158–169.

[21] Emmer, M. (1986). M.C. Escher: an interdisciplinary congress, *Structural Topology* **12**, pp. 61–65.

[22] Fáry, I. (1948). On straight line representation of planar graphs, *Acta Univ. Szeged. Sect. Sci. Math.* **11**, pp. 229–233.

[23] Filippov, A. F. (1950). An elementary proof of Jordan's theorem, *Uspehi Matem. Nauk (N.S.)* **5**, 5(39), pp. 173–176.

[24] Frederico, P. J. (1969). Enumeration of polyhedra: The number of 9-hedra, *J. Combinatorial Theory* **7**, pp. 155–161.

[25] Frederico, P. J. (1974/75). Polyhedra with 4 to 8 faces, *Geometriae Dedicata* **3**, pp. 469–481.

[26] Geelen, J. F., Gerards, A. M. H. and Kapoor, A. (2000). The excluded minors for GF(4)-representable matroids, *J. Combin. Theory Ser. B* **79**, 2, pp. 247–299.

[27] Geretschläger, R. (1995). Euclidean constructions and the geometry of origami, *Math. Mag.* **68**, 5, pp. 357–371.

[28] Goldberg, M. (1942). Polyhedral linkages, *Nat. Math. Mag.* **16**, pp. 323–332.

[29] Grashof, F. (1883). *Theoretische Maschinenlehre* (Verlag L. Voss, Leipzig).

[30] Grünbaum, B. (1975). Polygons, in *The geometry of metric and linear spaces (Proc. Conf., Michigan State Univ., East Lansing, Mich., 1974)* (Springer, Berlin), pp. 147–184. Lecture Notes in Math., Vol. 490.

[31] Grünbaum, B. and Shephard, G. C. (1988). Les charpentes de plaques rigides, *Structural Topology* , 14, pp. 1–8, dual French-English text.

[32] Hall, A. S., Jr. (1961). *Kinematics and linkage design* (Prentice-Hall Engineering Science Series. Prentice-Hall, Inc., Englewood Cliffs, N.J.).

[33] Hilton, P. and Pedersen, J. (1983). Approximating any regular polygon by folding paper, *Math. Mag.* **56**, 3, pp. 141–155.

[34] Hrones, J. A. and Nelson, G. L. (1951). *Analysis of the Four-Bar Linkage. Its Application to the Synthesis of Mechanisms* (The Technology Press of The Massachusetts Institute of Technology).

[35] Hughes Jones, R. (1982). Folding polyhedra, *Structural Topology* , 7, pp. 45–50, with a French translation.

[36] Hull, T. (1996). A note on "impossible" paper folding, *Amer. Math. Monthly* **103**, 3, pp. 240–241.

[37] Hunt, K. (1973). Constant-velocity shaft couplings: a general theory, *J. Engineering for Industry* , pp. 455–464.

[38] Hunt, K. H. (1978). *Kinematic geometry of mechanisms* (The Clarendon

Press Oxford University Press, New York), ISBN 0-19-856124-5, the Oxford Engineering Science Series.
[39] Julesz, B. (1960). Binocular depth perception of computer-generated patterns, *Bell Syst. Tech. J.* **39**, pp. 1125–1162.
[40] Julesz, B. (1986). Stereoscopic vision, *Vision Res.* **26**, pp. 1601–1612.
[41] Keable, J. (ed.) (1993). *Hodges, How the pyramids were built* (Warminster, Wiltshire: Aris and Phillips, Teddington House, Warminster, Wiltshire).
[42] Kempe, A. (1877). *How to draw a straight line; a lecture on linkages* (Macmillan, London).
[43] Kruskal, J. B., Jr. (1956). On the shortest spanning subtree of a graph and the traveling salesman problem, *Proc. Amer. Math. Soc.* **7**, pp. 48–50.
[44] Kubovy, M. (1986). *The psychology of perspective and renaissance art* (Cambridge University Press, Cambridge).
[45] Maeder, R. (1995). Single-image stereograms, *The Mathematica Journal* **5**, pp. 50–60.
[46] Mason, J. H. (1971). Geometrical realization of combinatorial geometries, *Proc. Amer. Math. Soc.* **30**, pp. 15–21.
[47] Mayhew, D. and Royle, G. (2011). Matroid seeker, Http://people.csse.uwa.edu.au/gordon/small-matroids.html.
[48] Montroll, J. (1979). *Origami for the enthusiast* (Dover Publications).
[49] Oxley, J. G. (1992). *Matroid theory*, Oxford Science Publications (The Clarendon Press Oxford University Press, New York), ISBN 0-19-853563-5.
[50] Penrose, L. and Penrose, R. (1958). Impossible objects - a special type of illusion, *Brit. J. Psych* **49**, pp. 31–33.
[51] Pirenne, M. (1970). *Optics, painting, and photography* (Cambridge University Press, Cambridge).
[52] Richter, J. (1970). *The literary works of Leonardo da Vinci* (Phaidon Press, London).
[53] Row, T. S. (1966). *Geometric exercises in paper folding* (Dover).
[54] Sertöz, S. (ed.) (1988). *The mathematical implications of Escher's prints, The World of M.C. Escher* (Abradale Press, New York).
[55] Servatius, B. (1995). Graphs, digraphs, and the rigidity of grids, *UMAP J.* **16**, 1, pp. 43–69.
[56] Seymour, P. D. (1979). Matroid representation over GF(3), *J. Combin. Theory Ser. B* **26**, 2, pp. 159–173.
[57] Tarnai, T. (1988). Les mécanismes finis et la charpente octogonale de la cathédrale d'Ely, *Structural Topology* , 14, pp. 9–20, dual French-English text, With an appendix by J. Eddie Baker.
[58] Trigg, C. (1978). Tetrahedral models from envelopes, *Mathematics Magazine* **51**, pp. 66–67.
[59] Veblen, O. and Young, J. W. (1965). *Projective geometry. Vol. 1* (Blaisdell Publishing Co. Ginn and Co. New York-Toronto-London).
[60] Wheatstone, C. (1838). Contributions to the physiology of vision. part the first, *Philosophical Trans. Royal Soc. London* **128**, pp. 371–394.
[61] Whiteley, W. (1979). Realizability of polyhedra, *Structural Topology* , 1, pp.

46–58, 73.

[62] Whitney, H. (1935). On the abstract properties of linear dependence, *Amer. J. Math.* **57**, pp. 509–33.

[63] Wilf, H. S. (1971). The friendship theorem, in *Combinatorial Mathematics and its Applications (Proc. Conf., Oxford, 1969)* (Academic Press, London), pp. 307–309.

[64] Williams, D. (1967). The dynamics of the golf swing, *Quarterly J. Appl. Maths.* **20**, pp. 247–264.

[65] Williams, V. (1968). A proof of sylvester's theorem on collinear points, *Amer. Math. Monthly* **75**, pp. 980–982.

[66] Ziegler, G. M. (1995). *Lectures on polytopes, Graduate Texts in Mathematics*, Vol. 152 (Springer-Verlag, New York), ISBN 0-387-94365-X.

Index